LINEAR PARAMETER-VARYING SYSTEM IDENTIFICATION

New Developments and Trends

ADVANCED SERIES IN ELECTRICAL AND COMPUTER ENGINEERING

Editor: W. K. Chen *(University of Illinois, Chicago, USA)*

Published:

Vol. 1: Broadband Matching: Theory and Implementations (Second Edition)
by W. K. Chen

Vol. 2: Active Network Analysis
by W. K. Chen

Vol. 3: Linear Networks and Systems: Algorithms and Computer-Aided Implementations (Second Edition) (Two Volumes)
by W. K. Chen

Vol. 4: Introduction to Engineering Electromagnetic Fields
by K. Umashankar

Vol. 5: Graph Theory and its Engineering Applications
by W. K. Chen

Vol. 6: Introductory Signal Processing
by R. Priemer

Vol. 7: Diagnostic Measurements in LSI/VLSI Integrated Circuits Production
by A. Jakubowski, W. Marciniak and H. Przewlocki

Vol. 8: An Introduction to Control Systems (Second Edition)
by K. Warwick

Vol. 9: Orthogonal Functions in Systems and Control
by K. B. Datta and B. M. Mohan

Vol. 10: Introduction to High Power Pulse Technology
by S. T. Pai and Q. Zhang

Vol. 11: Systems and Control: An Introduction to Linear, Sampled and Nonlinear Systems
by T. Dougherty

Vol. 12: Protocol Conformance Testing Using Unique Input/Output Sequences
by X. Sun, C. Feng, Y. Shen and F. Lombardi

Vol. 13: Semiconductor Manufacturing Technology
by C. S. Yoo

Vol. 14: Linear Parameter-Varying System Identification: New Developments and Trends
by Paulo Lopes dos Santos, Teresa Paula Azevedo Perdicoúlis, Carlo Novara, Jose A. Ramos and Daniel E. Rivera

Advanced Series in Electrical and Computer Engineering – Vol. 14

LINEAR PARAMETER-VARYING SYSTEM IDENTIFICATION

New Developments and Trends

Editors

Paulo Lopes dos Santos
Universidade do Porto, Portugal

Teresa Paula Azevedo Perdicoúlis
Universidade de Trás-os-Montes e Alto Douro, Portugal

Carlo Novara
Politecnico di Torino, Italy

Jose A Ramos
Nova Southeastern University, USA

Daniel E Rivera
Arizona State University, USA

World Scientific

NEW JERSEY • LONDON • SINGAPORE • BEIJING • SHANGHAI • HONG KONG • TAIPEI • CHENNAI

Published by

World Scientific Publishing Co. Pte. Ltd.
5 Toh Tuck Link, Singapore 596224
USA office: 27 Warren Street, Suite 401-402, Hackensack, NJ 07601
UK office: 57 Shelton Street, Covent Garden, London WC2H 9HE

British Library Cataloguing-in-Publication Data
A catalogue record for this book is available from the British Library.

Advanced Series in Electrical and Computer Engineering — Vol. 14
LINEAR PARAMETER-VARYING SYSTEM IDENTIFICATION
New Developments and Trends

ISBN-13 978-981-4355-44-5
ISBN-10 981-4355-44-5

Printed in Singapore by World Scientific Printers.

Preface

Increasing demands for high-performance control systems call for greater sophistication in dynamical modeling approaches based on system identification techniques. To this end, Linear Parameter Varying (LPV) models represent an important means to achieve advanced control of complex dynamical systems that can be implemented in practice.

This book is a review volume that reports on the state-of-the-art in LPV system identification, drawing from the work of some of the leading researchers in the field. Following an introductory chapter, twelve chapters focused on some of the most recent LPV identification methods for both discrete-time and continuous-time models are presented. Diverse approaches such as optimization methods for input/output LPV model identification, set membership methods, subspace methods for state-space LPV models, and identification using orthonormal basis functions methods are considered. Because of the strong connection between LPV systems, hybrid switching systems and piecewise affine models, identification of these types of systems are considered as well.

The impetus for this volume has been a series of invited sessions on LPV identification that have been organized at recent conferences in the control systems engineering field. These are the 2008 IEEE International Symposium on Computer-Aided Control System Design in San Antonio, Texas and the 2008 IEEE Conference on Decision and Control in Cancún, Mexico. The success of these sessions led to the development of a Special Issue in Advances in LPV Modeling and Identification Applications that was published in January 2011 in the IEEE Transactions on Control Systems Technology. These activities were conducted under the auspices of the IEEE-Control Systems Society Technical Committee on System Identification and Adaptive Control (TC-SIAC). Some of the book editors were editors of this Special Issue, and many authors of the chapters are also members of the TC-SIAC.

The book constitutes a means to assess the current state-of-the-art in system identification for this important class of models. We believe that a systematic and comprehensive presentation of the most important methods can be a valuable tool for both researchers and industrial practitioners, enabling simpler solutions to many real life problems using novel LPV and piecewise affine modeling frameworks. We anticipate the readership ranging from graduate students to seasoned research professionals, with relevance of this volume in the areas of process control, aeronautics and space, automotive and mechatronics, and robotics and transportation networks.

We acknowledge the following individuals who have been helpful in this effort: Prof. Marco Lovera from Politecnico di Milano and Dr Elena Nash from World Scientific. We are also grateful for the support given by our families.

Sincerely,

Paulo Lopes dos Santos

T-P Azevedo Perdicoúlis

C. Novara

J. A. Ramos

D. E. Rivera

Contents

Acronyms

Acronym	Definition
ACF	Autocorrelation function
AR	Autogressive
ARMA	Autoregressive moving average
ARMAX	Autoregressive moving average with exogenous inputs
ARX	Autoregressive with exogenous inputs
BB	Black-box
BJ	Box-Jenkings
BN	Bias norm
CA	Correlation analysis
CSMA	Carrier sense multiple access
CT	Continuous time
CVA	Covariance analysis
DOF	Degrees of freedom
DT	Discrete time
EA	Expectation minimization
EM	Expectation maximization
FKcM	Fuzzy Kolmogorov c-max
FFS	Feasible function set
FFT	Fast Fourier transform
FIR	Finite impulse response
FPS	Feasible parameter set
FRF	Frequency response function
GOBF	Generalized orthonormal basis functions
GPCA	Generalized principal component analysis
HLPV	Hybrid linear parameter varying
IO	Input-output
I/O	Input-output
IRR	Impulse response representation
IV	Instrumental variables
LFR	Linear fractional representation

LFT	Linear fractional transformation
LMI	Linear matrix inequality
LMS	Least mean square
LPV	Linear parameter varying
LS	Least squares
LS-SVM	Least squares support vector machines
LTI	Linear time-invariant
MA	Moving average
MCS	Monte Carlo simulation
MIMO	Multiple input and multiple output
MIQP	Mixed integer quadratic programming
MISO	Multiple input and single output
ML	Maximum likelihood
MOESP	MIMO output-error state space
N4SID	Numerical algorithms for subspace state space system identification
NARMAX	Non-linear auto-regressive moving average with exogenous input
NNG	Non-negative garrote
NSR	Noise-to-signal ratio
OBF	Orthonormal basis functions
OE	Output-error
PBSID	Predictor based subspace identification
PEM	Prediction error minimization
PO-MOESP	Past outputs MIMO output-error state space
PRBS	Pseudo-random binary sequence
PSE	Percentage simulation error
PWA	Piecewise affine systems
PWARX	Piecewise affine ARX
RIV	Refined instrumental variable
RIVC	Refined instrumental variables for continuous-time models
RLS	Recursive least squares
SA	Spectral analysis
SARX	Switched autoregressive moving average
SDP	Semi-definite programming
SI	System identification
SIDDS	Subspace identification downsampling

SISO	Single input and single output
SLPV	Switch linear parameter varying
SMILE	State-space model interpolation of local estimates
SM	Set membership
SMI	Subspace model identification
SNR	Signal-to-noise ratio
SOS	Sum of squares
SP	Stochastic processes
SRIV	Simplified refined instrumental variable
SRIVC	Simplified refined instrumental variable for continuous-time models
SS	State-space
SVD	Singular value decomposition
SLPV	Switching LPV
UBB	Unbounded but bounded
ZOH	Zero order hold

Chapter 1

Introduction

C. Novara[†], P. Lopes dos Santos[‡], T.-P. Azevedo Perdicoúlis[*], J. A. Ramos[♮] & D. E. Rivera[♯]

[†]*Dipartimento di Automatica e Informatica*
Politecnico di Torino, 10129 Torino, Italy
carlo.novara@polito.it

[‡]*Faculdade de Engenharia da Universidade do Porto*
Rua do Dr Roberto Frias s/n
4200-465 Porto, Portugal
pjsantos@fe.up.pt *and* jmartins@fe.up.pt

[*]*ISR—Coimbra & UTAD*
Departamento das Engenharias, UTAD,
5000-801 Vila Real, Portugal
tazevedo@utad.pt

[♮]*Farquhar College of Arts and Sciences*
Division of Mathematics, Science and Technology
Nova Southeastern University
3301 College Avenue, Fort Lauderdale, FL 33314, USA
jr1284@nova.edu

[♯]*School for Engineering of Matter, Transport, and Energy*
Mail Stop 876106, Arizona State University
Tempe, Arizona 85287-6106
daniel.rivera@asu.edu

The study of Linear Parameter Varying (LPV) systems was originally motivated by the control design methodology of gain scheduling [Rugh and Shamma (2000)]. The notion of LPV systems, first introduced in [Shamma and Athans (1990)], proved to be very relevant by allowing the automatic control community to overcome some limitations of the gain scheduling approach. Examples of such limitations are the large number of linear models often required to achieve a given performance, and the necessity for slow changes between two operating points. Currently, modeling, identification, and control design of LPV systems form an active research area that contributes to the development of gain scheduling controllers able to deal with fast variation of the operating points and with tight performance bounds.

By the end of the last decade, the increasing importance of LPV models in control system design has motivated the rise of a new identification problem. From the methodological point of view, several LPV identification methods have been proposed, including regression approaches [Bamieh and Giarré (2002); Hsu *et al.* (2008)], subspace techniques [Nemani *et al.* (1995); Verdult (2002); Lovera and Mercere (2007); Felici *et al.* (2007); Lopes dos Santos *et al.* (2008)], basis function expansions [Tóth (2008); Tóth *et al.* (2009)], linear matrix inequalities [Mazzaro and Sznaier (2001)], nonlinear programming [Lee and Poolla (1999)], successive approximations [Lopes dos Santos *et al.* (2008)], interpolation [Lovera and Mercere (2007); Steinbuch *et al.* (2003)], least squares [Bamieh and Giarré (2002); Lee and Poolla (1999)], and set membership approaches [Mazzaro and Sznaier (2001); Cerone and Regruto (2008); Novara *et al.* (2011)]. Strongly related to LPV methods, piecewise affine methods have also become an important tool for gain scheduling control, since they allow the user to capture many features of the real physical systems, and enable tractable mathematical analysis [Pepona *et al.* (2011)]. Despite the existence of several methods already available, systematic approaches able to deal with the complexity and the stringent specifications of advanced applications are still needed. From the application point of view, many real systems can be reasonably described by LPV models, yielding increasing interest for this identification framework within different areas, as for instance aerospace [Doll *et al.* (2008); Marcos and Balas (2004); Lu and Wu (2006); Lu *et al.* (2006); Lee *et al.* (2001); Ohara *et al.* (2001); Kron *et al.* (2005); Marcos *et al.* (2008); Ghersin and Pena (2002); Pfifer and Hecker (2011); Ferreres (2011); Szabó *et al.* (2011)], automotive [Chabaan (2003); Fialho and Balas (2000); Henry and Applebee (1996); Hingwe *et al.* (2002); Hong *et al.* (2002); Hunt *et al.* (2000); Onat *et al.* (2007); Tan (2003); Wei and del Re (2007); Novara *et al.* (2011)], motorbike control [Corno *et al.* (2011)], mechanics [Giarré *et al.* (2006)], mechatronics [Paijmans *et al.* (2006); Steinbuch *et al.* (2003); De Caigny *et al.* (2011)], traffic management [Luspay *et al.* (2011)], robotics [van Helvoort *et al.* (2004); Abbas and Werner (2011)], termodynamics [Mercè Andre *et al.* (2011)], bio-engineering [Previdi and Carpanzano (2003)], gas transportation networks [Lopes dos Santos *et al.* (2011)], web applications [Tanelli *et al.* (2011)], database [Szabó *et al.* (2011)], semiconductor manufacturing [Groot Wassink *et al.* (2005)] and computing systems [Qin and Wang (2007b,a); Tanelli *et al.* (2008)]. During the last biennium, invited sessions on LPV identification were organised for the 2008 IEEE International Symposium on Computer-Aided Control Systems and for the 2008 IEEE Conference on Decision and Control . Also, a Special Issue in Advances in LPV Modelling and Identification, Methods and Applications has been recently published in IEEE Transactions on Control Systems Technology [Lovera *et al.* (2011)]. However, only a limited number of works devoted to the application of the LPV identification techniques have been published in the relevant literature.

This book describes the state-of-the-art in LPV system identification and shows some recent advances in piecewise affine system identification. The presented methods provide an important step towards the systematization of several LPV system identification approaches and are valuable tools for both researchers and industry practitioners, allowing for the solution of many advanced real life problems. The book is organized as follows.

In Chapter 2, *Hybrid LPV modelling and identification* by Giarré, Falugi and Badalamenti, an overview of the existing literature for LPV and Hybrid Switching models for identification and control is provided. Starting from some motivating examples, the modeling of Hybrid LPV systems is introduced, together with some possible identification schemes. An application example is considered in details, with reference to the modeling of the traffic flow in wireless ad-hoc networks.

Chapter 3, *Set-membership identification of input-output LPV models with uncertain time-varying parameters* by Cerone, Piga and Regruto, considers the identification of LPV models when both the output and the time-varying parameter measurements are affected by bounded noise, and presents a relaxation technique to compute parameter uncertainty bounds by means of convex optimization.

In Chapter 4, *Set membership identification of state-space LPV systems* by Novara, a sparse set membership identification method for state-space LPV systems is proposed, allowing the identification of state-space

LPV models with worst-case optimality properties, and the evaluation of tight uncertainty bounds on the elements of the model parameter-varying matrices.

Chapter 5, *Identification of input/output LPV models* by Laurain, Tóth, Gilson and Garnier, presents a survey on identification of ARX, Output-Error and Box-Jenkins models for LPV systems, considering both the discrete-time and the continuous-time settings.

In Chapter 6, *Reducing the dimensions of data matrices involved in LPV subspace identification methods* by Verdult and Verhaegen, several LPV subspace methods are described, the construction of their data matrices is compared, and a method to reduce the dimensions of these matrices is discussed.

Chapter 7, *Subspace identification of MIMO LPV systems* by van Wingerden and Verhaegen, presents an LPV subspace identification algorithm able to deal with both open-loop and closed-loop data. The algorithm is based on the reconstruction of the LPV system matrices from an estimate of the state sequence.

Chapter 8, *Subspace identification of continuous-time state-space LPV models* by Bergamasco and Lovera, proposes an interpolation method for the estimation continuous-time state-space LPV models for systems operating in open-loop or closed-loop. In this method the identification of the local models is performed using recent developed subspace techniques using the prediction based framework [Bergamasco and Lovera (2010, 2011)], simultaneously enabling the straightforward treatment of MIMO and SISO modelling problems in state-space form and the ability to deal with data generated in closed-loop.

In Chapter 9, *Indirect continuous-time LPV system identification through a downsampled subspace approach* by Lopes dos Santos, Azevedo Perdicoúlis, Ramos and Martins de Carvalho, the successive approximations LPV subspace identification algorithm is modified to cope with continuous-time LPV state-space models. In particular, the LPV system of interest is discretized, and the obtained discrete-time model is identified by the successive approximations algorithm and then converted to a continuous-time model. Since affine parameter dependence is preserved only for fast sampling, a downsampled approach is used to estimate the sequence of the linear time invariant approximations to the LPV model.

In Chapter 10, *LPV system identification using series expansion models* by Tóth, Heuberger and Van den Hof, an Orthonormal Basis Functions model structure is proposed with advantageous properties with respect to many existing methods in terms of estimation and realization. A solid theoretic basis for the description of LPV systems in terms of this model structure is presented, together with a general approach to LPV identification.

In Chapter 11, *System identification of linear parameter varying state-space models* by Wills and Ninness, an expectation-maximisation algorithm is derived, allowing the estimation of multivariable linear models with time-varying parameters having affine dependence on a known signal.

Chapter 12, *Piecewise affine identification of interconnected systems with LFR structure* by Paoletti and Garulli, proposes a two-step iterative procedure with non-increasing cost for the identification of systems in linear fractional representation, i.e., systems composed of interconnected LTI systems and static nonlinearities.

In Chapter 13, *Identification and model (in)validation of switched ARX systems: A moment-based approach* by Feng, Ozay, Lagoa and Sznaier, it is first shown that computationally efficient algorithms can be obtained by recasting the identification problem into a polynomial optimization form and exploiting its inherent sparse structure. Then, the related problem of validating the identified models against (new) experimental data is addressed.

In summary, this book consists of twelve chapters, following this introduction, among which five mainly rely on regression representations (Chapters 2, 3, 5, 10, 13), six on state-space representations (Chapters 4, 6–9, 11), and one on the linear fractional representation (Chapter 12). Among the chapters relying on regression representations, several approaches are considered, such as basis function expansion (Chapters 2 and 10), set membership (Chapter 3), ARX, Output-Error and Box-Jenkins (Chapter 5), and sparsification (Chapter 13). Among the chapters relying on state-space representations, four consider subspace methods (Chapters 6–9), one on a sparse set membership approach (Chapter 4,), and one on a maximum-likelihood criterion (Chapter 11). It can be observed that both, statistical noise assumptions (Chapters 2, 5–11) and unknown-but-bounded noise assumptions (Chapters 2–4, 12, 13), are extensively dealt with in the book.

References

Abbas, H. and Werner, H. (2011). Frequency-weighted discrete-time LPV model reduction using structurally balanced truncation, *IEEE Transactions on Control Systems Technology* **19**, 1, pp. 140–147.

Bamieh, B. and Giarré, L. (2002). Identification of linear parameter varying models, *International Journal of Robust and Nonlinear Control* **12**, 9, pp. 841–853.

Bergamasco, M. and Lovera, M. (2010). Continuous-time subspace identification in closed-loop, in *Proceedings of the International Symposium on Mathematical Theory of Networks and Systems* (Budapest, Hungary).

Bergamasco, M. and Lovera, M. (2011). Continuous-time predictor-based subspace identification using Laguerre filters, *IET Control Theory and Applications* Special issue on Continuous-time Model Identification (to appear).

Cerone, V. and Regruto, D. (2008). Set-membership identification of LPV models with uncertain measurements of the time-varying parameter, in *Proceedings of the Conference on Decision and Control* (Cancún, Mexico).

Chabaan, R. (2003). H-infinity control and gain scheduling method for electric power assist steering system, *US Patent 6651771* .

Corno, M., Tanelli, M., Savaresi, S. and Fabbri, L. (2011). Design and validation of a gain-scheduled controller for the electronic throttle body in ride-by-wire racing motorcycles, *IEEE Transactions on Control Systems Technology* **19**, 1, pp. 18–30.

De Caigny, J., Camino, J. and Swevers, J. (2011). Interpolation-based modeling of MIMO LPV systems, *IEEE Transactions on Control Systems Technology* **19**, 1, pp. 46–63.

Doll, C., Chiappa, C. and Biannic, J.-M. (2008). LFT modelling of the 2-dof longitudinal nonlinear aircraft behaviour, in *Proceedings of the IEEE Multiconference on Systems and Control* (San Antonio, TX, USA).

Felici, F., van Wingerden, J. W. and Verhaegen, M. (2007). Subspace identification of MIMO LPV systems using a periodic scheduling sequence, *Automatica* **43**, 10, pp. 1684–1697.

Ferreres, G. (2011). Computation of a flexible aircraft LPV/LFT model using interpolation, *IEEE Transactions on Control Systems Technology* **19**, 1, pp. 132–139.

Fialho, I. J. and Balas, G. (2000). Design of nonlinear controllers for active vehicle suspensions using parameter-varying control synthesis, *Vehicle Systems Dynamics* **33**, 5, pp. 351–370.

Ghersin, A. and Pena, R. S. (2002). LPV control of a 6-dof vehicle, *IEEE Transactions on Control Systems Technology* **10**, 6, pp. 883–887.

Giarré, L., Bauso, D., Falugi, P. and Bamieh, B. (2006). LPV model identification for gain scheduling control: An application to rotating stall and surge control problem, *Control Engineering Practice* **14**, 4, pp. 351–361.

Groot Wassink, M., van del Wal, M., Scheerer, C. and Bosgra, O. (2005). LPV control for a wafer stage: beyond the theoretical solution, *Control Engineering Practice* **13**, 2, pp. 231–245.

Henry, R. and Applebee, M. (1996). Vehicle suspension system with gain scheduling, *US Patent 5497324*.

Hingwe, P., tan, H., Packard, A. and Tomizuka, M. (2002). Linear parameter varying controller for automated lane guidance: experimental study on tractor-trailers, *IEEE Transactions on Control System Technology* **10**, 6, pp. 793–806.

Hong, K., Sohn, H. and Hedrick, J. (2002). Modified skyhook control of semi-active suspensions: A new model, gain scheduling, and hardware-in-the-loop tuning, *Journal of Dynamic Systems, Measurement, and Control* **124**, 1, pp. 158–167.

Hsu, K., Vincent, T. and Poolla, K. (2008). Nonparametric methods for the identification of linear parameter varying systems, in *Proceedings of the IEEE International Conference on Computer-Aided Control Systems* (San Antonio, TX, USA).

Hunt, K., Johansen, T., Kalkkuhl, J., Fritz, H. and Gottsche, T. (2000). Speed control design for an experimental vehicle using a generalized gain scheduling approach, *IEEE Transactions on Control Systems Technology* **8**, 3, pp. 381–395.

Kron, A., de Lafontaine, J. and Peuvédic, C. L. (2005). Mars entry and aerocapture robust control using static output feedback and LPV techniques, in *Proceedings of the International ESA Conference on Guidance, Navigation and Control Systems* (Loutraki, Greece).

Lee, C., Shin, M. and Chung, M. (2001). A design of gain-scheduled control for a linear parameter varying system: an application to flight control, *Control Engineering Practice* **9**, 1, pp. 11–21.

Lee, L. and Poolla, K. (1999). Identification of linear parameter-varying systems using nonlinear programming, *Journal of Dynamic Systems Measurement and Control* **121**, 1, pp. 71–78.

Lopes dos Santos, P., Azevedo Perdicoúlis, T.-P., Ramos, J. A., Martins de Carvalho, J. L., Jank, G. and Milhinhos, J. (2011). An LPV modeling and identification approach to leakage detection in high pressure natural gas transportation networks, *IEEE Transactions on Control Systems Technology* **19**, 1, pp. 77–92.

Lopes dos Santos, P., Ramos, J. A. and Martins de Carvalho, J. L. (2008). Identification of LPV systems using successive approximations, in *Proceedings of the IEEE Conference on Decision and Control* (Cancún, Mexico).

Lovera, M. and Mercere, G. (2007). Identification for gain-scheduling: a balanced subspace approach, in *Proceedings of the American Control Conference* (New York City, NY, USA).

Lovera, M., Novara, C., dos Santos, P. L. and Rivera, D. (2011). Guest editorial special issue on applied LPV modeling and identification, *IEEE Transactions on Control Systems Technology* **19**, 1, pp. 1–4.

Lu, B. and Wu, F. (2006). Probabilistic robust linear parameter-varying control of an F-16 aircraft, *Journal of Guidance, Control and Dynamics* **29**, 6, pp. 1454–1460.

Lu, B., Wu, F. and Kim, S. (2006). Switching LPV control of an F-16 aircraft via controller state reset, *IEEE Transactions on Control Systems Technology* **14**, 2, pp. 267–277.

Luspay, T., Kulcsár, B., van Wingerden, J. W., Verhagen, M. and Bokor, J. (2011). Linear parameter varying identification of freeway traffic models, *IEEE Transactions on Control Systems Technology* **19**, 1, pp. 31–45.

Marcos, A. and Balas, G. (2004). Development of linear-parameter-varying models for aircraft, *Journal of Guidance, Control and Dynamics* **27**, 2, pp. 218–228.

Marcos, A., Penin, L. and Sotto, E. D. (2008). LFT modelling for the analysis of relative motion controllers in eccentric orbits, in *Proceedings of the IEEE Multi-conference on Systems and Control* (San Antonio, TX, USA).

Mazzaro, C. and Sznaier, M. (2001). An LMI approach to model (in) validation of LPV systems, in *Proceedings of The American Control Conference*, Vol. 6 (IEEE, Arlington, VA, USA).

Mercè Andre, G., Palsson, H. and Poinot, T. (2011). Continuous-time linear parameter-varying identification of a cross flow heat exchanger: A local approach, *IEEE Transactions on Control Systems Technology* **19**, 1, pp. 64–76.

Nemani, M., Ravikanth, R. and Bamieh, B. (1995). Identification of linear parametrically varying systems, in *Proceedings of the IEEE Conference on Decision and Control* (New Orleans, LA, USA).

Novara, C., Ruiz, F. and Milanese, M. (2011). Direct identification of optimal SM-LPV filters and application to vehicle yaw rate estimation, *IEEE Transactions on Control Systems Technology* **19**, 1, pp. 5–17.

Ohara, A., Yamaguchi, Y. and Morito, T. (2001). LPV modeling and gain scheduled control of re-entry vehicle in approach and landing phase, in *Proceedings of the AIAA Guidance, Navigation, and Control Conference and Exhibit* (Montreal, Canada).

Onat, C., Kucukdemiral, I. B., Sivrioglu, S. and Yuksek, I. (2007). LPV model based gain-scheduling controller for a full vehicle active suspension system, *Journal of Vibration and Control* **13**, 11, pp. 1629–1666.

Paijmans, B., Symens, W., Van Brussel, H. and Swevers, J. (2006). A gain-scheduling-control technique for mechatronic systems with position-dependent dynamics, in *Proceedings of the American Control Conference* (Minneapolis, MN, USA).

Pepona, E., Paoletti, S., Garulli, A. and Date, P. (2011). Identification of piecewise affine LFR models of interconnected systems, *IEEE Transactions on Control Systems Technology* **19**, 1, pp. 148–155.

Pfifer, H. and Hecker, S. (2011). Generation of optimal linear parametric models for LFT-based robust stability analysis and control design, *IEEE Transactions on Control Systems Technology* **19**, 1, pp. 118–131.

Previdi, F. and Carpanzano, E. (2003). Design of a gain scheduling controller for knee-joint angle control by using functional electrical stimulation, *IEEE Transactions on Control Systems Technology* **11**, 3, pp. 310–324.

Qin, W. and Wang, Q. (2007a). An LPV approximation for admission control of an internet web server: identification and control, *Control Engineering*

Practice **15**, 12, pp. 1457–1467.

Qin, W. and Wang, Q. (2007b). Modeling and control design for performance management of web servers via an LPV approach, *IEEE Transactions on Control Systems Technology* **15**, 2, pp. 259–275.

Rugh, W. and Shamma, J. (2000). Research on gain scheduling, *Automatica* **36**, pp. 1401–1425.

Shamma, J. and Athans, M. (1990). Analysis of gain scheduled control for nonlinear plants, *IEEE Transactions on Automatic Control* **35**, 8, pp. 898–907.

Steinbuch, M., van de Molengraft, R. and van der Voort, A. (2003). Experimental modelling and LPV control of a motion system, in *Proceedings of the American Control Conference* (Denver, CO, USA).

Szabó, Z., Marcos, A., Mostaza Prieto, D., Kerr, M. L., Rodonyi, G., Bokor, J. and Bennani, S. (2011). Development of an integrated LPV/LFT framework: Modeling and data-based validation tool, *IEEE Transactions on Control Systems Technology* **19**, 1, pp. 104–117.

Tan, H. (2003). Estimating vehicle velocities using linear-parameter-varying and gain varying scheduling theories, *US Patent 6618651* .

Tanelli, M., Ardagna, D. and Lovera, M. (2008). LPV model identification for power management of web service systems, in *Proceedings of the IEEE International Symposium on CACSD* (San Antonio, TX, USA).

Tanelli, M., Ardagna, D. and Lovera, M. (2011). Identification of LPV state space models for autonomic web service systems, *IEEE Transactions on Control Systems Technology* **19**, 1, pp. 93–103.

Tóth, R. (2008). *Modeling and identification of linear parameter-varying systems*, Ph.D. thesis, Technische Universiteit Delft, The Netherlands.

Tóth, R., Heuberger, P. and Van den Hof, P. (2009). Asymptotically optimal orthonormal basis functions for LPV system identification, *Automatica* **45**, 6, pp. 1359–1370.

van Helvoort, J. J. M., Steinbuch, M., Lambrechts, P. F. and van Molengraft, R. (2004). Analytical and experimental modelling for gain-scheduling of a double scara robot, in *Proceedings of the IFAC Symposium on Mechatronic Systems* (Sydney, Australia).

Verdult, V. (2002). *Nonlinear system identification: a state space approach*, Ph.D. thesis, University of Twente, The Netherlands.

Wei, X. and del Re, L. (2007). Gain scheduled H_∞ control for air path systems of diesel engines using LPV techniques, *IEEE Transactions on Control Systems Technology* **15**, 3, pp. 406–415.

Chapter 2

Hybrid LPV modeling and identification

L. Giarré[†], P. Falugi[‡] & R. Badalamenti[†]

[†]*DIETT, Università di Palermo, 90128 Palermo, Italy*
giarre@unipa.it
romina.badalamenti@unipa.it

[‡]*Imperial College, London, SW7 2AZ, U.K.*
paola.falugi@gmail.com

In the last two decades, LPV and Hybrid Switching models have been extensively studied and identification and control design techniques have been devised. In the present chapter, after an overview of the existing literature, starting from some motivating examples, we introduce the modeling of Hybrid LPV systems. An application example is considered in detail, with reference to the modeling of the traffic flow in wireless ad-hoc networks, together with some possible identification schemes. Then, we highlight some open problems.

1. Introduction

Linear Parameter Varying (LPV), Hybrid Switching and Piece-Wise Affine (PWA) models arise from different applications and have been studied in connection to various control schemes. Linear models allow for efficient design and software tools but often fail to provide a faithful description of the real systems. The following three categories of systems, whose study hinges upon the underlying linear sub-components, appear to be adequate in many cases:

- LPV systems: the parameters are unknown but measurable.
- Switching linear systems: the parameters can take any value in a given finite set.

- PWA systems: the linear model depends on the region to which the system variables belong.

We can trace back LPV modeling origin to the gain scheduling techniques for control design and to aerospace applications. In particular, the use of LPV models as an approximation of nonlinear models became popular in the nineties [Shamma and Athans (1991a)]. In the literature, a large amount of work on LPV control design and methods to obtain a set of stabilizing controllers proliferated since then. Actually, many industrial and advanced applications benefit from the use of such techniques. Identification of LPV models has been approached since then in various ways, by using State Space (SS) [Lovera *et al.* (1998)], [Nemani *et al.* (1995)], [Lee and Poolla (1996)], Input/Output (I/O) [Bamieh and Giarré (1999)], [Previdi and Lovera (1999)], and set membership (SM) [Mazzaro *et al.* (1999)] approaches.

On the other hand, since the beginning of this century, the importance of switching and hybrid dynamic systems [Goebel *et al.* (2009)] has been recognized, relying also upon the fact that robust stabilizing control can be designed, and new identification techniques have been extensively considered in the literature, [Juloski *et al.* (2005)], [Bemporad *et al.* (2005)]. Considering these models, we ask to ourselves some questions: What are the major differences among the three models? Can one find a unified model that comprises all three? Can they be considered as a set of Linear Time Invariant (LTI) models with parameters varying in a set to be treated with standard robust control techniques or seen as simple LTI models and controlled with adaptive controllers? What are the peculiarities of such models? They are in a very strict connection with more sophisticated controllers, since they can guarantee less conservative performance than a unique worst-case robust controller and since modern control theory allows for the simultaneous achievement of system robustness, constraints fulfillment and (sub)optimal performance. Moreover, hybrid systems are dynamic systems consisting of components with continuous and discrete behavior and thus they require a new hybrid systems and control theory that has been developed in the past years and is still an active research topic.

Any identification scheme has the purpose not only to give the best approximating model, but also to provide the best model that is useful for the design of the controller. This is the reason why, when the control design theory advances, a parallel advance is also needed in the identification theory. In the present chapter, we investigate the above questions, presenting a review of the existing LPV identification methods, together with a motivating example of a hybrid LPV (HLPV) model. The main question we are focusing on is the following: is it possible to introduce a unified framework comprising existing identification methodologies or a new specific theory is required?

Since the topic of the present book is specifically to investigate LPV identification, we omit in the present chapter a review of existing methods for identification of hybrid and piece-wise affine models. A survey of the principal identification methods for hybrid systems can be found in [Paoletti *et al.* (2007)], [Juloski *et al.* (2005)] and [Juloski *et al.* (2006)].

The chapter is organized as follows. In Section 2, we present a literature review of LPV identification approaches, with particular attention to the I/O one. In Section 3, we formulate the problem for a HLPV model, with some motivations and attention to I/O models and its equivalence with SS representations. In Section 4, the HLPV model of the traffic flow in wireless ad-hoc networks is presented. Finally, in Section 5, we make some remarks and discuss open problems.

2. Literature review on LPV identification

In the following review, we are not deliberately listing all the LPV control papers that have appeared in the literature. Having in mind that any different identification methodology can be suitable to a particular control scheme, or can be useful to validate a possible model, we focus our attention only on the identification papers. Nevertheless, to understand the importance of such an approach, we recall the papers [Shamma and Athans (1991b)], [Rugh and Shamma (2000)], where the paradigm of LPV systems was introduced for gain scheduling. As reported in [Bamieh and Giarré (2001)], in many practical situations, a control system is subject to transitions among different conditions associated with variations of some fundamental parameters:

- physical modifications of part of the plant;

- modeling problems, i.e., when the linearization is applied to a system which may work in different operating points.

One approach to deal with is the *gain scheduling*, that is when the compensator parameters are chosen on-line as function of the plant parameters. A major drawback of linearization is that the control design based on the linearized dynamics does not necessarily exhibit good performance or even stability when operating away from the equilibrium. One possibility is to linearize along a trajectory which is not restricted to a local operating region. However, this trajectory must be known in advance in order to perform the control design, and such advance knowledge is often not available. *Gain Scheduling* is then a two-step procedure where: 1) design local linear controllers based on linearization of the nonlinear system at several different equilibria (operating conditions); 2) a global nonlinear controller for the nonlinear system is obtained by *interpolating* or *scheduling* among the local operations points. A useful paradigm to study gain scheduling is then the one of LPV systems. The terminology Linear Parameter time-Varying systems has been introduced in [Shamma and Athans (1991b)]. A discrete LPV system is represented in SS as

$$\begin{aligned} x_{k+1} &= A(p(k))x_k + B(p(k))u_k \\ y_k &= C(p(k))x_k + D(p(k))u_k \end{aligned}$$

The exogenous parameter $p(k)$ is assumed unknown *a priori*. However, it can be measured or estimated upon operation of the system. The question is then why distinguish LPV from Linear Time Varying (LTV) systems? In LTV systems, the time variations are known beforehand (as in periodic systems). In LPV systems, *a priori* assumptions on the scheduling parameter are bounds on its magnitude or rate of change. Rather than modeling the dynamic evolution of a particular variable, one can treat it as an exogenous independent parameter [Shamma and Athans (1991a)], [Shamma (1996)], and the survey [Rugh and Shamma (2000)], as well as references therein.

Since the beginning of the nineties, the literature on LPV Control and LPV Identification has started growing almost in parallel. The first attempt to solve the problem of LPV identification dates back to the paper of [Nemani *et al.* (1995)], where the Linear Fractional Transformation (LFT) LPV system identification problem is solved having a single time-varying block and SS measurements. The problem was shown to be equivalent to a linear regression, and certain conservative conditions for persistency of excitation were given. Then, the papers [Lee and Poolla (1996)] and [Lee and Poolla (1999)]; [Bamieh and Giarré (1999)] and [Bamieh and Giarré (2000)]; [Lovera *et al.* (1998)] and [Previdi and Lovera (1999)]; [Mazzaro *et al.* (1999)] have addressed the problem under different points of view. In [Lovera *et al.* (1998)], linear SS measurements were considered for LFT LPV system having a single time-varying block. In particular, the work of [Lee and Poolla (1996)] and [Lee and Poolla (1999)], address the problem of identifying a discrete LPV model with an LFT structure, that is $y_k = P(p_k, \theta)u_k + v_k$, where p_k is the measured time-varying parameter, θ the vector of parameters to be identified and an identification problem based on a prediction-error cost is solved via nonlinear programming. In [Previdi and Lovera (1999)], LPV models are identified with a hybrid linear/nonlinear procedure. The nonlinear part is identified through neural networks and the linear parametric part through a Least Squares (LS) algorithm. Finally, the approach in [Mazzaro *et al.* (1999)] describes a robust, worst-case identification approach. The proposed technique was also improved with an iterative procedure implementing a validation of the identified controlled model.

In [Bamieh and Giarré (1999)], [Bamieh and Giarré (2000)] and [Bamieh and Giarré (2002)], the problem of identifying I/O discrete-time LPV models of nonlinear or time varying systems has been solved, assuming that inputs, outputs and the scheduling parameters are measured, and a form of the functional dependence of the coefficients on the parameters is known. In those paper, the authors show how the identification problem can be reduced to a linear regression so that a Least Mean Square (LMS) and Recursive Least Squares (RLS) identification algorithm can be used. Conditions on the persistency of excitation in terms of the inputs and of the parameter trajectories are given to ensure the consistency of the algorithms when the functional dependence is of polynomial type. We report in details this contribution in Subsection 2.

Since the end of the last century, the literature on LPV has tremendously been spread and it is almost impossible to exhaustively list all the contributions in the present chapter, but the main methodologies can be clustered according to various paradigms and criteria.

According to [Casella and Lovera (2008)] and [Lovera (2008)], the literature can be divided in two main groups: identification methods based on SS vs. I/O models, referring to the structure of the identified model. When data are collected in a single experiment, where the parameter is also excited, the approach is referred to as *global*, whereas when data are obtained from multiple experiments with constant parameter values and the many LTI models obtained have to be interpolated is called *local.* Most of the SS approaches are based on the subspace identification method, [Viberg (1995)], [van Overschee and de Moor (1996)], because increasing demand for Multi Input Multi Output (MIMO) LTI and LPV models have also called for deeper study of SS and LFT modeling.

In the first group, SS, we can list: [Lovera and Mercere (2007)], where the subspace identification idea is combined with suitable properties of balanced SS realizations to identify MIMO parameter dependent model data generated by local identification experiments; [Verdult and Verhaegen (2002)], where a subspace identification for MIMO LPV systems with affine parameter-dependency is presented; [Verdult and Verhaegen (2005)], where an attempt to improve the curse of dimensionality is proposed; [van Wingerden and Verhaegen (2009)] where subspace identification for bilinear and LPV systems with affine parameter dependence operating under open- and closed-loop conditions is shown; [Felici *et al.* (2007)], where a new subspace identification approach, using a periodic scheduling sequence for the varying parameter, was suggested; [Lee and Poolla (1996)], [Lee and Poolla (1999)], where the parameter dependence of the MIMO LPV is rewritten as an LFT and the prediction-error cost is minimized via nonlinear programming; [DeCaigny *et al.* (2009)], where *local* LTI models are simply constructed interpolating SISO LPV SS models.

In the second group, I/O, we can list: [Bamieh and Giarré (1999)], [Bamieh and Giarré (2002)] and [Giarré *et al.* (2006)], where, as described in the next section, the LPV I/O system is recast in a pseudo-regressor form that can be identified via standard LMS algorithms; [Previdi and Lovera (2003)],[Previdi and Lovera (2004)], where a Nonlinear LPV (NLPV) model class is taken into account in analogy with *local* model networks and a separable least squares estimation algorithm is presented. Other more recent I/O approaches are: [Tóth *et al.* (2009a)], where I/O identification of a Single Input Single Output (SISO) model with *global* and *local* experiments based on orthonormal basis function approximation is formulated; [Laurain *et al.* (2010)], where an instrumental variable approach is used; [Tóth *et al.* (2010a)], just to cite one, and for an extensive reference of all contributions, [Tóth (2010)] and references therein, where the problem of system representation, equivalence among I/O and SS, possible different basis have been chosen and discretization problems have been analyzed in details. In particular, we recall [Tóth *et al.* (2007)], where possible pitfalls in representation are investigated: [Abbas *et al.* (2010)], where the SS realization of an LPV I/O model is described and [Tóth *et al.* (2009b)], where the behavioral approach has been introduced to construct a representation-free description of LPV systems.

Another possible categorization is continuous vs. discrete time identification, referring to the type of considered data, or statistical vs. deterministic noise. In this last group, we need to consider the Unknown But Bounded (UBB) framework, and the Set Membership approaches. In the latter group, UBB and SM, we list: [Cerone and Regruto (2008)], where polytopic outer approximation of the feasible parameter set and the parameters uncertainty intervals of the LPV model are computed solving an Error In Variable problem; [Belforte and Gay (2004)] where optimal worst case input sequences for LPV Finite Impulse Response (FIR) models under bounded disturbances have been determined; [Sznaier and Mazzaro (2003)], where a control-oriented identification framework for a class of LPV systems is studied, together with a methodology for validating the model, that is solved in terms of linear matrix inequality feasibility problems; [Ruiz *et al.* (2010)], where a new approach to filter design for LPV systems is proposed, based on the direct design of the filter from data, avoiding the intermediate step of model identification, within a SM framework. As [Ruiz *et al.* (2010)], also other papers have considered the nonparametric approach to LPV identification. For example, [Hsu *et al.* (2008)], where the identification of LPV systems is solved by taking a nonparametric approach, and not imposing any *a priori* parameterizations on the model structure, nor assuming that a natural parameterization is suggested from an analytical understanding of the underlying process; [Estrada *et al.* (2008)], where a new identification approach is proposed, whose main advantage lies in the fact that it presents a compact and precise nonlinear model with a small number of parameters, yielding to the time evolution of the scheduling variable that explains the nonlinear behavior of the system; [Bianchi and Sanchez-Pena (2010)], [Mazzaro *et al.* (1999)], where a robust LPV identification/invalidation method is presented in SS. Validation of an LPV/LFT framework is also considered in [Szabó *et al.* (2011)], where a data-based tool is presented as a step towards industrialization of these techniques.

Finally, we can cite another important set of contributions as application oriented identification methods or as identification for control. The application fields are so many that a complete list is not viable, so we recall here only some of the most recent ones. The list, that is far to be exhaustive, underlines how many different fields can benefit from the use of LPV model description: performance of web-services [Qin and Wang (2007b)], [Qin and Wang (2007a)]; inverted pendulum [Boonto and Werner (2008)]; turbo charged diesel engine [Salcedo and Martinez (2008)]; open-channel irrigation [Duviella *et al.* (2010)], [Blesa *et al.* (2010)]; air path systems of diesel engines [Wei and del Re (2007)]; environmental modeling [Beflorte *et al.* (2005)]; glass manufacturing [Wattamwar *et al.* (2010)]; heated and cooled iron solid square plate [Bos *et al.* (2009)]; continuous stirred tank reactor [Tóth *et al.* (2010b)]; motion systems [Steinbuch *et al.* (2003)]; high pressure natural gas transportation networks [Lopes dos Santos *et al.* (2011)]; stall and surge phenomena of compressors [Giarré *et al.* (2006)]; and many others fields such as vehicles, aircrafts, motorcycles, web-services, highway-traffic as described in the special issue [Lovera *et al.* (2011)] and articles therein.

2.1. *I/O LPV Identification [Bamieh and Giarré (2002)]*

We briefly recall now the I/O method presented in [Bamieh and Giarré (2002)]. The class of discrete-time LPV models is parameterized as follows

$$M(\delta,p)=\mathscr{A}(\delta,p)^{-1}\mathscr{B}(\delta,p), \tag{1}$$

where δ is the delay operator i.e. $(\delta\ y)_k\ :=\ y_{(k-1)}$, and

$$\begin{aligned}\mathscr{B}(\delta,p) &:= b_0(p)+b_1(p)\delta+\ldots+b_{n_b}(p)\delta^{n_b}\\ \mathscr{A}(\delta,p) &:= I+a_1(p)\delta+\ldots+a_{n_a}(p)\delta^{n_a},\end{aligned} \tag{2}$$

$n=n_a+n_b+1$ is the number of parametric functions to be identified and the varying parameter p is a function of discrete time ($p=p(k)$). The fractional representation in (1) is short hand for the operational notation

$$\mathscr{A}(\delta,p)\ y(k)\ =\ \mathscr{B}(\delta,p)\ u(k), \tag{3}$$

where y is the output and u is the input, and we assume, for the sake of simplicity, that $y\in\mathbb{R}$ and $u\in\mathbb{R}$. We assume that the functions $\{a_i\}$, $\{b_j\}$ in (2) are *linear combinations* of a set of known fixed basis functions $\{f_1,\ldots,f_N\}$, e.g.

$$a_1(p)=a_1^1f_1(p)+\ldots+a_1^Nf_N(p), \tag{4}$$

where the constants $a_1^\ell,\quad \ell=1,\ldots,N$ are real numbers. Thus any particular model in our class is completely characterized by the real numbers $\{a_i^\ell\}$ and $\{b_j^\ell\}$. The goal of a parametric identification scheme is then to find these constants from data. The right basis functions may not be known, however, in [Bamieh and Giarré (2002)] the polynomial dependence was considered as an example, but other choices have been presented and used in the literature, i.e., the functions $f_l(p)$ are powers of p

$$f_l(p) = p^{l-1} \quad l = 1, \ldots, N.$$

In this case, the coefficient functions become

$$\begin{aligned} a_i(p) &= a_i^1 + a_i^2 p + \ldots + a_i^N p^{N-1} \\ b_j(p) &= b_j^1 + b_j^2 p + \ldots + b_j^N p^{N-1}. \end{aligned} \tag{5}$$

Defining an $n \times N$ matrix Θ which contains all the coefficients to be identified:

$$\Theta := \begin{bmatrix} a_1^1 & \ldots & a_1^N \\ a_2^1 & \ldots & a_2^N \\ \vdots & \vdots & \vdots \\ a_{na}^1 & \ldots & a_{na}^N \\ b_0^1 & \ldots & b_0^N \\ \vdots & \vdots & \vdots \\ b_{nb}^1 & \ldots & b_{nb}^N \end{bmatrix}$$

and the *extended regressor* Ψ, which will be made up of past I/O data and parameter trajectories

$$\Psi_k := \phi_k \pi_k := \begin{bmatrix} -y_{k-1} \\ \vdots \\ -y_{k-n_a} \\ u_k \\ \vdots \\ u_{k-n_b} \end{bmatrix} \left[f_1(p_k) \; f_2(p_k) \; \ldots \; f_N(p_k) \right],$$

where the subscript for the time index is used (e.g. $p_k := p(k)$), and with other definitions given in [Bamieh and Giarré (2002)], the model (2)-(4) can now be rewritten as

$$y_k = \langle \Theta, \Psi_k \rangle , \tag{6}$$

Now, the problem has been recast in a pseudo-regressor form and standard LMS can solve the related identification problem, as reported in the referred paper, together with persistency of excitation conditions. Other persistency conditions are given in [Wei and del Re (2006)]. We now ask ourselves if such a framework can be extended and used also for hybrid systems. First we notice that in the above model no *a priori* assumption on the variation of p is stated, thus allowing us to use the same regressor also for switching or even stochastically switching p. Second, the considered discrete model was not derived by a continuous discretized one, or from a SS model, but directly from a discrete one. However, equivalence among SS and I/O and proper discretization have been extensively discussed in the literature, [Tóth (2010)]. Thus we notice that it is also easy to extend the above regressor to include the past terms of p_k, such as $p_{k-1}, p_{k-2}, ..$, as required.

3. HLPV modeling: Problem formulation

LPV models are an appealing class of models because they can capture the behavior of complex systems allowing the use of powerful LPV control synthesis tools. In Section 2 a detailed literature review on the identification and modeling problem related to LPV models have been presented. In addition to what we reported, we recall that another reason of interest in LPV models is the possibility of allowing a linear-like geometric theory of control systems [Bokor and Balas (2005); Tóth (2010)]. In [Bokor and Balas (2005)] it is shown that, using affine LPV representations, various analysis and control problems can be solved by linear algebraic manipulations similarly to LTI systems, provided the scheduling variables satisfy some particular condition. Recently a rigorous and extensive analysis of a broader class of LPV systems [Tóth (2010)] has been carried out showing that many system theoretic concepts can be characterized by simple algebraic manipulations. For these reasons, rewriting the original nonlinear system of equations into LPV form is particularly desirable in many application areas. Many input affine nonlinear systems and switched linear systems can also be rewritten in the form of qLPV systems once a suitable unknown but measurable vector parameter has been defined (for instance as a portion of the state and/or input variables). In [Lim and Chan (2003)] and [Zhang *et al.* (2008)] the analysis of stability of HLPV systems and respectively, the model reduction of Switched Linear Parameter-Varying (SLPV) systems is presented. Such models are represented in a SS formalism as follows:

$$\begin{aligned} x_{k+1} &= A_{\sigma_k}(p_k)x_k + B_{\sigma_k}(p_k)u_k + w_k \\ y_k &= C_{\sigma_k}(p_k)x_k + D_{\sigma_k}(p_k)u_k + v_k \\ \sigma_k &\in \mathscr{S}, \end{aligned} \tag{7}$$

with state variable $x \in \mathbb{R}^{n_x}$, input $u \subset \mathbb{R}^{n_u}$, output $y \in \mathbb{R}^{n_y}$ and scheduling parameter $p \in \mathbb{R}^{n_p}$. In the above model the switching signal σ_k is an arbitrary sequence taking values in $\mathscr{S} \triangleq \{1,\ldots,s\}$, where s is the number of LPV models and solutions that can be related to the input u_k or the state x_k or some other external input. Correspondingly, we get the continuous matrix-valued functions $A_\sigma(p) : \mathbb{R}^{n_p} \to \mathbb{R}^{n_x \times n_x}$, $B_\sigma(p) : \mathbb{R}^{n_p} \to \mathbb{R}^{n_x \times n_u}$, $C_\sigma(p) : \mathbb{R}^{n_p} \to \mathbb{R}^{n_y \times n_x}$ or $D_\sigma(p) : \mathbb{R}^{n_p} \to \mathbb{R}^{n_y \times n_u}$, where $\sigma \in \mathscr{S}$.

In the hybrid systems literature (see [Lunze (2009)] and references therein) the evolution of the discrete state has been described in various ways, and σ_k can be an arbitrary, deterministic and finite-valued exogenous input (Jump Linear models (JL)) or a stochastic signal generated by an irreducible Markov chain, (Markov Jump Linear models (MJL)).
In PWA, [Sontag (1981)], σ_k is given by

$$\sigma_k = i \quad \text{iff} \begin{bmatrix} x_k \\ u_k \end{bmatrix} \in \Omega_i, \quad i = 1,\ldots,s, \tag{8}$$

where Ω_i is a complete partition of the state input-domain.
In MJL models the evolution of the discrete state is commonly described by a first order Markov chain as follows for all $k \geq 0$

$$\begin{aligned} \pi_{j,i}(k+1,k,\gamma) &= \mathbb{P}\left[\sigma_{k+1} = j | \sigma_k = i, \gamma\right] \\ \rho_i(0,\gamma) &= \mathbb{P}\left[\sigma_0 = i \mid \gamma\right] \end{aligned} \tag{9}$$

where the initial probability and transition probabilities are known functions of a (potentially unknown) parameter vector $\gamma \in \mathbb{R}^s$.

3.1. *Hybrid LPV motivation*

These systems are motivated by practical applications such as active magnetic bearing system [Lu and Wu (2004)], missile autopilot system [Lim and How (2003)], F-16 aircraft system [Lu *et al.* (2006)]. In [Bianchi and Sanchez-Pena (2010)] a novel design procedure for switched LPV controllers has been presented, with applications to magnetic bearing systems. In [Preitl *et al.* (2008)] a PWA quasi LPV model of a Hybrid Solar Vehicle is developed, with an analysis of the controllability and stability. In [LeBel and Rodrigues (2008)], a controller for PWA Parameter Varying (PWAPV) systems was applied to the problem of path following of a wheeled mobile robot.

The interest in Switching LPV (SLPV) models is strongly justified from the control design point of view to improve drastically the performance. Indeed SLPV models give a more accurate description of the nonlinear dynamics (see for instance [Lim and How (2003)]) than the LFT or affine LPV models, suitable for advanced nonsmooth design techniques of gain scheduled controllers. These advanced techniques automatically select a parameter dependent Lyapunov function (conversely to the standard approaches used in LPV design methods), by considering the parameter-dependency of the system dynamics and giving rise to a problem with a finite number of Linear Matrix Inequalities (LMIs). Hence this modeling approach generates models presenting computational and performance advantages in terms of control design. Moreover, as noticed in [Farhood and Feron (2009)], HLPV systems are related to non stationary LPV systems. The reference trajectories do not have to be pre-specified as long the reference states and controls are within the covered SS region.

3.2. *I/O and SS model representations*

Both in the Hybrid systems and LPV literature the problem of equivalence among I/O and SS representation, the notion of minimality and the notion of observability have been investigated. In particular, in [Paoletti *et al.* (2010)] PWA SS models and PWA Auto Regressive eXogenus (PWARX) models were analyzed and their equivalence studied; in [Paoletti *et al.* (2008)], the equivalence between Switched Affine State Space (SWA) models and Switched affine ARX (SARX) was discussed, while in [Tóth *et al.* (2007); Tóth (2010)] LPV_{IO} I/O representation and LPV_{SS} SS representation was studied.

Following the discussion in [Paoletti *et al.* (2008, 2010); Tóth *et al.* (2007); Tóth (2010)] a meaningful I/O representation of an HLPV model is the following one:

Definition $\mathbf{HLPV}_{IO}$

$$y(k) = -\sum_{i=1}^{n_a}(a_i \diamond \rho_k)y(k-i) + \sum_{j=0}^{n_b}(b_j \diamond \rho_k)u(k-j-d) + g(\rho_k), \qquad (10)$$

where $\rho_k = [p_k, \sigma_k]$, and $\diamond$ denotes the evaluation of a function f over the trajectory of ρ_k, i.e. $f \diamond \rho_k = f(p_k, p_{k-1}, p_{k-2}, \dots p_{k-\hat{n}}, \sigma_{k-\hat{n}}, \dots, \sigma_k)$ (see [Tóth (2010)]) with $\hat{n} \triangleq \max\{n_a, n_b\}$ and g a function taking into account the affine terms. In some cases, the selected I/O description can be simplified considering special forms of dynamic dependence on p, [Abbas *et al.* (2010)] allowing SS realizations without introducing dynamic dependence on p. The selected I/O description is appealing to describe the behavior of many practical applications and to derive a corresponding SS representation, if any. The importance of subsequently determining a suitable SS representation is due to the fact that many efficient control techniques are formulated using SS information. The advantage of a parameter varying representation with respect to artificial neural networks or NARMAX is given by the possibility to interpret the behavior of a complex system as the effect of a linear model and a scheduling variable describing nonlinearity. Allowing switching and $\mathscr{C}^1$ scheduling variables gives the opportunity to describe a broader class of systems than classical LPV and the property of PWA maps to be a universal approximator is inherited and strengthen by the model (10). Moreover the description (10) can be useful to perform closed loop identification whenever a switching controller is exploited. The main drawback is the possibly high number of models involved in (10). Thus, an important and interesting question in the solution of the identification problem is how to design classes of inputs allowing to handle a low number of modes, [Paoletti *et al.* (2010)].

An LPV model is characterized by the measurability of the scheduling variables. Then, under the assumption of measurability of ρ_k, it would be possible to proceed in the solution of the identification problem. Indeed, for several applications it is possible to include additional information concerning the scheduling variables. Such additional information can be fruitfully used in order to achieve a good approximating structure maintaining its simplicity, [Estrada *et al.* (2008); Previdi and Lovera (2004); Cinquemani *et al.* (2007)]. For instance, in [Estrada *et al.* (2008)] the time evolution of the scheduling variable p_k is described in terms of the measured output and input of the system and explains the nonlinear behavior of the systems in a compact form.

4. A motivating example: Traffic modeling in wireless ad-hoc networks

Wireless ad-hoc networks consist of a number of untethered nodes able to communicate with each other by means of intermediate nodes, collaboratively forwarding ongoing traffic. Because of the nature of the wireless medium, data communications in ad-hoc networks are intrinsically broadcast, so that links exist between any pairs of nodes that are within the transmission range of each other. These features make ad-hoc networks easy to be deployed and suitable for a large number of applications, spanning from low-range sensor networks targeted to distributed monitoring, to high-range mesh networks targeted to build infrastructure-less transport networks.

In a recent paper [Giarré *et al.* (2011)], a single channel radio network has been studied, made of a set, V, of nodes distributed uniformly over a given area, and some form of coordination among nodes has been introduced to significantly improve the transport capacity of the network, with respect to the standard simple uncoordinated Carrier Sense Multiple Access (CSMA) protocols. A simple form of coordination can be provided by pre-allocating temporal intervals in which different sets of nodes are allowed to access the shared wireless medium. Indeed, this mechanism avoids collisions due to hidden-nodes, thus reducing channel and energy wastes. The approach is based on distributed coloring algorithms introducing cooperation mechanisms among the nodes and can be customized as a function of the network topology and traffic load. Synchronization among nodes is also studied in [Tinnirello *et al.* (2010)]. Each node $i \in V$ can communicate only with a subset of adjacent nodes V_i. We say that i is (radio) *visible* only to the nodes in V_i. Differently, i is *hidden* to the remaining nodes in $V \setminus V_i$.

The network structure can be represented through an edge labeled graph $G = (V, E)$. Specifically, the node set V includes all the nodes i of the network and the edge set E includes all the pairs of adjacent nodes (i, j) that are in radio visibility of each other. We call each edge $(i, j) \in E$ as link ℓ, that is labeled with its maximum transmission rate B_{max} reached only when we are in the best condition. $B^\ell \leq B_{max}$ is the effective transmission rate, that is time variable and depends on nearby radio conditions, and among the other factors, is a function of the distance between nodes i and j and of the possible presence of obstacles. The channel traffic load can be modeled as follows, taking into account and extending the model originally proposed by [Lee *et al.* (2007)]. For a given topology, some of the nodes are considered sources and some are considered the final receivers. The routing of the flows has been set, so that for each flow the path is pre-assigned. A possible example is depicted in Figure 1, where nodes 1, 2 and 3 are the sources and nodes 24, 29 and 30 are the final receivers.

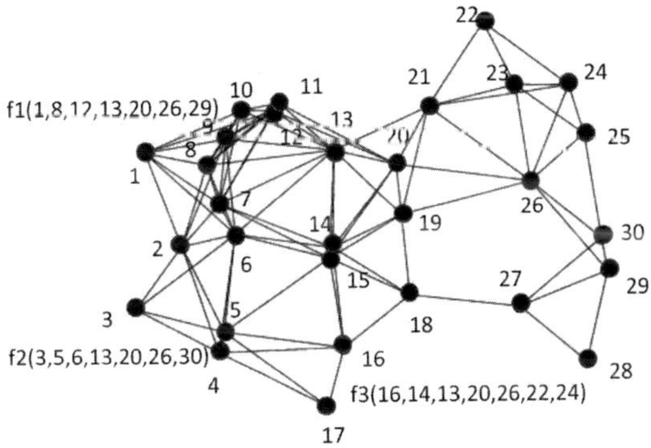

Fig. 1. An example of a network with 30 nodes and 3 flows (nodes 1, 2 and 3 are the sources, and nodes 24, 29 and 30 are the final receivers).

In this communication network we consider $\ell = 1, \ldots n_\ell$ links crossed by n flows $f_1, f_2, \ldots, f_n$. We assume all links to be unidirectional. Every link has a queue that, when necessary, holds packets before transmission. If bytes in the queue reach a maximum value (that may be different for each link), incoming packets will be dropped. Queue dynamics in a link ℓ in the path of a generic flow f can be represented by:

$$\dot{q}_f^\ell = s_f^\ell - d_f^\ell - r_f^\ell,$$

where r_f^ℓ is the flow transmission rate through the link ℓ. s_f^ℓ is the ℓ link/f flow arrival rate that (i.e. the rate of the flow packets arriving at the node j for the link $\ell = (j, k)$). We assume the hypothesis of flow conservation in which we neglect drops that occur in transmission medium, so s_f^ℓ is equal to the transmission rate of the flow at the previous link, or to flow sending rate if the flow starts at link ℓ. d_f^ℓ denotes the flow drop rate at link ℓ. For each link we define vectors whose components refer to each flow in the link:

$$\begin{aligned}
q^\ell &= [q_{f1}^\ell \; q_{f2}^\ell \; \cdots \; q_{fn}^\ell]' \\
s^\ell &= [s_{f1}^\ell \; s_{f2}^\ell \; \cdots \; s_{fn}^\ell]' \\
d^\ell &= [d_{f1}^\ell \; d_{f2}^\ell \; \cdots \; d_{fn}^\ell]' \\
r^\ell &= [r_{f1}^\ell \; r_{f2}^\ell \; \cdots \; r_{fn}^\ell]'.
\end{aligned}$$

In this model the input variable is s^ℓ, the state variable q^ℓ, and the output is the transmission rate at the link ℓ, $y = r^\ell$. We rewrite the queue dynamics model as:

$$\dot{q}^\ell = s^\ell - h(s^\ell) - k(s^\ell, q^\ell)$$
$$y = \left[r_f^\ell\right]'$$

where $d^\ell = h(s^\ell)$ and $r_f^\ell = k(s^\ell, q^\ell)$.

Queue dynamics can be described with a hybrid system in which, according to [Lee *et al.* (2007)], three different models can be separately modeled, depending of the value of $\sum_{i=1}^n q_{fi}^\ell$ and $\sum_{i=1}^n s_{fi}^l$. These are the parameters that allow us to transit from a state to another. In each case the expressions for h and k will be different:

- $\sum_{i=1}^n q_{fi}^\ell = 0$ - **Queue Empty**: in this case there are no drops and all incoming packets are transmitted. In this case we have

  $$\dot{q}^\ell = s^\ell - k(s^\ell),$$

  where

  $$k_{fi}(s^\ell) = \begin{cases} s_{fi}^\ell, & \sum_{i=1}^n s_{fi}^\ell \le B^\ell \\ B^\ell \frac{s_{fi}^\ell}{\sum_{i=1}^n s_{fi}^\ell}, & \sum_{i=1}^n s_{fi}^\ell > B^\ell; \end{cases}$$

- $0 < \sum_{i=1}^n q_{fi}^\ell < q_{max}^\ell$ or $\sum_{i=1}^n q_{fi}^\ell = q_{max}^\ell$ and $\sum_{i=1}^n s_{fi}^\ell \le B^\ell$ - **Queue neither empty nor full**: also in this case there are no drops but the packets are transmitted at the head of the queue (B^ℓ bytes per unit of times). In this case we can rewrite the dynamics as:

  $$\dot{q}^\ell = s^\ell - k(q^\ell),$$

  where

  $$k_{fi}(q^\ell) = B^\ell \frac{q_{fi}^\ell}{\sum_{i=1}^n q_{fi}^\ell}$$

- $\sum_{i=1}^n q_{fi}^\ell \ge q_{max}^\ell$ and $\sum_{i=1}^n s_{fi}^\ell > B^\ell$ - **Queue full**: in this case the packets are transmitted at the head of the queue as in the previous case, but all the packets that exceed the maximum value of the queue will be dropped. In this case:

  $$\dot{q}^\ell = s^\ell - h(s^\ell) - k(q^\ell),$$

where

$$h(s^\ell) = s^\ell - B^\ell \frac{s_{fi}^\ell}{\sum_{i=1}^n s_{fi}^\ell}$$

$$k_{fi}(q^\ell) = B^\ell \frac{q_{fi}^\ell}{\sum_{i=1}^n q_{fi}^\ell}.$$

Interest in the identification and validation of this model derives from the importance of on-line estimating the traffic load and the effective value of the bandwidth in the network, to implement a distributed optimization enhancing performance in multi-hop transmissions.

The aim of this paper is to show how to simply recast such a hybrid model as an HLPV. Note that in this example we describe a simple deterministic hybrid model, while in the literature stochastic hybrid systems with applications to communication networks have been also introduced, [Hespanha (2005)] in which transitions between discrete modes are triggered by stochastic events. In future works, we will focus our attention also on stochastic HLPV systems.

Hereafter, we consider a single link ℓ and we assume that the number n of data flows is known and fixed to its maximum, which is equal to the number N_s of sources. Note that this is a restrictive assumption because in general n can also be time-varying, $n \in [0, N_s]$. We drop the superscript ℓ for the sake of clarity. Such a hybrid model can be recast as a HLPV model, by defining $p1 := \sum_{i=1}^n q_{fi}$ and $p2 := \sum_{i=1}^n s_{fi}$ and assuming that the parameter $p = [p1 \quad p2]'$ is the measurable varying parameter. This parameter takes into account an aggregated value of the queue and the total incoming flow rate to the link. For each link the input is the n-dimensional vector $u = s$, the state is the n-dimensional $x = q$ and the measured output is given by the n-dimensional vector $z = [z_i]$ where

$$z_i = y_i + v, \quad i = 1, \dots, n,$$

and v is a measuring noise equally distributed on each flow f_i. Moreover, we assume that the theoretical value of the band B_{max} and the maximum value of the queue q_{max} are known. The effective value assumed by the band is instead assumed unknown and we refer to it as $b_w \in [B_{min}, B_{max}]$. The condition of an empty queue corresponds to $p1 < \epsilon$, with ϵ sufficiently small. Clearly, the parameters p need to satisfy some constraints as follows:

C.1 $p2 \geq p1 + B_{max}$, if $p1 \geq \epsilon$

C.2 $p2 \leq B_{min}$, if $p1 < \epsilon$.

Numerical simulations have been carried out to gather data with $B_{min} = 6Mbit$, $B_{max} = 54Mbit$, $\epsilon = 0.0001$, and $q_{max} = 0.03Mbit$. The considered network is given in Figure 1, with $n = 3$ and the traffic flow in the link $\ell = (13, 20)$ is modeled. The system is simulated in continuous time, but parameters are kept constant during sampling time and data have been collected at each sampling time interval. The parameter p has been chosen to excite all possible states and satisfy the above constraints C.1 and C.2,

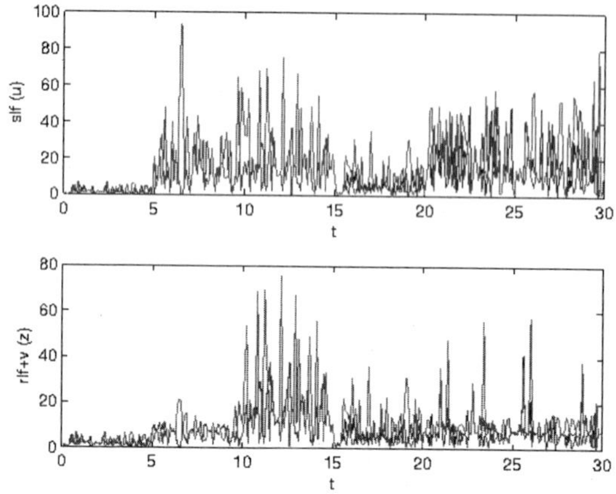

Fig. 2. I/O data over an interval of $30s$ and sampling time T=0.1s for each flow $i = 1, 2, 3$.

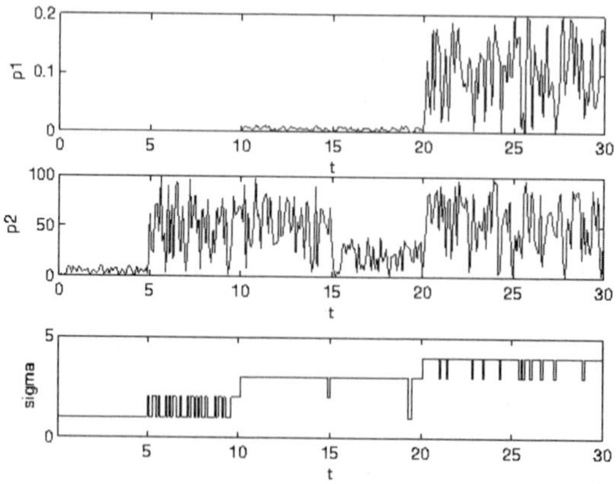

Fig. 3. Parameters $p1$, $p2$ and σ over an interval of $30s$ and sampling time $T = 0.1s$.

as shown in Figure 3. The corresponding simulated data are shown in Figure 2, for each flow ($i = 1, 2, 3$). The measured output is obtained by adding a uniform random noise v with unitary variance to the output y. The simulation duration is $30s$ and the sampling time is $T = 0.1s$.

The above hybrid model assume the following $HLPV_{SS}$ form:

$$\begin{aligned} \dot{x} &= A(\rho)x + B(\rho)u \\ y &= C(\rho)x + D(\rho)u \\ z &= y + v \end{aligned} \tag{11}$$

where $\rho = [p' \quad \sigma]'$ is the parameter vector and the switching parameter σ is defined as follow:

$$\sigma = \sigma_k = j, \quad j = \{1, 2, 3, 4\} \iff p \in \Omega_j$$

$$\Omega = \begin{cases} \Omega_1 \text{ if } (p1 \leq \epsilon \quad \text{AND} \quad p2 \leq b_w) \\ \Omega_2 \text{ if } (p1 \leq \epsilon \quad \text{AND} \quad p2 > b_w) \\ \Omega_3 \text{ if } (\epsilon < p1 < q_{max}) \qquad \text{OR} \quad (p1 = q_{max} \quad \text{AND} \quad p2 \leq b_w) \\ \Omega_4 \text{ if } (p1 > q_{max}) \qquad \text{OR} \quad (p1 = q_{max} \quad \text{AND} \quad p2 > b_w). \end{cases} \tag{12}$$

Note that the above regions are depending on the unknown parameter b_w. A possible first approximation is the following where the known maximum value of the band B_{max}, instead of the effective value b_w is used.

$$\Omega = \begin{cases} \Omega_1 \text{ if } (p1 \leq \epsilon \quad \text{AND} \quad p2 \leq B_{max}) \\ \Omega_2 \text{ if } (p1 \leq \epsilon \quad \text{AND} \quad p2 > B_{max}) \\ \Omega_3 \text{ if } (\epsilon < p1 < q_{max}) \quad \text{OR} \quad (p1 = q_{max} \quad \text{AND} \quad p2 \leq B_{max}) \\ \Omega_4 \text{ if } (p1 > q_{max}) \quad \text{OR} \quad (p1 = q_{max} \quad \text{AND} \quad p2 > B_{max}) \end{cases} \tag{13}$$

With these definitions, the effective Ω_1 and Ω_3 in (12) are bigger than the approximating regions given in (13) and the Ω_2 and Ω_4 are smaller, depending on the realization of the parameter b_w. There are other possible modeling choices, varying also on how much *a priori* information we want to use in the identification method, as we will discuss soon.

Notice that all the matrices are diagonal matrices with the same entries for $i = 1, \ldots, n$: $A = diag(a_i) = diag(a)$, $B = diag(b_i) = diag(b)$, $C =$

$diag(c_i) = diag(c)$ and $D = diag(d_i) = diag(d)$ with:

$$a(\rho) = \begin{cases} 0 & \text{if } j = 1,2 \\ -\frac{b_w}{p1} & \text{if } j = 3,4 \end{cases}$$

$$b(\rho) = \begin{cases} 0 & \text{if } j = 1 \\ 1 - \frac{b_w}{p2} & \text{if } j = 2 \\ 1 & \text{if } j = 3 \\ \frac{b_w}{p2} & \text{if } j = 4 \end{cases}$$

$$c(\rho) = \begin{cases} 0 & \text{if } j = 1,2 \\ \frac{b_w}{p1} & \text{if } j = 3,4 \end{cases}$$

$$d(\rho) = \begin{cases} 1 & \text{if } j = 1 \\ \frac{b_w}{p2} & \text{if } j = 2 \\ 0 & \text{if } j = 3,4. \end{cases}$$

Because the above SS model is diagonal and the entry terms are all the same, a first order scalar SS model for each single flow, $\Sigma = \{a(\rho), b(\rho), c(\rho), d(\rho)\}$ can be taken into account by itself.

As mentioned before in the chapter, both the discretization and the problem of equivalence of representations, for both Hybrid and LPV systems, is a topic of research addressed in various papers, [Tóth (2010)], [Paoletti *et al.* (2010)], and for HLPV models is a key problem that needs more investigation and will be argument for future research. For the present example, the parameters are kept constant during the sampling time interval, so that the switching can occur only at the sampling time instant and then, both the I/O representation and the discrete-time model can be easily derived. The corresponding contonuous time input-output model is obtained simply by deriving the output expression and substituting the derivative term $\dot{x}$:

$$\gamma(\rho)\dot{y}_i(t) = \alpha(\rho, \dot{\rho})y_i(t) + (\beta(\rho))u_i(t)$$

where

$$\gamma(\rho) = \begin{cases} 0 \text{ if } j = 1,2 \\ 1 \text{ if } j = 3,4 \end{cases}$$

$$\alpha(\rho, \dot{\rho}) = \begin{cases} c(\rho) & \text{if } j = 1,2 \\ a(\rho) - \frac{\dot{p1}}{p1} & \text{if } j = 3,4 \end{cases}$$

and

$$\beta(\rho) = \begin{cases} d(\rho) & \text{if } j = 1, 2 \\ c(\rho)b(\rho) & \text{if } j = 3, 4 \end{cases}$$

The discretization of the former model, via a backward Euler approximation with a ZOH, gives rise to a simple HLPV$_{IO}$:

$$\mathscr{A}(\delta, \rho_k, \rho_{k-1})y_i(k) = \mathscr{B}(\delta, \rho_k)u_i(k), \tag{14}$$

where $\mathscr{A}(\delta, \rho_k, \rho_{k-1}) = a_0(\rho_k, \rho_{k-1}) + a_1(\rho_k)\delta$ and $\mathscr{B}(\delta, \rho_k) = b_0(\rho_k)$ and the discrete time parameters are related to the continuous time parameters as follows:

$$a_0(\rho_k, \rho_{k-1}) = \begin{cases} 1 & \text{if } j = 1, 2 \\ (2 - Ta(\rho_k))p1_k - p1_{k-1} & \text{if } j = 3, 4 \end{cases}$$

$$a_1(\rho_k) = \begin{cases} 0 & \text{if } j = 1, 2 \\ p1_k & \text{if } j = 3, 4 \end{cases}$$

$$b_0(\rho_k) = \begin{cases} d(\rho_k) & \text{if } j = 1, 2 \\ Tc(\rho_k)b(\rho_k)p1_k & \text{if } j = 3, 4 \end{cases}$$

Clearly, the functional dependence on ρ_k, ρ_{k-1} is unknown and is one of the object of the identification, as described in Section 2. Moreover, model (14) can be easily recast with the structure given in (10) as

$$y(k) = (\alpha_1 \diamond \rho_k)y(k-1) + (\alpha_0 \diamond \rho_k)u(k),$$

with $(\alpha_1 \diamond \rho_k) = \alpha_1(p1_k, p1_{k-1}, \sigma_k)$ and $(\beta_0 \diamond \rho_k) = \beta_0(p1_k, p2_k, p1_{k-1}, \sigma_k)$.

Depending on the available *a priori* information on the model, we can suggest now some possible different identification solutions.

1) *Off line identification.* Assume that the regions are known. Design the identification experiment choosing the input at the sources, measuring $p2$ at each link and selecting the variation of $p1$ satisfying the constraints $C.1$ and $C.2$ and the persistence of excitations conditions as in [Bamieh and Giarré (2002)]. Divide the data according to each region j and use the standard I/O LPV identification method, recasting model (14) in a pseudo-regressor form as described in Subsection 2.1 and identify each LPV model corresponding to a value of j.

2) *On line identification.*

a) Assume that the regions are known. Design the identification experiment choosing the input at the sources, measuring $p2$ at each link, and selecting the variation of $p1$ satisfying the constraints $C.1$ and $C.2$ and the persistence of excitations conditions as in [Bamieh and Giarré (2002)], and, at the same time, guaranteeing that the parameter σ is kept constant in a sufficiently long time to guarantee convergence of the algorithm. Then, use the standard I/O LPV identification method, recasting model (14) in a pseudo-regressor form as described in Subsection 2.1 and identify each LPV model.

b) Assume that the regions are not known, but the number $s = 4$ is known. Design an identification experiment as in 2a). *A posteriori*, separate each LPV model, from the identified one.

3) *Two step identification.* Assume that the approximating regions need to be validated. First, perform a validation of the *a priori* information on p, to find approximated regions and validate them. Second, run one of the previous identification scheme 1) or 2). This solution is still under investigation.

4) *Global experiment.* In case we would like to perform an on line identification and the assumption that σ is kept constant in a sufficiently long time cannot be guaranteed, or if the model is more general than the one obtained in the present example, then a global experiment for the model given in (10) needs to be performed. A suitable identification algorithm determining in one-shot both the functional dependence on p and on σ needs to be performed. This general solution is still under investigation.

5. Final remarks and open problems

In the previous sections the definition of an HLPV model has been illustrated. HLPV models inherit properties of LPV and PWA affine systems with the addition of some knowledge on the scheduling variables. This kind of models are motivated by recent works as discussed in Subsection 3.1. In Section 4, we have highlighted some undergoing research topics to set and solve well posed and significant identification problems. However, there are still other important issues and open problems that need to be faced.

- The selection of an appropriate model structure describing a physical system is currently a relevant challenging problem in the modeling and identification field. Once the choice of using an HLPV has been performed, the selection of the structure, based on *a priori* information and data-based tests, becomes an open problem. Moreover, it is of great interest to investigate which are the advantages and disadvantages of exploiting an HLPV model with respect to other model classes describing nonlinear behaviors. In this context, the study of HLPV approximating properties is of interest.
- The equivalence transformations between SS and I/O models and the geometric properties for the PWA model and LPV models are still an active area of research. For HLPV models the previous properties need to be further investigated, enhancing the role of the additional available information.
- Data experiment design to guarantee identifiability and persistent excitation conditions is a fundamental issue in model identification and for complex systems is a challenging problem. Here, it becomes of important to investigate how to extend and generalize the conditions proposed in [Bamieh and Giarré (2002); Wei and del Re (2006); Khalate *et al.* (2009); Vidal and Anderson (2004)].
- A rigorous discretization theory is a critical issue for well posed identification problems. Devising efficient identification techniques for HLPV needs to be widely explored under different assumptions on the model parameters. In particular, the stochastic setting appears to be appealing in application contexts [Lygeros and Prandini (2010); Cinquemani *et al.* (2007); Hespanha (2005)].

Acknowledgments

We acknowledge B. Bamieh for useful discussions on the use of LPV identification techniques for hybrid systems and for his contribution in developing the LPV I/O identification algorithms and I. Tinnirello for useful discussions on wireless ad-hoc networks.

References

Abbas, H., Tóth, R. and Werner, H. (2010). State-space realization of LPV input-output models: practical methods for the user, in *Proceedings of the American Control Conference* (Baltimore, MD, USA).

Bamieh, B. and Giarré, L. (1999). Identification for linear parameter varying models, in *Proceedings of the IEEE Conference on Decision and Control* (Phoenix, AZ, USA).

Bamieh, B. and Giarré, L. (2000). Identification for a general class of LPV models, in *Proceedings of the IFAC Symposium on System Identification* (Santa Barbara, CA, USA).

Bamieh, B. and Giarré, L. (2001). LPV models: identification for gain-scheduling control, in *Proceedings of the European Control Conference* (Porto, Portugal).

Bamieh, B. and Giarré, L. (2002). Identification of linear parameter varying models, *International Journal of Robust and Nonlinear Control* **12**, 9, pp. 841–853.

Beflorte, G., Dabbene, F. and Gay, P. (2005). LPV approximation of distributed parameter systems in environmental modelling, *Environmental Modelling & Software* **20**, pp. 1063–1070.

Belforte, G. and Gay, P. (2004). Optimal worst case estimation for LPV/FIR models with bounded errors, *Systems & Control Letters* **53**, pp. 259–268.

Bemporad, A., Garulli, A., Paoletti, S. and Vicino, A. (2005). A bounded-error approach to piecewise affine system identification, *IEEE Transactions on Automatic Control* **50**, 10, pp. 1567–1580.

Bianchi, F. and Sanchez-Pena, R. (2010). A novel approach for switched LPV controllers, *International Journal of Control* **83**, 8, pp. 1710–1717.

Blesa, J., Puig, V. and Bolea, Y. (2010). Fault detection using interval LPV models in an open-flow canal, *Control Engineering Practice* **18**, pp. 460–470.

Bokor, J. and Balas, G. (2005). Linear parameter varying systems: A geometric theory and applications, in *Proceedings of the IFAC World Congress* (Prague, Czech Republic).

Boonto, S. and Werner, H. (2008). Closed-loop system identification of LPV input-output models - application to an arm-driven inverted pendulum, in *Proceedings of the IEEE Conference on Decision and Control* (Cancún, Mexico).

Bos, R., Bombois, X. and Van den Hof, P. (2009). Accelerating simulations of computationally intensive first principle models using accurate quasi-linear parameter varying models, *Journal of Process Control* **19**, pp. 1601–1609.

Casella, F. and Lovera, M. (2008). LPV/LFT modelling and identification: Overview, synergies and a case study, in *Proceedings of the IEEE Multiconference on Systems and Control* (San Antonio, TX, USA).

Cerone, V. and Regruto, D. (2008). Set-membership identification of LPV models with uncertain measurements of the time-varying parameter, in *Proceedings of the IEEE Conference on Decision and Control* (Cancún, Mexico).

Cinquemani, E., Porreca, R., Ferrari-Trecate, G. and Lygeros, J. (2007). A general framework for the identification of Jump Markov Linear Systems, in *Proceedings of the IEEE Conference on Decision and Control* (New Orleans, LA, USA).

DeCaigny, J., Camino, J. and Swevers, J. (2009). Interpolating model identification for SISO linear parameter-varying systems, *Mechanical Systems and Signal Processing* **23**, pp. 2395–2417.

Duviella, E., Puig, V., Charbonnaud, P., Escobet, T., Carrillo, F. J. and Quevedo, J. (2010). Supervised gain-scheduling multimodel versus linear parameter varying internal model control of open-channel systems for large operating conditions, *Jourmal of Irrigation and Drainage Engineering* , pp. 543–552.

Estrada, G., Kirchsteiger, H. and del Re, L. (2008). Compact representation of a nonlinear system using the NLPV approach, in *Proceedings of the IEEE Multi-conference on Systems and Control* (San Antonio, TX, USA).

Farhood, M. and Feron, E. (2009). A parameter-dependent Lyapunov approach for the control of nonstationary and hybrid LPV systems, in *Proceedings of the American Control Conference* (St. Louis, MO, USA).

Felici, F., van Wingerden, J. W. and Verhaegen, M. (2007). Subspace identification of MIMO LPV systems using a periodic scheduling sequence, *Automatica* **43**, 10, pp. 1684–1697.

Giarré, L., Bauso, D., Falugi, P. and Bamieh, B. (2006). LPV model identification for gain scheduling control: An application to rotating stall and surge control problem, *Control Engineering Practice* **14**, 4, pp. 351–361.

Giarré, L., Rosa, F. G. L., Pesenti, R. and Tinnirello, I. (2011). Coloring-based resource allocation in ad-hoc wireless networks, in *Proceedings of the IFIP Annual Mediterranean Ad Hoc Networking Workshop* (Favignana Island, Sicily, Italy).

Goebel, R., Sanfelice, R. G. and Teel, A. R. (2009). Hybrid dynamical systems, *IEEE Control Systems Magazine* **29**, 2, pp. 28–93.

Hespanha, J. (2005). A model for stochastic hybrid systems with application to communication networks, *Nonlinear Analysis Special Issue on Hybrid Systems* **62**, 8, pp. 1353–1383.

Hsu, K., Vincent, T. and Poolla, K. (2008). Nonparametric methods for the identification of linear parameter varying systems, in *Proceedings of the IEEE Multi-conference on Systems and Control* (San Antonio, TX, USA).

Juloski, A., Heemels, W., Ferrari-Trecate, G., Vidal, R., Paoletti, S. and Niessen, J. H. G. (2005). Comparison of four procedures for the identification of hybrid systems, in *Proceedings of the International Conference on Hybrid Systems: Computation and Control*, Vol. 3414 (Springer-Verlag, Zurich, Switzerland).

Juloski, A., Paoletti, S. and Roll, J. (2006). Recent techniques for the identification of piecewise affine and hybrid systems, in L. Z. e. C. A. L. Menini (ed.), *Current trends in nonlinear systems and control* (CRC Press, Boston, MA, USA), pp. 77–97.

Khalate, A. A., Bombois, X., Tóth, R. and Babuska, R. (2009). Optimal experimental design for LPV identification using a local approach, in *Proceedings of the IFAC Symposium on System Identification* (Saint-Malo, France).

Laurain, V., Gilson, M., Tóth, R. and Garnier, H. (2010). Refined instrumental variable methods for identification of LPV Box-Jenkins models, *Automatica* **46**, pp. 959–967.

LeBel, S. and Rodrigues, L. (2008). Piecewise-Affine parameter-varying control of wheeled mobile robots, in *Proceedings of the American Control Conference* (Seattle, WA, USA).

Lee, J., Bohacek, S., Hespanha, J. P. and Obraczka., K. (2007). Modeling communication networks with hybrid systems, *IEEE/ACM Transactions on Networking* **15**, pp. 630–643.

Lee, L. and Poolla, K. (1999). Identification of linear parameter-varying systems using nonlinear programming, *Journal of Dynamic Systems, Measurement and Control* **121**, 1, pp. 71–78.

Lee, L. H. and Poolla, K. (1996). Identification of linear parameter-varying systems using nonlinear programming, in *Proceedings of the IEEE Conference on Decision and Control* (Kobe, Japan).

Lim, S. and Chan, K. (2003). Analysis of hybrid linear parameter-varying systems, in *Proceedings of the American Control Conference* (Denver, CO, USA).

Lim, S. and How, J. P. (2003). Modeling and $\mathscr{H}_\infty$ control for switched linear parameter-varying missile autopilot, *IEEE Transactions on Control Systems Technology* **11**, 6, pp. 830–838.

Lopes dos Santos, P., Azevedo Perdicoúlis, T.-P., Ramos, J. A., Martins de Carvalho, J. L., Jank, G. and Milhinhos, J. (2011). An LPV modelling and identification approach to leakage detection in high pressure natural gas transportation networks, *IEEE Transactions on Control Systems Technology* , pp. 77–92.

Lovera, M. (2008). *LPV model identification: overview and perspectives*, URL `ftp://ftp.elet.polimi.it/users/Marco.Lovera/PRES/GT_Ident_Mosar.pdf`, available on line.

Lovera, M. and Mercere, G. (2007). Identification for gain-scheduling: a balanced subspace approach, in *Proceedings of the American Control Conference* (New York, NY, USA).

Lovera, M., Novara, C., Lopes dos Santos, P. and Rivera, D. (2011). Special issue on applied LPV modeling and identification, *IEEE Transactions on Control Systems Technology* **19**, 1.

Lovera, M., Verhaegen, M. and Chou, C. (1998). State-space identification of mimo linear parameter varying models, in *Proceedings of Mathematical Theory of Networks and Systems* (Padova, Italy).

Lu, B. and Wu, F. (2004). Switching LPV control designs using multiple parameter-dependent lyapunov functions, *Automatica* **40**, pp. 1973–1980.

Lu, B., Wu, F. and Kim, S. (2006). Switching LPV control of an F-16 Aircraft via controller state reset, *IEEE Transactions on Control Systems Technology* **14**, pp. 267–277.

Lunze, J. (ed.) (2009). *Handbook of Hybrid Systems Control Theory, Tools, Applications* (Cambridge University press, Cambridge, UK).

Lygeros, J. and Prandini, M. (2010). Stochastic hybrid systems: A powerful framework for complex, large scale applications, *European Journal of Control* **16**, pp. 583–594.

Mazzaro, M., Movsichoff, E. and Pena, R. (1999). Robust identification of linear parameter varying systems, in *Proceedings of the American Control Conference* (San Diego, CA, USA).

Nemani, M., Ravikanth, R. and Bamieh, B. (1995). Identification of linear parametrically varying systems, in *Proceedings of the IEEE Conference on Decision and Control* (New Orleans, LA, USA).

Paoletti, S., Garulli, A., Roll, J. and Vicino, A. (2008). A necessary and sufficient condition for input-output realization of switched affine state space models, in *Proceedings of the IEEE Conference on Decision and Control* (Cancún, Mexico).

Paoletti, S., Juloski, A., Ferrari-Trecate, G. and Vidal, R. (2007). Identification of hybrid systems: a tutorial, *European Journal of Control* **13**, pp. 242–260.

Paoletti, S., Roll, J., Garulli, A. and Vicino, A. (2010). On the input-output representation of piecewise affine state space models, *IEEE Transactions on Automatic Control* **55**, 1, pp. 60–73.

Preitl, Z., Kulcsar, B. and Bokor, J. (2008). Piecewise linear parameter varying mathematical model of a hybrid solar vehicle, in *Proceedings of the IEEE Intelligent Vehicles Symposium* (Eindhoven, the Netherlands).

Previdi, F. and Lovera, M. (1999). Identification of a class of linear models with nonlinearly varying parameters, in *Proceedings of European Control Conference* (Bretthauer Karlsruhe, Germany).

Previdi, F. and Lovera, M. (2003). Identification of a class of nonlinear parametrically varying models, *International Journal on Adaptive Control and Signal Processing* **17**, pp. 33–50.

Previdi, F. and Lovera, M. (2004). Identification of non-linear parametrically varying models using separable least squares, *International Journal of Control* **77**, pp. 1382–1392.

Qin, W. and Wang, Q. (2007a). An LPV approximation for admission control of an internet web server: Identification and control, *Control Engineering Practice* **15**, pp. 1457–1467.

Qin, W. and Wang, Q. (2007b). Modeling and control design for performance management of web servers via an LPV approach, *IEEE Transactions on Control Systems Technology* **15**, 2, pp. 259–275.

Rugh, W. and Shamma, J. (2000). Research on gain scheduling, *Automatica* **36**, 10, pp. 1401–1425.

Ruiz, F., Novara, C. and M., M. (2010). Direct design from data of optimal filters for LPV systems, *Systems & Control Letters* **59**, pp. 1–8.

Salcedo, J. and Martinez, M. (2008). LPV identification of a turbocharged diesel engine, *Applied Numerical Mathematics* **58**, pp. 1553–1571.

Shamma, J. (1996). Linearization and gain scheduling, in W. Levine (ed.), *Control Handbook* (CRC Press, Boca Raton, FL, USA), pp. 388–396.

Shamma, J. and Athans, M. (1991a). Guaranteed properties of gain scheduled control for linear parameter varying plants, *Automatica* **27**, pp. 559–564.

Shamma, J. and Athans, M. (1991b). Guaranteed properties of gain scheduled control for linear parameter varying plants, *Automatica* **27**, 3, pp. 559–564.

Sontag, E. (1981). Nonlinear regulation: The piecewise linear approach, *IEEE Trans. on Automatic Control* **26**, 2, pp. 346–358.

Steinbuch, M., van de Molengraft, R. and van der Voort, A. (2003). Experimental modelling and LPV control of a motion system, in *Proceedings of American Control Conference* (Denver, CO, USA).

Szabó, Z., Marcos, A., Prieto, D., Kerr, M., Rodonyi, G., Bokor, J. and Bennani, S. (2011). Development of an integrated LPV/LFT framework: Modeling and data-based validation tool, *IEEE Transactions on Control Systems Technology* **19**, 1, pp. 104–117.

Sznaier, M. and Mazzaro, C. (2003). An LMI approach to control-oriented identification and model (in)validation of LPV systems, *IEEE Transactions on Automatic Control* **48**, pp. 1619–1624.

Tinnirello, I., Giarré, L. and Pesenti, R. (2010). Decentralized synchronization for zigbee wireless sensor networks in Multi-Hop Topology, in *Proceedings of the IFAC Workshop on Distributed Estimation and Control in Networked Systems* (Annecy, France).

Tóth, R. (2010). *Modeling and Identification of Linear Parameter-Varying Systems*, Lecture Notes in Control and Information Sciences, 403 (Spinger-Verlag, Berlin/Heidelberg).

Tóth, R., den Hof, P. M. J. V., Ludlage, J. H. A. and Heuberger, P. S. C. (2010a). Identification of nonlinear process models in an LPV framework, in *Proceedings of the IFAC International Symposyum on Dynamics and Control of Process Systems - DYCOPS 2010* (Leuven, Belgium).

Tóth, R., Felici, F., Heuberger, P. S. C. and den Hof, P. M. J. V. (2007). Discrete time LPV I/O and state space representations, difference of behavior and pitfalls in interpolation, in *Proceedings of the European Control Conference* (Kos, Greece).

Tóth, R., Heuberger, P. S. and den Hof, P. M. V. (2009a). Asymptotically optimal orthonormal basis functions for LPV system identification, *Automatica* **45**, pp. 1359–1370.

Tóth, R., Van den Hof, P., Ludlage, J. and Heuberger, P. (2010b). Identification of nonlinear process models in an LPV framework, in *Proceedings of the IFAC International Symposium on Dynamics and Control of Process Systems - DYCOPS 2010* (Leuven, Belgium).

Tóth, R., Willems, J. C., Heuberger, P. S. C. and Van den Hof, P. (2009b). The behavioral approach to linear parameter–varying systems, in *Proceedings of the European Control Conference* (Budapest, Hungary).

van Overschee, P. and de Moor, B. (1996). *Subspace identification for linear systems theory, implementation, applications* (Kluwer Academic Publishers, Boston, MA, USA).

van Wingerden, J. and Verhaegen, M. (2009). Subspace identification of bilinear and LPV systems for open- and closed-loop data, *Automatica* **45**, pp. 372–381.

Verdult, V. and Verhaegen, M. (2002). Subspace identification of multivariable linear parameter-varying systems, *Automatica* **38**, 5, pp. 805–814.

Verdult, V. and Verhaegen, M. (2005). Kernel method for subspace identification of multivariable LPV and bilinear systems, *Automatica* **41**, 9, pp. 1557–1565.

Viberg, M. (1995). Subspace-based methods for the identification of linear time-invariant systems, *Automatica* **31**, pp. 1835–1852.

Vidal, R. and Anderson, B. (2004). Recursive identification of switched ARX hybrid models: Exponential convergence and persistence of excitation, in *Proceedings of the IEEE Conference on Decision and Control* (Atlantis, Paradise Islands, Bahamas).

Wattamwar, S. W., Weiland, S. and Backx, T. (2010). Identification of low-order parameter-varying models for large-scale systems, *Journal of Process Control* **20**, pp. 158–172.

Wei, X. and del Re, L. (2006). On persistent excitation for parameter estimation of quasi-LPV systems and its application in modeling of diesel engine torque, in *Proceedings of the IFAC Symposium on System Identification* (Newcastle, Australia).

Wei, X. and del Re, L. (2007). Gain scheduled H_∞ control for air path systems of diesel engines using LPV techniques, *IEEE Transactions on Control Systems Technology* **15**, 3, pp. 406–415.

Zhang, L., Shi, P. and Basin, M. (2008). Model reduction for switched linear parameter varying systems with average dwell time, in *Proceedings of the American Control Conference* (Seattle, WA, USA).

Chapter 3

Set-membership identification of input-output LPV models with uncertain time-varying parameters

V. Cerone, D. Piga & D. Regruto

Dipartimento di Automatica e Informatica
Politecnico di Torino, 10123 Torino, Italy
vito.cerone@polito.it
dario.piga@polito.it
diego.regruto@polito.it

In this chapter, we consider the identification of single-input single-output linear-parameter-varying models when both the output and the time-varying parameter measurements are affected by bounded noise. First, the problem of computing exact parameter uncertainty intervals is formulated in terms of semialgebraic optimization. Then, a suitable relaxation technique is presented to compute parameter bounds by means of convex optimization. Advantages of the presented approach with respect to previously published results are discussed.

1. Introduction

The linear-parameter-varying (LPV) modeling approach has received major attention from the identification and control community in recent years and can now be regarded as one of the most popular tools to derive mathematical description of nonlinear/time-varying phenomena. LPV models belong to the more general class of linear time-varying models and, roughly speaking, they can be defined as linear systems where either the matrices of the state equations or the coefficients of the difference equation relating the input and the output signals depend on one or more time varying parameters whose real-time measurements are assumed to be available. According to [Rugh and Shamma (2000)], a LPV model of a nonlinear plant can be obtained from the physical equations by means of two different approaches. The first one, called *linearization scheduling*, is based on Jacobian linearization of the nonlinear system around a family of equilibrium points parameterized by a set of measurable parameters (the so called scheduling variables). Such a linearization leads to a family of linear models, parameterized by the scheduling variables, which actually represent an LPV model of the nonlinear plant. The second approach, called *quasi-LPV scheduling*, is based on the idea of hiding nonlinearities in the system equations by suitably redefining some measurable state variables as time-varying parameters (see [Rugh and Shamma (2000)] for details). As to the identification of LPV models, since the work by Nemani, Ravikanth and Bamieh [Nemani *et al.* (1995)] which seems to be the first paper addressing the problem, a relevant number of approaches has appeared in the literature, which consider LPV models with a single time-varying parameter and assume that the measurements of all the state variables are available. A parameter estimation scheme based on the minimization of a prediction error cost function by means of nonconvex optimization is proposed in [Lee and Poolla (1999)] where LPV models with multiple time-varying parameters are considered under the assumption of linear fractional transformation (LFT) parameter dependence. Least mean squares and recursive least squares algorithms are proposed in [Bamieh and Giarré (2002)] to solve the identification of LPV input-output models assuming that measurements of input, output and scheduling parameters are available. Persistency of excitation conditions in terms of inputs and scheduling parameters trajectories are also derived. Subspace identification of multiple-input multiple-output (MIMO) LPV models with affine parameters dependence is considered in a number of papers. In [Verdult and Verhaegen (2002)] it is shown that standard subspace algorithms cannot be used in practice to identify LPV models since the dimensions of the data matrices grow exponentially with the system order. Significant improvements over the method proposed in [Verdult and Verhaegen (2002)] are presented in [Verdult and Verhaegen (2005)], exploiting kernel methods, and in [Felici *et al.* (2007)] where an instrumental variable approach is considered and the positive effect of using periodic scheduling sequences is highlighted. Iterative subspace system identification is considered by Lopes dos Santos and co-workers in [Lopes dos Santos *et al.* (2008)]. Application of LPV subspace identification algorithms to both periodic and nonlinear systems are discussed in [Verdult *et al.* (2004)] where the proposed approach is applied to the modeling of the helicopter rotor dynamics. Separable least squares are exploited in [Previdi and Lovera (2004)] to derive a novel algorithm for a class of nonlinear parameter-varying models represented in the form of a LFT, while an approach based on orthonormal basis functions is presented in [Tóth *et al.* (2010)]. A detailed overview of the available LPV modeling and identification approaches can be found in the book [Tóth (2010)], while a good collection of challenging real-world applications of such techniques can be found in the special issue [Lovera *et al.* (2011)]. In all the papers mentioned above, the authors assume that the measurement error is statistically described. An alternative to the stochastic description of measurement errors is the bounded-errors or set-membership characterization, where uncertainties are assumed to belong to a given set (see, e.g., [Milanese *et al.* (1996)]). In this context, all parameter vectors belonging to the *feasible parameter set* (*FPS*), i.e., parameters consistent with measurements, error bounds and the assumed model structure, are feasible solutions to the identification problem. To the authors best knowledge, only few contributions can be found in the literature addressing the identification of LPV models when the measurement errors are supposed to be bounded. In [Sznaier and Mazzarro (2003); Bianchi and Sanchez-Pena (2010)] identification and model invalidation of LPV systems in the presence of bounded noise and a possible nonparametric part is considered. The obtained results are applied to a practical problem arising in the context of active vision showing the effectiveness of the proposed approach. Belforte and Gay [Belforte and Gay (2004)] consider the identification of discrete-time LPV models with finite impulse response structure and output measurements affected by bounded noise. The problem of the optimal worst-case experiment design is addressed.

In this chapter, a procedure for set-membership identification of single-input single-output (SISO) discrete-time LPV models when both the output and the time-varying parameters measurements are affected by bounded noise is considered. Preliminary results on this problem are presented in [Cerone and Regruto (2008)] where a polytopic outer approximation of the nonconvex exact FPS is proposed. Results from [Cerone and Regruto (2008)] are successfully applied to the problem of deriving an LPV model of the vehicle lateral dynamics in [Cerone *et al.* (2011)]. In this chapter, a new relaxation approach is presented. The parameter uncertainty intervals computed by means of such an approach are guaranteed to be tighter than those obtained though the algorithm proposed in [Cerone and Regruto (2008)]. The chapter is organized as follows. Section 2 is devoted to the problem formulation. In Section 3 we show that computation of tight parameter bounds requires the solution to polynomial optimization problems. The new relaxation procedure is discussed in Section 4 and a detailed analysis of its properties is reported in Section 5. A simulated example is reported in Section 6 in order to show the effectiveness of the proposed approach and to highlight the improvement of the presented procedure in the evaluation of uncertainty intervals with respect to the algorithm proposed in [Cerone and Regruto (2008)].

2. Problem formulation

Consider the SISO discrete-time LPV model described in terms of the linear difference equation

$$\mathscr{A}(q^{-1},\lambda(k))w(k)=\mathscr{B}(q^{-1},\lambda(k))u(k), \tag{1}$$

where $u(k)$ and $w(k)$ are the input and the output signals, respectively, while $\lambda(k)\in\mathbb{R}^{\mu}$, $\lambda(k)=[\lambda_1(k)\lambda_2(k)\ldots\lambda_{\mu}(k)]^{\mathrm{T}}$ is a vector of time-varying parameters which, according to the LPV modeling and control literature (see, e.g., [Rugh and Shamma (2000)]) are assumed to be measurable. $\mathscr{A}(\cdot)$ and $\mathscr{B}(\cdot)$ are polynomials in the backward shift operator q^{-1},

$$\mathscr{A}(q^{-1},\lambda(k))=1+a_1(\lambda(k))q^{-1}+\ldots+a_{na}(\lambda(k))q^{-na}, \tag{2}$$

$$\mathscr{B}(q^{-1},\lambda(k))=b_0(\lambda(k))+b_1(\lambda(k))q^{-1}+\ldots+b_{nb}(\lambda(k))q^{-nb}, \tag{3}$$

where $na\geq nb$ and the coefficients a_i and b_j are assumed to be nonlinear memoryless mappings of parameters $\lambda(k)$. In order to lighten the mathematical notation we will use the symbol f_k to denote $f(k)$ from now on. According to such a simplified notation, the coefficients a_i and b_j are written as

$$a_i(\lambda_k)=\sum_{r=1}^{n_i}a_{i,r}\phi_{i,r}(\lambda_k), \tag{4}$$

$$b_j(\lambda_k)=\sum_{h=0}^{m_j}b_{j,h}\psi_{j,h}(\lambda_k), \tag{5}$$

where $\phi_{i,r}(\cdot)$ and $\psi_{j,h}(\cdot)$ are known nonlinear basis functions. In our work we assume that $\phi_{i,r}(\cdot)$ and $\psi_{j,h}(\cdot)$ belong to the canonical polynomial basis in the parameters λ_k. Let y_k and z_k be the noise-corrupted measurements of w_k and λ_k, respectively, i.e.,

$$y_k = w_k + v_k, \tag{6}$$

$$z_k = \lambda_k + \varepsilon_k, \tag{7}$$

where $\varepsilon_k = [\varepsilon_{1_k}\varepsilon_{2_k}\dots\varepsilon_{\mu_k}]^{\mathrm{T}}$. Measurements uncertainties v_k and ε_{s_k} are known to be within given bounds Δv_k and $\Delta\varepsilon_{s_k}$, more precisely

$$|v_k| \le \Delta v_k, \tag{8}$$

and

$$\varepsilon_k \in \mathscr{E}_k = \{\varepsilon_k \in \mathbb{R}^{\mu} : |\varepsilon_{k_s}| \le \Delta\varepsilon_{k_s}, s = 1, 2, \dots, \mu\}. \tag{9}$$

Thus, the noise free signals w_k and λ_k are such that

$$y_k - \Delta v_k = \underline{w}_k \le w_k \le \overline{w}_k = y_k + \Delta v_k, \tag{10}$$

and

$$\begin{aligned} \lambda_k \in \Lambda_k = \{\lambda_k \in \mathbb{R}^{\mu} : \\ \lambda_{k_s} - \Delta\varepsilon_{k_s} = \underline{\lambda}_{k_s} \le \lambda_{k_s} \le \overline{\lambda}_{k_s} = \lambda_{k_s} + \Delta\varepsilon_{k_s}, s = 1, 2, \dots, \mu\}. \end{aligned} \tag{11}$$

The unknown parameter vector $\theta \in \mathbb{R}^{n_\theta}$ to be estimated is defined as

$$\begin{aligned} \theta^{\mathrm{T}} = [&a_{1,1}\dots a_{1,n_1}\dots a_{na,1}\dots a_{na,n_{na}} \\ &b_{0,0}\dots b_{0,m_1}\dots b_{nb,1}\dots b_{nb,m_{nb}}], \end{aligned} \tag{12}$$

with $n_\theta = \sum_{i=1}^{na} n_i + \sum_{j=0}^{nb} m_j$. In this chapter, we address the problem of deriving uncertainty intervals on the θ. For the sake of simplicity and without loss of generality, in the rest of the chapter we consider the case of a scalar scheduling parameter λ_k, that is $\lambda_k \in \mathbb{R}$.

3. Evaluation of tight parameter bounds

The set $\mathscr{D}$ of all the LPV system parameters θ and the noise-free signals λ_k and w_k consistent with the measurement data sequence, the assumed model structure and the error bounds are described by equations (1) - (9), i.e.,

$$\begin{aligned}\mathscr{D} = \Big\{&(\theta, \lambda, w) \in \mathbb{R}^{n_\theta + N + (N - na)} : \\ &\mathscr{A}(q^{-1}, \lambda_k) w_k = \mathscr{B}(q^{-1}, \lambda_k) u_k, \\ &\underline{\lambda}_k \leq \lambda_k \leq \overline{\lambda}_{s_k}, \;\; \underline{w}_r \leq w_r \leq \overline{w}_r, \\ &k = na + 1, \ldots, N; \; r = 1, \ldots, N\Big\},\end{aligned} \tag{13}$$

with $\lambda = [\lambda_{na+1}, \ldots, \lambda_N]^{\mathrm{T}}$ and $w = [w_1, \ldots, w_N]^{\mathrm{T}}$. Thus, for $j = 1, \ldots, n_\theta$, tight bounds on the parameter θ_j can be computed by solving the constrained optimization problems

$$\underline{\theta}_j = \min_{(\theta,\lambda,w)\in\mathscr{D}} \theta_j, \quad \overline{\theta}_j = \max_{(\theta,\lambda,w)\in\mathscr{D}} \theta_j. \tag{14}$$

Parameter uncertainty intervals on θ_j are defined as

$$PUI_j = \left[\underline{\theta}_j; \, \overline{\theta}_j\right]. \tag{15}$$

Because of polynomial constraints $\mathscr{A}(q^{-1}, \lambda_k) w_k = \mathscr{B}(q^{-1}, \lambda_k) u_k$ defining the feasible region $\mathscr{D}$, problems (14) are nonconvex. Therefore, standard nonlinear optimization tools (gradient method, Newton method, etc.) can not be used because they can be trapped in local minima/maxima. As a consequence, the PUI_j obtained using these tools is not guaranteed to contain the true unknown parameter θ_j, which is a key requirement of any set-membership identification method. One possible solution to overcome such a problem is to relax the identification problems (14) into convex optimization problems in order to numerically compute lower bounds of $\underline{\theta}_j$ as well as upper bounds of $\overline{\theta}_j$. It can be shown that (14) are semialgebraic optimization problems with inherent structured sparsity. Approximate solutions of $\underline{\theta}_j$ and $\overline{\theta}_j$ can be computed through a direct implementation of the sparse linear-matrix-inequality (LMI)-based relaxation proposed in [Waki *et al.* (2006); Lasserre (2006)]. Unfortunately, due to high memory usage, the relaxation order δ has to be quite low to implement such an identification procedure in a commercial workstation. Roughly speaking, δ should not be greater than 2 when the number of parameters θ is 6 and the number N of measurements is 30. In order to deal with problems with a larger number of measurements and parameters, a relaxation method for evaluating parameter bounds of LPV systems in the set-membership context is proposed in [Cerone and Regruto (2008)] and successfully applied to the problem of vehicle lateral dynamics modeling [Cerone *et al.* (2011)].

In this chapter, we propose a new approach to compute guaranteed bounds on parameters θ that generalizes and improves the algorithm proposed in [Cerone and Regruto (2008)]. Such a method, which is referred to as semi-static LPV relaxation, provides parameter bounds which are guaranteed to be tighter than the ones obtained in [Cerone and Regruto (2008)].

4. Semi-static LPV relaxation

For the sake of clarity, a general overview of the proposed method is first presented in Section 4.1. Then, detailed technical results are provided in Section 4.2.

4.1. *Overview of the relaxation procedure*

The key idea to relax the identification problems (14) is to construct an outer-bound of the feasible set $\mathscr{D}$ by considering the noise-free signals λ_k and w_k, appearing in the definition of $\mathscr{D}$, independent of each other. Indeed, this could introduce conservativeness in evaluating parameter bounds since the correlation between consecutive measurements is lost. In order to reduce the conservativeness of such an approach, suitable equality constraints are added in the definition of the feasible set $\mathscr{D}$ as follows:

(i) given a fixed integer n, we consider n consecutive equality constraints in the description of $\mathscr{D}$ in (13);

(ii) we substitute the first constraint in the second one obtaining a new equality that retains correlations among the two; then, the newly obtained equation is substituted into the third constraint. The procedure is repeated until all the n consecutive constraints selected in (i) are nested;

(iii) the new equations obtained in (ii) are added in the description of $\mathscr{D}$.

Then, a set $\mathscr{D}^{(ss,n)}$, outer-bounding $\mathscr{D}$, is sought by considering the noise-free signals λ_k and w_k appearing in the equations obtained in (ii) independent of each other. In such a way, the correlation between n consecutive measurements is not completely lost since it is kept by the nested substitutions in (ii). Thanks to the structure of $\mathscr{D}^{(ss,n)}$, parameter bounds can be evaluated through the solution to polynomial problems with n_θ variables (the unknown LPV system parameter θ). It is worth noting that the number of variables involved in the solution of the new problems does not depend on the measurement size N and, as a consequence, it is significantly smaller than $n_\theta + 2N - na$, which is the number of variables involved in (14).

4.2. *Technical results*

Technical details of the proposed method are here discussed. As stated in the previous section, we first rewrite the feasible parameter set $\mathscr{D}$ in (13) in an alternative way, as described in the following. For a given integer $n \in [1, N - na]$ and $s = na + 1, \ldots, N$, let us define the set $\mathscr{S}_s^{(n)}$ as:

$$\begin{aligned} \mathscr{S}_s^{(n)} =& \{(\theta, \lambda, w) \in \mathbb{R}^{n_\theta + 2N - na} : \mathscr{A}(q^{-1}, \lambda_k) w_k = \mathscr{B}(q^{-1}, \lambda_k) u_k, \\ & \underline{\lambda}_k \leq \lambda_k \leq \overline{\lambda}_k, \quad \underline{w}_r \leq w_r \leq \overline{w}_k, \\ & k = s, \ldots, \min\{s + n - 1, N\}; \\ & r = s - na, s - na + 1, \ldots, \min\{s + n - 1, N\}\}. \end{aligned} \tag{16}$$

The set $\mathscr{S}_s^{(n)}$ in (16) is defined by the constraints describing $\mathscr{D}$ in (13) obtained by at most n consecutive measurements. For instance, $\mathscr{S}_{na+1}^{(n)}$ is only defined by the constraints obtained by the measurements from time $k = na + 1$ up to time $k = na + n$, $\mathscr{S}_{na+2}^{(n)}$ is defined by the constraints obtained by the measurements from time $t = na+2$ up to time $t = na+n+1$, and so on, up to $\mathscr{S}_N^{(n)}$, that will be defined just by the constraint obtained by the measurement at time N. Thus, the FPS $\mathscr{D}$ can be also written as

$$\mathscr{D} = \bigcap_{s=na+1}^{N} \mathscr{S}_s^{(n)}. \tag{17}$$

In this work, the integer n is referred to as the dynamic-horizon. We now rewrite the constraint $\mathscr{A}(q^{-1}, \lambda_k) w_k = \mathscr{B}(q^{-1}, \lambda_k) u_k$ in the matrix form

$$\underline{y}_k = \mathbf{A}(\lambda_k) \underline{w}_{k-1} + \mathbf{B}(\lambda_k) \underline{u}_k + \underline{v}_k, \tag{18}$$

where

$$\underline{y}_k = \begin{bmatrix} y_k \\ y_{k-1} \\ \vdots \\ y_{k-na+1} \end{bmatrix}, \quad \underline{v}_k = \begin{bmatrix} v_k \\ v_{k-1} \\ \vdots \\ v_{k-na+1} \end{bmatrix}, \quad \underline{u}_k = \begin{bmatrix} u_k \\ u_{k-1} \\ \vdots \\ u_{k-nb} \end{bmatrix}, \underline{w}_k = \underline{y}_k - \underline{v}_k. \tag{19}$$

$$\mathbf{A}(\lambda_k) \in \mathbb{R}^{na,na} : \mathbf{A}(\lambda_k) = \begin{bmatrix} -a_1(\lambda_k) & -a_2(\lambda_k) & \dots & -a_{na-1}(\lambda_k) & -a_{na}(\lambda_k) \\ 1 & 0 & \dots & 0 & 0 \\ 0 & 1 & \dots & 0 & 0 \\ \vdots & . & . & . & \vdots \\ 0 & 0 & \dots & 1 & 0 \end{bmatrix} \tag{20}$$

$$\mathbf{B}(\lambda_k) \in \mathbb{R}^{na,nb+1} : \mathbf{B}(\lambda_k) = \begin{bmatrix} b_0(\lambda_k) & b_1(\lambda_k) & \dots & b_{nb}(\lambda_k) \\ 0 & 0 & \dots & 0 \\ \vdots & . & . & \vdots \\ 0 & 0 & \dots & 0 \end{bmatrix}. \tag{21}$$

From nested substitutions of the equations defining $\mathscr{S}_s^{(n)}$ in (16), the set $\mathscr{S}_s^{(n)}$ can be written in terms of matrices $\mathbf{A}$ and $\mathbf{B}$ as

$$\begin{aligned} \mathscr{S}_s^{(n)} =& \big\{(\theta, \lambda, w) \in \mathbb{R}^{n_\theta + 2N - na} : \\ & y_{s+k} = [1, 0, \dots, 0] \left[\left[\prod_{i=0}^{k} \mathbf{A}(\lambda_{s+i}) \right] w_{s-i} + \right. \\ & \left. + \sum_{j=0}^{k} \left[\prod_{r=0}^{j-1} \mathbf{A}(\lambda_{s+k-r}) \right] \mathbf{B}(\lambda_{s+k-j}) \underline{u}_{s+k-j} \right] + v_{s+k} \\ & \underline{\lambda}_{s+k} \le \lambda_{s+k} \le \overline{\lambda}_{s+k}, \ \ \underline{w}_{s-1} \le w_{s-1} \le \overline{w}_{s-1}, \ \ |v_{s+k}| \le \Delta v_{s+k} \\ & k = 0, \dots, \min\{n-1, N-s\}\big\}, \end{aligned} \tag{22}$$

or equivalently as

$$\begin{aligned} \mathscr{S}_s^{(n)} =& \big\{(\theta, \lambda, v) \in \mathbb{R}^{n_\theta + 2N - na} : \\ & y_{s+k} = \sum_{i=1}^{na} a_i^{(s+k)}(\theta, \lambda)[y_{s-i} - v_{s-i}] + \\ & + \sum_{j=0}^{k} \sum_{i=0}^{nb} b_{ji}^{(s+k)}(\theta, \lambda) u_{k+s-j-i} + v_{s+k} \\ & \underline{\lambda}_{s+k} \le \lambda_{s+k} \le \overline{\lambda}_{s+k}, \ \ \underline{w}_{s-1} \le w_{s-1} \le \overline{w}_{s-1}, \ \ |v_{s+k}| \le \Delta v_{s+k} \\ & k = 0, \dots, \min\{n-1, N-s\}\big\}, \end{aligned} \tag{23}$$

where $\underline{w}_{s-1} = \left[\underline{w}_{s-1}, \ldots, \underline{w}_{s-na}\right]^T$, $\overline{w}_{s-1} = \left[\overline{w}_{s-1}, \ldots, \overline{w}_{s-na}\right]^T$, while $a_i^{(s+k)}(\theta, \lambda)$, $b_j^{(s+k)}(\theta, \lambda)$ are polynomial functions of the system parameters θ and noise-free signals $\lambda_s, \lambda_{s+1}, \ldots, \lambda_{s+k}$. In particular, $a_i^{(s+k)} = \left[\prod_{i=0}^{k} \mathbf{A}(\lambda_{s+i})\right](1,i)$ and $b_{ji}^{(s+k)} = \left[\left[\prod_{r=0}^{j-1} \mathbf{A}(\lambda_{s+k-r})\right] \mathbf{B}(\lambda_{s+k-j})\right](1,i)$, where $[R](1,i)$ denotes the entry in the first row and i-th column of a generic matrix R.

Let $m_i^{(s+k)}$ be the number of monomials of the polynomial $a_i^{(s+k)}(\theta, \lambda)$ and $f_q^{a_i^{(s+k)}}(\theta) \cdot r_q^{a_i^{(s+k)}}(\lambda)$, with $q = 1, \ldots, m_i^{(s+k)}$, the q-th monomial of $a_i^{(s+k)}(\theta, \lambda)$, where $f_q^{a_i^{(s+k)}}(\theta)$ is a homogeneous monomial depending only on the parameters θ, while $r_q^{a_i^{(s+k)}}(\lambda)$ depends only on the uncertain scheduling parameters $\lambda_s, \lambda_{s+1}, \ldots, \lambda_{s+k}$. For example, if $a_i^{(s+k)} = a_{11}^2 \phi_{11}(\lambda_{k+1})\phi_{11}(\lambda_k) + a_{11}a_{12}\phi_{11}(\lambda_{k+1})\phi_{12}(\lambda_k)$ then $m_i^{(s+k)} = 2$, $f_1^{a_i^{(s+k)}}(\theta) = a_{11}^2$, $q_1^{a_i^{(s+k)}}(\lambda) = \phi_{11}(\lambda_{k+1})\phi_{11}(\lambda_k)$, $f_2^{a_i^{(s+k)}}(\theta) = a_{11}a_{12}$ and $q_2^{a_i^{(s+k)}}(\lambda) = \phi_{11}(\lambda_{k+1})\phi_{12}(\lambda_k)$. In the same way, let $n_{ji}^{(s+k)}$ be the number of monomials of the polynomial $b_{ji}^{(s+k)}(\theta, \lambda)$ and $f_q^{b_{ji}^{(s+k)}}(\theta) \cdot r_q^{b_{ji}^{(s+k)}}(\lambda_s, \lambda_{s+1}, \ldots, \lambda_{s+k})$, with $q = 1, \ldots, n_{ji}^{(s+k)}$, the q-th monomial of $b_{ji}^{(k)}(\theta, \lambda)$, where $f_q^{b_{ji}^{(k)}}(\theta)$ is a homogeneous monomial depending only on the system parameters θ. In such a way, the set $\mathscr{S}_s^{(n)}$ in (23) can be written as

$$\mathscr{S}_s^{(n)} = \Bigg\{ (\theta, \lambda, w) \in \mathbb{R}^{n_\theta + 2N - na} :$$

$$y_{s+k} = \sum_{i=1}^{na} \sum_{q=1}^{m_i^{(s+k)}} f_q^{a_i^{(s+k)}}(\theta) \cdot r_q^{a_i^{(s+k)}}(\lambda) \cdot w_{s-i} +$$

$$+ \sum_{j=0}^{k} \sum_{i=0}^{nb} \sum_{q=1}^{n_{ji}^{(s+k)}} f_q^{b_{ji}^{(s+k)}}(\theta) \cdot r_q^{b_{ji}^{(s+k)}}(\lambda) \cdot u_{k+s-j-i} + v_{s+k}$$

$$\underline{\lambda}_{s+k} \leq \lambda_{s+k} \leq \overline{\lambda}_{s+k}, \ \underline{w}_{s-1} \leq w_{s-1} \leq \overline{w}_{s-1}, \ |v_{s+k}| \leq \Delta v_{s+k} \Bigg\}. \tag{24}$$

Remark 3.1. The maximum degree of the monomials $f_q^{a_i^{(s+k)}}(\theta)$ and $f_q^{b_{ji}^{(s+k)}}(\theta)$ is equal to n.

Let us define

$$\rho_q^{a_i^{(s+k)}}(\lambda, w_{s-i}) = r_q^{a_i^{(k+s)}}(\lambda) \cdot w_{s-i}, \tag{25}$$

$$\rho_q^{b_{ji}^{(s+k)}}(\lambda) = r_q^{b_{ji}^{(s+k)}}(\lambda) \cdot u_{k+s-j-i}. \tag{26}$$

Because of the uncertainty affecting λ, $r_q^{a_i^{(k+s)}}(\lambda)$ and $r_q^{b_{ji}^{(s+k)}}(\lambda)$ are uncertain variables belonging to the intervals

$$\underline{r}_q^{a_i^{(k+s)}} \leq r_q^{a_i^{(k+s)}}(\lambda) \leq \overline{r}_q^{a_i^{(k+s)}}, \tag{27}$$

$$\underline{r}_q^{b_{ji}^{(s+k)}} \leq r_q^{b_{ji}^{(s+k)}}(\lambda) \leq \overline{r}_q^{b_{ji}^{(s+k)}}, \tag{28}$$

where

$$\underline{r}_q^{a_i^{(k+s)}} = \min_{\lambda \in \Lambda} r_q^{a_i^{(k+s)}}(\lambda); \quad \overline{r}_q^{a_i^{(k+s)}} = \max_{\lambda \in \Lambda} r_q^{a_i^{(k+s)}}(\lambda), \tag{29}$$

$$\underline{r}_q^{b_{ji}^{(s+k)}} = \min_{\lambda \in \Lambda} r_q^{b_{ji}^{(s+k)}}(\lambda); \quad \overline{r}_q^{b_{ji}^{(s+k)}} = \max_{\lambda \in \Lambda} r_q^{b_{ji}^{(s+k)}}(\lambda), \tag{30}$$

with $\Lambda = \{\lambda \in \mathbb{R}^N : \underline{\lambda} \leq \lambda \leq \overline{\lambda}\}$.

Solving problems (29) and (30) requires the computation of the global minimum/maximum of monomials $r_q^{a_i^{(k+s)}}(\lambda)$ and $r_q^{b_{ji}^{(s+k)}}(\lambda)$. Since the feasible set Λ is compact, the Weierstrass theorem guarantees that $r_q^{a_i^{(k+s)}}(\lambda)$ and $r_q^{b_{ji}^{(s+k)}}(\lambda)$ achieve their global minimum and maximum on the feasible set Λ. Such global minimum and maximum must either be stationary points or lie on the boundary of Λ. It must be pointed out that the computation of stationary points is straightforward since $r_q^{a_i^{(k+s)}}(\lambda)$ and $r_q^{b_{ji}^{(s+k)}}(\lambda)$ are monomials. Besides, their behavior on the boundary of Λ can be easily analyzed since the set Λ is an axis-aligned box in the scheduling parameter space.

Given bounds $\underline{r}_q^{a_i^{(k+s)}}$, $\overline{r}_q^{a_i^{(k+s)}}$, $\underline{r}_q^{b_{ji}^{(s+k)}}$ and $\overline{r}_q^{b_{ji}^{(s+k)}}$, extreme values of the interval variables $\rho_q^{a_i^{(s+k)}}(\lambda, w_{s-i})$ and $\rho_q^{b_{ji}^{(s+k)}}(\lambda)$, defined as

$$\underline{\rho}_q^{a_i^{(s+k)}} = \min_{\lambda \in \Lambda, \underline{w}_{s-i} \leq w_{s-i} \leq \overline{w}_{s-i}} \rho_q^{a_i^{(s+k)}}(\lambda, w_{s-i}); \tag{31a}$$

$$\overline{\rho}_q^{a_i^{(s+k)}} = \max_{\lambda \in \Lambda, \underline{w}_{s-i} \leq w_{s-i} \leq \overline{w}_{s-i}} \rho_q^{a_i^{(s+k)}}(\lambda, w_{s-i}); \tag{31b}$$

$$\underline{\rho}_q^{b_{ji}^{(s+k)}} = \min_{\lambda\in\Lambda} \rho_q^{b_{ji}^{(s+k)}}(\lambda); \tag{32a}$$

$$\overline{\rho}_q^{b_{ji}^{(s+k)}} = \max_{\lambda\in\Lambda} \rho_q^{b_{ji}^{(s+k)}}(\lambda); \tag{32b}$$

can be computed through basic results on interval arithmetic applied to equations (25)-(26).

On the basis of the computed bounds $\underline{r}_q^{a_i^{(k+s)}}$, $\overline{r}_q^{a_i^{(k+s)}}$, $\underline{r}_q^{b_{ji}^{(s+k)}}$ and $\overline{r}_q^{b_{ji}^{(s+k)}}$, a set $\mathscr{S}_s^{(ss,n)}$ outerbounding $\mathscr{S}_s^{(n)}$ can be constructed as described in the following proposition.

Proposition 3.1. *Let us define* $\mathscr{S}_s^{(ss,n)}$ *as*

$$\begin{aligned}
\mathscr{S}_s^{(ss,n)} = \Big\{\theta \in \mathbb{R}^{n_\theta} : \\
\Big|y_{s+k} - \sum_{i=1}^{na}\sum_{q=1}^{m_i^{(k)}} f_q^{a_i^{(k)}}(\theta)\cdot\rho_q^{a_i^{(s+k)}}(c) - \sum_{j=0}^{k}\sum_{i=0}^{nb}\sum_{q=1}^{n_{ji}^{(s+k)}} f_q^{b_{ji}^{(k)}}(\theta)\cdot\rho_q^{b_{ji}^{(s+k)}}(c)\Big| \leq \\
\leq \sum_{i=1}^{na}\sum_{q=1}^{m_i^{(s+k)}} f_q^{a_i^{(k)}}(\theta)\cdot sign(f_q^{a_i^{(s+k)}}(\theta))\Delta\rho_q^{a_i^{(s+k)}} + \\
+ \sum_{j=0}^{k}\sum_{i=0}^{nb}\sum_{q=1}^{n_{ji}^{(k)}} f_q^{b_{ji}^{(k)}}(\theta)\cdot sign(f_q^{b_{ji}^{(s+k)}}(\theta))\Delta\rho_q^{b_{ji}^{(s+k)}} + \Delta v_{s+k}\Big\},
\end{aligned} \tag{33}$$

where $\rho_q^{a_i^{(s+k)}}(c)$ *and* $\rho_q^{b_{ji}^{(s+k)}}(c)$ *are the central value of* $\rho_q^{a_i^{(s+k)}}(\lambda, w_{s-i})$ *and* $\rho_q^{b_{ji}^{(s+k)}}(\lambda)$*, respectively, i.e.,*

$$\rho_q^{a_i^{(s+k)}}(c) = \frac{\overline{\rho}_q^{a_i^{(s+k)}} + \underline{\rho}_q^{a_i^{(s+k)}}}{2}, \quad \rho_q^{b_{ji}^{(s+k)}}(c) = \frac{\overline{\rho}_q^{b_{ji}^{(s+k)}} + \underline{\rho}_q^{b_{ji}^{(s+k)}}}{2}, \tag{34}$$

while $\Delta\rho_q^{a_i^{(s+k)}}$ *and* $\Delta\rho_q^{b_{ji}^{(s+k)}}$ *are the uncertainty bounds of* $\rho_q^{a_i^{(s+k)}}(\lambda, w_{s-i})$ *and* $\rho_q^{b_{ji}^{(s+k)}}(\lambda)$*, respectively, i.e.,*

$$\Delta\rho_q^{a_i^{(s+k)}} = \frac{\overline{\rho}_q^{a_i^{(s+k)}} - \underline{\rho}_q^{a_i^{(s+k)}}}{2}, \quad \Delta\rho_q^{b_{ji}^{(s+k)}} = \frac{\overline{\rho}_q^{b_{ji}^{(s+k)}} - \underline{\rho}_q^{b_{ji}^{(s+k)}}}{2}. \tag{35}$$

Then, the set $\mathscr{S}_s^{(ss,n)}$ *is an outer bound of* $\mathscr{S}_s^{(n)}$*, i.e.,* $\mathscr{S}_s^{(n)} \subseteq \mathscr{S}_s^{(ss,n)}$.

Proof. Let us rewrite the equality constraints defining $\mathscr{S}_s^{(n)}$ in (24) in terms of $\rho_q^{a_i^{(s+k)}}$ and $\rho_q^{b_{ji}^{(s+k)}}$, i.e.,

$$y_{s+k} = \sum_{i=1}^{na} \sum_{q=1}^{m_i^{(k)}} f_q^{a_i^{(k)}}(\theta) \cdot \rho_q^{a_i^{(s+k)}} + \sum_{j=0}^{k} \sum_{i=0}^{nb} \sum_{q=1}^{n_{ji}^{(k)}} f_q^{b_{ji}^{(k)}}(\theta) \cdot \rho_q^{b_{ji}^{(s+k)}} + v_{s+k}. \tag{36}$$

Then, let us write $\rho_q^{a_i^{(s+k)}}(\lambda, v_{s-i})$ and $\rho_q^{b_{ji}^{(s+k)}}(\lambda)$ in terms of their central values $\rho_q^{a_i^{(s+k)}}(c)$, $\rho_q^{b_{ji}^{(s+k)}}(c)$ and their perturbations $\delta\rho_q^{a_i^{(s+k)}}$, $\delta\rho_q^{b_{ji}^{(s+k)}}$, where

$$\rho_q^{a_i^{(s+k)}}(\lambda, w_{s-i}) = \rho_q^{a_i^{(s+k)}}(c) + \delta\rho_q^{a_i^{(s+k)}}, \tag{37}$$

$$\rho_q^{b_{ji}^{(s+k)}}(\lambda) = \rho_q^{b_{ji}^{(s+k)}}(c) + \delta\rho_q^{b_{ji}^{(s+k)}}, \tag{38}$$

where

$$|\delta\rho_q^{a_i^{(s+k)}}| \leq \Delta\rho_q^{a_i^{(s+k)}}, \quad |\delta\rho_q^{b_{ji}^{(s+k)}}| \leq \Delta\rho_q^{b_{ji}^{(s+k)}}. \tag{39}$$

By substituting eqs. (37) and (38) into (36) we get:

$$\begin{aligned} y_{s+k} - \sum_{i=1}^{na} \sum_{q=1}^{m_i^{(k)}} f_q^{a_i^{(k)}}(\theta) \cdot \rho_q^{a_i^{(s+k)}}(c) - \sum_{j=0}^{k} \sum_{i=0}^{nb} \sum_{q=1}^{n_{ji}^{(k)}} f_q^{b_{ji}^{(k)}}(\theta) \cdot \rho_q^{b_{ji}^{(s+k)}}(c) = \\ = \sum_{i=1}^{na} \sum_{q=1}^{m_i^{(k)}} f_q^{a_i^{(k)}}(\theta) \cdot \delta\rho_q^{a_i^{(s+k)}} + \sum_{j=0}^{k} \sum_{i=0}^{nb} \sum_{q=1}^{n_{ji}^{(k)}} f_q^{b_{ji}^{(k)}}(\theta) \cdot \delta\rho_q^{b_{ji}^{(s+k)}} + v_{s+k}. \end{aligned} \tag{40}$$

By taking the absolute value in eq. (40), using the triangle inequality and conditions (39) and $|v_{s+k}| \leq \Delta v_{s+k}$, the following inequalities hold:

$$
\begin{aligned}
&|y_{s+k} - \sum_{i=1}^{na}\sum_{q=1}^{m_i^{(k)}} f_q^{a_i^{(k)}}(\theta)\cdot \rho_q^{a_i^{(s+k)}}(c) - \sum_{j=0}^{k}\sum_{i=0}^{nb}\sum_{q=1}^{n_{ji}^{(k)}} f_q^{b_{ji}^{(k)}}(\theta)\cdot \rho_q^{b_{ji}^{(s+k)}}(c)| = \\
&= |\sum_{i=1}^{na}\sum_{q=1}^{m_i^{(k)}} f_q^{a_i^{(k)}}(\theta)\cdot \delta\rho_q^{a_i^{(s+k)}} + \sum_{j=0}^{k}\sum_{i=0}^{nb}\sum_{q=1}^{n_{ji}^{(k)}} f_q^{b_{ji}^{(k)}}(\theta)\cdot \delta\rho_q^{b_{ji}^{(s+k)}} + v_{s+k}| \leq \\
&\leq \sum_{i=1}^{na}\sum_{q=1}^{m_i^{(k)}} |f_q^{a_i^{(k)}}(\theta)|\cdot |\delta\rho_q^{a_i^{(s+k)}}| + \sum_{j=0}^{k}\sum_{i=0}^{nb}\sum_{q=1}^{n_{ji}^{(k)}} |f_q^{b_{ji}^{(k)}}(\theta)|\cdot |\delta\rho_q^{b_{ji}^{(s+k)}}| + |v_{s+k}| \leq \\
&\leq \sum_{i=1}^{na}\sum_{q=1}^{m_i^{(k)}} f_q^{a_i^{(k)}}(\theta)\cdot sign(f_q^{a_i^{(k)}}(\theta))\Delta\rho_q^{a_i^{(s+k)}} + \\
&+ \sum_{j=0}^{k}\sum_{i=0}^{nb}\sum_{q=1}^{n_{ji}^{(k)}} f_q^{b_{ji}^{(k)}}(\theta)\cdot sign(f_q^{b_{ji}^{(k)}}(\theta))\Delta\rho_q^{b_{ji}^{(s+k)}} + \Delta v_{s+k}.
\end{aligned} \tag{41}
$$

This means that when constraints defining $\mathscr{S}_s^{(n)}$ in (24) are satisfied, then inequalities (41), which in turn define $\mathscr{S}_s^{(ss,n)}$, hold. $\mathscr{S}_s^{(n)} \subseteq \mathscr{S}_s^{(ss,n)}$. □

On the basis of Proposition 3.1, an outer bound of the FPS $\mathscr{D}$ can be constructed as described by the following result.

Proposition 3.2. *Let us define the set* $\mathscr{D}^{(ss,n)}$ *as:*

$$\mathscr{D}^{(ss,n)} = \bigcap_{s=na+1}^{N} S_s^{(ss,n)}. \tag{42}$$

The set $\mathscr{D}^{(ss,n)}$ *is an outer bound of* $\mathscr{D}$, *that is:*

$$\mathscr{D}^{(ss,n)} \supseteq \mathscr{D}. \tag{43}$$

Proof. From Proposition 3.1, $\mathscr{S}_s^{(ss,n)} \supseteq \mathscr{S}_s^{(n)}$, $s = na+1,\ldots,N$,

$$\mathscr{D}^{(ss,n)} = \bigcap_{s=na+1}^{N} \mathscr{S}_s^{(ss,n)} \supseteq \bigcap_{s=na+1}^{N} \mathscr{S}_s^{(n)} = \mathscr{D}, \tag{44}$$

that proves (43). □

For a given dynamic horizon n, let us define the parameter uncertainty interval $PUI_j^{(n)} = \left[\underline{\theta}_j^{(n)}; \overline{\theta}_j^{(n)}\right]$, with $j = 1, \ldots, n_\theta$, where

$$\underline{\theta}_j^{(n)} = \min_{\theta \in \mathscr{D}^{(ss,n)}} \theta_j; \quad \overline{\theta}_j^{(n)} - \max_{\theta \in \mathscr{D}^{(ss,n)}} \theta_j. \tag{45}$$

Remark 3.2. Only the unknown LPV system parameters θ are optimization variables for problems (45); on the contrary, in the original identification problems (14), the noise-free signals λ and w were also treated as decision variables.

Evaluation of the intervals $PUI_j^{(n)}$ requires the solution to the optimization problems (45) over the nonconvex feasible region $\mathscr{D}^{(ss,n)}$. In the following we describe how to solve numerically such problems by exploiting the particular structure of $\mathscr{D}^{(ss,n)}$.

Proposition 3.3. *Set $\mathscr{D}^{(ss,n)}$ is a union of at most 2^{n_θ} semialgebraic sets $\mathscr{D}_l^{(ss,n)}$ in the parameter space $\mathbb{R}^{n_\theta}$, i.e.,*

$$\mathscr{D}^{(ss,n)} = \bigcup_{l=1}^{2^{n_\theta}} \mathscr{D}_l^{(ss,n)}, \tag{46}$$

where $\mathscr{D}_l^{(ss,n)}$ is the intersection between $\mathscr{D}^{(ss,n)}$ and the l-th orthant $\mathscr{O}_l$ of the system parameter space $\mathbb{R}^{n_\theta}$, i.e.,

$$\mathscr{D}_l^{(ss,n)} = \mathscr{D}^{(ss,n)} \cap \mathscr{O}_l. \tag{47}$$

Furthermore, the degree of polynomial inequality constraints defining $\mathscr{D}_l^{(ss,n)}$ is less than or equal to n.

Proof. Indeed,

$$\mathbb{R}^{n_\theta} = \bigcup_{l=1}^{2^{n_\theta}} \mathscr{O}_l. \tag{48}$$

Then, from (48) and the definition of $\mathscr{D}_l^{(ss,n)}$ in (47), condition (46) follows. In fact,

$$\bigcup_{l=1}^{2^{n_\theta}} \mathscr{D}_l^{(ss,n)} = \bigcup_{l=1}^{2^{n_\theta}} \mathscr{D}^{(ss,n)} \cap \mathscr{O}_l = \mathscr{D}^{(ss,n)} \cap \bigcup_{l=1}^{2^{n_\theta}} \mathscr{O}_l = \mathscr{D}^{(ss,n)} \cap \mathbb{R}^{n_\theta} = \mathscr{D}^{(ss,n)}. \tag{49}$$

We are left to prove that $\mathscr{D}_l^{(ss,n)}$ is a semialgebraic set. Since $f_q^{a_i^{(s+k)}}(\theta)$ and $f_q^{b_{ji}^{(s+k)}}(\theta)$ are monomials of the system parameters θ, in the orthant $\mathscr{O}_l$ the sign of $f_q^{a_i^{(s+k)}}(\theta)$ and $f_q^{b_{ji}^{(s+k)}}(\theta)$ is known. In the orthant $\mathscr{O}_l$, inequalities defining $\mathscr{S}_s^{(ss,n)}$ in (33) (and so defining also $\mathscr{D}_l^{(ss,n)}$) are polynomial functions of θ, thus $\mathscr{D}_l^{(ss,n)}$ is a semialgebraic set. Besides, from Remark 3.1, the degree of polynomials $f_q^{a_i^{(s+k)}}(\theta)$ and $f_q^{b_{ji}^{(s+k)}}(\theta)$ is less than or equal to n, then the degree of polynomials defining $\mathscr{D}_l^{(ss,n)}$ is at most n. □

Thanks to the structure of the set $\mathscr{D}^{(ss,n)}$ highlighted by Proposition 3.3, problems (45) can be decomposed into a collection of 2^{n_θ} polynomial optimization problems. i.e.,

$$\underline{\theta}_j^{(n)} = \min_{l=1,\ldots,2^{n_\theta}} \underline{\theta}_{jl}^{(n)}, \quad \overline{\theta}_j^{(n)} = \max_{l=1,\ldots,2^{n_\theta}} \overline{\theta}_{jl}^{(n)}, \tag{50}$$

where

$$\underline{\theta}_{jl}^{(n)} = \min_{\theta \in \mathscr{D}_l^{(ss,n)}} \theta_j, \quad \overline{\theta}_{jl}^{(n)} = \max_{\theta \in \mathscr{D}_l^{(ss,n)}} \theta_j. \tag{51}$$

Approximate solutions of (51) can be computed through the LMI-relaxation techniques proposed in [Lasserre (2001)], [Parrillo (2003)] and [Chesi *et al.* (2003)] since the variables involved in that optimization problem are only the unknown LPV system parameters θ. In this work we exploit the approach proposed by [Lasserre (2001)], which results on the theory of moments to construct a hierarchy of SDP problems, whose solutions converge to the global optimum of the original polynomial problem as the number of successive relaxations, the relaxation order δ, increases. In particular, for a given relaxation order δ, application of the theory-of-moments relaxation to problems (51) leads to the approximate solutions $\underline{\theta}_{jl}^{(n,\delta)}$ and $\overline{\theta}_{jl}^{(n,\delta)}$ computed by solving the convex SDP problems

$$\underline{\theta}_{jl}^{(n,\delta)} = \min_{p \in \mathscr{D}_l^{(ss,n,\delta)}} f_j(p); \; \overline{\theta}_{jl}^{(n,\delta)} = \max_{p \in \mathscr{D}_l^{(ss,n,\delta)}} f_j(p), \tag{52}$$

where p is the decision variable vector of dimension $\binom{n_\theta + 2\delta}{2\delta}$. The objective function $f_j(p)$ in (52) is linear in the variables p and the feasible region $\mathscr{D}_l^{(ss,n,\delta)}$ is a convex set defined by LMIs taking into account the constraints that characterize the semialgebraic set $\mathscr{D}_l^{(ss,n,\delta)}$. It must be pointed out that the minimum allowed value $\underline{\delta}$ of the LMI relaxation order has to be equal to $\left\lceil \frac{\chi\left(\mathscr{D}_l^{(ss,n)}\right)}{2} \right\rceil$, where $\chi\left(\mathscr{D}_l^{(ss,n)}\right)$ is the maximum degree of the polynomial constraints defining $\mathscr{D}_l^{(ss,n)}$ and $\lceil \cdot \rceil$ denotes the ceiling operator. From Proposition 3.3, the maximum degree of the polynomial inequalities that describe $\mathscr{D}_l^{(ss,n)}$ is always less than or equal to n, therefore, for the relaxation order $\delta \geq \underline{\delta} = \lceil \frac{n}{2} \rceil$, problems (52) are well defined. See [Lasserre (2001)] for technical details on the relaxation of semialgebraic optimization problems to convex SDP problems though the theory of moments.

Let us define the δ-relaxed uncertainty interval $PUI_j^{(n,\delta)}$ as:

$$PUI_j^{(n,\delta)} = \left[\underline{\theta}_j^{(n,\delta)};\ \overline{\theta}_j^{(n,\delta)}\right], \tag{53}$$

where

$$\underline{\theta}_j^{(n,\delta)} = \min_{l=1,\dots,2^{n_\theta}} \underline{\theta}_{jl}^{(n,\delta)}, \quad \overline{\theta}_j^{(n,\delta)} = \max_{l=1,\dots,2^{n_\theta}} \overline{\theta}_{jl}^{(n,\delta)}. \tag{54}$$

In the next section, the main features enjoyed by the computed parameter uncertainty intervals $PUI_j^{(n,\delta)}$ are described.

Remark 3.3. The semi-static LPV relaxation presented here is equivalent to the algorithm proposed in [Cerone and Regruto (2008)] when the dynamic-horizon n equals one.

5. Properties of the computed parameter uncertainty intervals $PUI_j^{(n,\delta)}$

A number of interesting features enjoyed by the computed parameter uncertainty intervals $PUI_j^{(n,\delta)}$ are reported in the following propositions.

Proposition 3.4. *For all $n = 1, \dots, N - na$ and $\delta \geq \underline{\delta}$, the true unknown parameter θ_j^o to be estimated belong to the computed interval $PUI_j^{(n,\delta)}$, i.e.,*

$$\theta_j^o \in PUI_j^{(n,\delta)}, \quad j = 1, \dots, n_\theta. \tag{55}$$

Proof. By definition of tight parameter bounds $\underline{\theta}_j$ and $\overline{\theta}_j$ in (14), the true system parameter θ_j^o belong to PUI_j, i.e.,

$$\theta_j^o \in PUI_j, \quad j = 1, \dots, n_\theta. \tag{56}$$

Furthermore, since $\mathscr{D}^{(ss,n)} \supseteq \mathscr{D}$, then $\underline{\theta}_j^{(n)}$ and $\overline{\theta}_j^{(n)}$ computed in eqs. (45) are such that

$$\underline{\theta}_j^{(n)} \leq \underline{\theta}_j, \quad \text{for all } j = 1, \ldots, n_\theta, \tag{57}$$

$$\overline{\theta}_j^{(n)} \geq \overline{\theta}_j, \quad \text{for all } j = 1, \ldots, n_\theta, \tag{58}$$

that is,

$$PUI_j \subseteq PUI_j^{(n)}, \quad \text{for all } j = 1, \ldots, n_\theta. \tag{59}$$

Besides, from properties of LMI relaxation techniques [Lasserre (2001)], the following conditions hold

$$\underline{\theta}_{jl}^{(n,\delta)} \leq \underline{\theta}_{jl}^{(n)}, \quad \text{for all } j = 1, \ldots, n_\theta,\ l = 1, \ldots, 2^{n_\theta},\ \delta \geq \underline{\delta} \tag{60}$$

$$\overline{\theta}_{jl}^{(n,\delta)} \geq \overline{\theta}_{jl}^{(n)}, \quad \text{for all } j = 1, \ldots, n_\theta,\ l = 1, \ldots, 2^{n_\theta},\ \delta \geq \underline{\delta}. \tag{61}$$

From conditions (60)-(61) and the definitions of $\underline{\theta}_j^{(n)}$-$\overline{\theta}_j^{(n)}$ in (45) and $\underline{\theta}_j^{(n,\delta)}$-$\overline{\theta}_j^{(n,\delta)}$ in (52), the following inequalities follow straightforwardly:

$$\underline{\theta}_j^{(n,\delta)} \leq \underline{\theta}_j^{(n)}, \quad \text{for all } j = 1, \ldots, n_\theta,\ \delta \geq \underline{\delta}, \tag{62}$$

$$\overline{\theta}_j^{(n,\delta)} \geq \overline{\theta}_j^{(n)}, \quad \text{for all } j = 1, \ldots, n_\theta,\ \delta \geq \underline{\delta}, \tag{63}$$

that is,

$$PUI_j^{(n)} \subseteq PUI_j^{(n,\delta)}, \quad \text{for all } j = 1, \ldots, n_\theta. \tag{64}$$

By combining conditions (56), (59) and (64), it follows that

$$\theta_j^o \in PUI_j \subseteq PUI_j^{(n)} \subseteq PUI_j^{(n,\delta)} \tag{65}$$

as stated in the proposition. □

Proposition 3.5. *For all $n = 1, \ldots, N - na$ and $\delta \geq \underline{\delta}$, the uncertainty interval $PUI_j^{(n,\delta)}$ becomes tighter as the LMI-relaxation order δ increases, that is:*

$$PUI_j^{(n,\delta+1)} \subseteq PUI_j^{(n,\delta)}. \tag{66}$$

Furthermore, the δ-relaxed uncertainty intervals $PUI_j^{(n,\delta)}$ converge to the intervals $PUI_j^{(n)}$ as far as the LMI relaxation order goes to infinity, that is:

$$\lim_{\delta\to\infty} PUI_j^{(n,\delta)} = PUI_j^{(n)}, \tag{67}$$

where eq. (67) *means that:*

$$\lim_{\delta\to\infty} \underline{\theta}_j^{(n,\delta)} = \underline{\theta}_j^{(n)}, \quad \lim_{\delta\to\infty} \overline{\theta}_j^{(n,\delta)} = \overline{\theta}_j^{(n)}. \tag{68}$$

Proof. From properties of monotone convergence of LMI relaxation techniques [Lasserre (2001)] applied to problems (51) and the corresponding SDP-relaxed problems (52), the following conditions hold:

$$\underline{\theta}_{jl}^{(n,\delta)} \le \underline{\theta}_{jl}^{(n,\delta+1)}, \quad \text{for all } j = 1,\ldots,n_\theta,\ l = 1,\ldots,2^{n_\theta},\ \delta \ge \underline{\delta}, \tag{69}$$

$$\overline{\theta}_{jl}^{(n,\delta)} \ge \overline{\theta}_{jl}^{(n,\delta+1)}, \quad \text{for all } j = 1,\ldots,n_\theta,\ l = 1,\ldots,2^{n_\theta},\ \delta \ge \underline{\delta}. \tag{70}$$

Then

$$\underline{\theta}_j^{(n,\delta)} \le \underline{\theta}_j^{(n,\delta+1)}, \quad \text{for all } j = 1,\ldots,n_\theta,\ \delta \ge \underline{\delta}, \tag{71}$$

$$\overline{\theta}_j^{(n,\delta)} \ge \overline{\theta}_j^{(n,\delta+1)}, \quad \text{for all } j = 1,\ldots,n_\theta,\ \delta \ge \underline{\delta}. \tag{72}$$

Likewise, from (71) and (72) and the definition of $PUI_j^{(n,\delta)}$, condition (66) holds. Conditions (68) follows from convergency properties of the theory-of-moment relaxation. □

Proposition 3.6. *For all $n = 1,\ldots,N-na-1$ and $\delta \ge \underline{\delta}$, the uncertainty interval $PUI_j^{(n,\delta)}$ becomes tighter as the dynamic-horizon n increases, that is:*

$$PUI_j^{(n+1,\delta)} \subseteq PUI_j^{(n,\delta)}. \tag{73}$$

Proof. Since all the constraints defining $\mathscr{S}_s^{(ss,n)}$ in (33) also define $\mathscr{S}_s^{(ss,n+1)}$, then all the constraints describing $\mathscr{D}_l^{(ss,n)}$ also describe $\mathscr{D}_l^{(ss,n+1)}$. Therefore, from properties of LMI relaxation techniques [Lasserre (2001)], all the LMIs defining $\mathscr{D}_l^{(ss,n,\delta)}$ describe $\mathscr{D}_l^{(ss,n+1,\delta)}$, too.

This means that

$$\mathscr{D}_l^{(ss,n)} \supseteq \mathscr{D}_l^{(ss,n+1,\delta)}, \quad \text{for all } l = 1,\ldots,2^{n_\theta},\ n = 1,\ldots,N-na-1,\ \delta \geq \underline{\delta} \tag{74}$$

As a consequence,

$$\underline{\theta}_j^{(n,\delta)} \leq \underline{\theta}_j^{(n+1,\delta)}, \quad \text{for all } n = 1,\ldots,N-na-1,\ \delta \geq \underline{\delta} \tag{75}$$

$$\overline{\theta}_j^{(n,\delta)} \geq \overline{\theta}_j^{(n+1,\delta)}, \quad \text{for all } n = 1,\ldots,N-na-1,\ \delta \geq \underline{\delta} \tag{76}$$

that is,

$$PUI_j^{(n+1,\delta)} \subseteq PUI_j^{(n,\delta)}. \tag{77}$$

□

It is worth remarking that, from Proposition (3.6), the semi-static LPV relaxation technique carried out with a dynamic-horizon n greater than 1, always provides tighter parameter bounds than the ones computed through the algorithm proposed by [Cerone and Regruto (2008)], which in turn is equivalent to the semi-static relaxation for $n = 1$ as highlighted by Remark 3.3.

6. Simulated example

In this section a simulated example is presented in order to show the effectiveness of the presented identification procedure. The LPV system considered here is described by (1) with $\mathscr{A}(q^{-1},\lambda(k)) = 1+0.5\lambda(k)q^{-1}+(-0.2+0.3\lambda(k)^2)q^{-2}$ and $\mathscr{B}(q^{-1},\lambda(k)) = 0.2q^{-1} + (1\lambda(k) + 0.2\lambda(k)^2)q^{-2}$. Therefore, the true parameters vector is $\theta = [a_{1,1},\ a_{2,1},\ a_{2,2},\ b_{1,1},\ b_{2,1},\ b_{2,2}]^{\mathrm{T}} = [0.5,\ -0.2,\ 0.3,\ 0.2,\ 1,\ 0.2]^{\mathrm{T}}$ and the functions $\phi_{i,r}$ and $\psi_{j,h}$ in (4) and (5), depending on the scheduling parameter $\lambda(k)$, are $\phi_{1,1} = \lambda(k)$, $\phi_{2,1} = 1$, $\phi_{2,2} = \lambda(k)^2$, $\psi_{1,1} = 1$, $\psi_{2,1} = \lambda(k)$ and $\psi_{2,2} = \lambda(k)^2$. The input sequence $\{u(k)\}$ is a random uniform distributed signal which takes values in the interval $[-2,\ 2]$, while $\lambda(k) = 2\cos(0.2k/N)$. The output $w(k)$ and the scheduling parameter $\lambda(k)$ are corrupted by random additive noises $v(k)$ and $\varepsilon(k)$, respectively, uniformly distributed in $[-\Delta v_k,\ +\Delta v_k]$ and $[-\Delta\varepsilon_k,\ +\Delta\varepsilon_k]$. The chosen error bounds Δv_k and $\Delta\varepsilon_k$ are such that the signal to noise ratios on the output SNR_w and on the scheduling parameter SNR_λ, defined as:

$$SNR_w = 10\log\left\{\frac{\sum_{k=1}^{N} w(k)^2}{\sum_{k=1}^{N} v(k)^2}\right\};\ SNR_\lambda = 10\log\left\{\frac{\sum_{k=1}^{N} \lambda(k)^2}{\sum_{k=1}^{N} \varepsilon(k)^2}\right\},$$

are equal to 29 db and 28 db, respectively. The length of the data sequence is $N = 500$. Bounds on the parameters are evaluated by solving (52) for a relaxation order $\delta = 2$. The open source freeware software Gloptipoly by [Henrion and Lasserre (2003)], has been used to convert the original identification problems (51) into the corresponding LMI relaxed problems (52), which are numerically solved by the SDP solver SeDuMi [Sturm (1999)]. First, parameter bounds are evaluated with the algorithm proposed in [Cerone and Regruto (2008)], which is equivalent to the LPV semi-static relaxation when $n = 1$. Then, in order to show the improvement on the parameter bounds evaluation of the semi-static relaxation approach with respect to the method in [Cerone and Regruto (2008)], the dynamic-horizon n is set equal to 2. The obtained parameter bounds $\underline{\theta}_j^{(n,\delta)}$ and $\overline{\theta}_j^{(n,\delta)}$, the central estimate $\theta_j^{c(n,\delta)}$, the parameter uncertainty bounds $\Delta\theta_j^{(n,\delta)}$ and the relative parameter uncertainty bounds $\%\Delta\theta_j^{(n,\delta)}$, defined as

$$\theta_j^{c(n,\delta)} = \frac{\overline{\theta}_j^{(n,\delta)} + \underline{\theta}_j^{(n,\delta)}}{2},\ \Delta\theta_j^{(n,\delta)} = \frac{\overline{\theta}_j^{(n,\delta)} - \underline{\theta}_j^{(n,\delta)}}{2},\ \%\Delta\theta_j^{(n,\delta)} = \frac{\Delta\theta_j^{(n,\delta)}}{|\theta_j^{c(n,\delta)}|}100.$$

are reported in Table 1. The reported results show that the true value of the parameters is always included in the computed uncertainty intervals. A comparison between the results reported in Table 1, shows that the semi-static LPV relaxation with a dynamic-horizon $n = 2$ provides parameter bounds tighter than those obtained with the algorithm proposed in [Cerone and Regruto (2008)], that is, when $n = 1$. We point out that, although the values of dynamic horizon n and relaxation order δ are fairly small, the presented procedure provides satisfactory uncertainty intervals, with relative parameter uncertainty bounds $\%\Delta\theta_j^{(n,\delta)}$ smaller than 30% for dynamic-horizon $n = 2$ and signal to noise ratios $SNR_w = 29$ db and $SNR_\lambda = 28$ db. Furthermore, the central estimates $\theta_j^{c(n,\delta)}$ are close to the true value of the system parameters θ_j.

Table 1. Parameter central estimates ($\theta_j^{c(n,\delta)}$), parameter bounds ($\underline{\theta}_j^{(n,\delta)}$, $\overline{\theta}_j^{(n,\delta)}$) and parameter uncertainties $\Delta\theta_j^{(n,\delta)}$, $\%\Delta\theta_j^{(n,\delta)}$ obtained for a relaxation order $\delta = 2$ and dynamic-horizon $n = 1$ and $n = 2$.

Dynamic horizon	Parameter	True Value	$\underline{\theta}_j^{(n,\delta)}$	$\theta_j^{c(n,\delta)}$	$\overline{\theta}_j^{(n,\delta)}$	$\Delta\theta_j^{(n,\delta)}$	$\%\Delta\theta_j^{(n,\delta)}$
$n = 1$	$a_{1,1}$	0.500	0.366	0.531	0.696	0.165	31 %
	$a_{2,1}$	-0.200	-0.263	-0.206	-0.149	0.057	28 %
	$a_{2,2}$	0.300	0.213	0.327	0.442	0.115	35 %
	$b_{1,1}$	0.200	0.173	0.201	0.229	0.028	14 %
	$b_{2,1}$	1.000	0.791	1.027	1.263	0.236	23 %
	$b_{2,2}$	0.200	0.168	0.245	0.321	0.077	31 %
$n = 2$	$a_{1,1}$	0.500	0.386	0.522	0.657	0.136	26 %
	$a_{2,1}$	-0.200	-0.245	-0.206	-0.167	0.039	19 %
	$a_{2,2}$	0.300	0.226	0.321	0.417	0.096	30 %
	$b_{1,1}$	0.200	0.183	0.199	0.216	0.017	8 %
	$b_{2,1}$	1.000	0.838	1.016	1.193	0.178	18 %
	$b_{2,2}$	0.200	0.178	0.241	0.303	0.063	26 %

7. Conclusion

A new technique to evaluate parameter uncertainty intervals for LPV systems when both the output and the scheduling parameter measurements are affected by bounded noise is presented in this chapter. First, it is shown that tight parameter bounds evaluation requires the solution to a set of nonconvex optimization problems where the number of decision variables increases with the number of measured data. In order to reduce the computational complexity of the identification problem, the original feasible parameter set is approximated by the union of semialgebraic regions described by polynomial inequalities depending only on the unknown system parameters. Then, evaluation of parameter bounds is formulated as a set of semialgebraic optimization problems where the decision variables are the only parameters of the system. Guaranteed parameter uncertainty intervals are numerically computed by means of recently proposed LMI-relaxation techniques for polynomial problems. The capability of the proposed identification technique to provide less conservative parameter bounds with respect to the previously published results is shown both theoretically and by means of a numerical example.

References

Bamieh, B. A. and Giarré, L. (2002). Identification of linear parameter varying models, *International Journal of Robust Nonlinear Control* **12**, 9, pp. 841–853.

Belforte, G. and Gay, P. (2004). Optimal worst case estimation for LPV-FIR models with bounded errors, *Systems & Control letters* **53**, pp. 259–268.

Bianchi, F. and Sanchez-Pena, R. S. (2010). Robust identification/invalidation in an LPV framework, *International Journal of Robust and Nonlinear Control* **20**, 3, pp. 301–312.

Cerone, V., Piga, D. and Regruto, D. (2011). Set-membership LPV model identification of vehicle lateral dynamics, *Automatica* **47**, 8, pp. 1794–1799.

Cerone, V. and Regruto, D. (2008). Set-membership identification of LPV models with uncertain measurements of the time-varying parameter, in *Proceedings of the IEEE Conference on Decision and Control* (Cancún, Mexico).

Chesi, G., Garulli, A., Tesi, A. and Vicino, A. (2003). Solving quadratic distance problems: an LMI-based approach, *IEEE Transactions on Automatic Control* **48**, 2, pp. 200–212.

Felici, F., van Wingerden, J. W. and Verhaegen, M. (2007). Subspace identification of MIMO LPV systems using a periodic scheduling sequence, *Automatica* **43**, pp. 1684–1697.

Henrion, D. and Lasserre, J. B. (2003). Gloptipoly: global optimization over polynomials with Matlab and SeDuMi, *ACM Transactions on Mathematical Software* **29**, pp. 165–194.

Lasserre, J. (2006). Convergent semidefinite relaxations in polynomial optimization with sparsity, *SIAM Journal on Optimization* **17**, 1, pp. 822–843.

Lasserre, J. B. (2001). Global optimization with polynomials and the problem of moments, *SIAM Journal on Optimization* **11**, pp. 796–817.

Lee, L. H. and Poolla, K. (1999). Identification of linear parameter-varying systems using nonlinear programming, *ASME Journal of Dynamic Systems, Measurement, and Control* **121**, pp. 71–78.

Lopes dos Santos, P., Ramos, J. A. and Martins de Carvalho, J. L. (2008). Subspace identification of linear parameter-varying systems with innovation-type noise models driven by general inputs and a measurable white noise time-varying parameter vector, *International Journal of Systems Science* **39**, 9, pp. 897–911.

Lovera, M., Novara, C., Lopes dos Santos, P. and Rivera, D. (2011). Guest editorial special issue on applied LPV modeling and identification, *IEEE Transactions on Control Systems Technology* **19**, 1.

Milanese, M., Norton, J., Piet-Lahanier, H. and Walter, E. (eds.) (1996). *Bounding approaches to system identification* (Plenum Press, New York City, NY, USA).

Nemani, M., Ravikanth, R. and Bamieh, B. A. (1995). Identification of linear parametrically varying sistems, in *Proceedings of the IEEE Conference on Decision and Control* (New Orleans, LA, USA).

Parrillo, P. (2003). Semidefinite programming relaxations for semialgebraic problems, *Mathematical Programming* **96**, pp. 293–320.

Previdi, F. and Lovera, M. (2004). Identification of non-linear parametrically varying models using separable least squares, *International Journal of Control* **77**, 16, pp. 1382–1392.

Rugh, W. J. and Shamma, J. S. (2000). Research on gain scheduling, *Automatica* **36**, 10, pp. 1401–1425.

Sturm, J. F. (1999). Using SeDuMi 1.02, a MATLAB Toolbox for optimization over symmetric cones, *Optimization Methods and Software* **11**, 12, pp. 625–653.

Sznaier, M. and Mazzarro, C. (2003). An LMI approach to control-oriented identification and model (in)validation of LPV systems, *IEEE Transactions on Automatic Control* **48**, 9, pp. 1619–1624.

Tóth, R., Heuberger, P. S. C. and den Hof, P. V. (2010). Asymptotically optimal orthonormal basis functions for LPV system identification, *Automatica* **45**, 6, pp. 1359–1370.

Tóth, R. (2010). *Modeling and identification of linear parameter-varying systems* (Springer-Verlag, Berlin, Germany).

Verdult, V., Lovera, M. and Verhaegen, M. (2004). Identification of linear parameter-varying state-space models with application to helicopter rotor dynamics, *International Journal of Control* **77**, 13, pp. 1149–1159.

Verdult, V. and Verhaegen, M. (2002). Subspace identification of multivariable linear parameter-varying systems, *Automatica* **38**, pp. 805–814.

Verdult, V. and Verhaegen, M. (2005). Kernel methods for subspace identification of multivariable LPV and bilinear systems, *Automatica* **41**, pp. 1557–1565.

Waki, H., Kim, S., Kojima, M. and Muramatsu, M. (2006). Sums of squares and semidefinite programming relaxations for polynomial optimization problems with structured sparsity, *SIAM Journal on Optimization* **17**, 1, pp. 218–242.

Chapter 4

Set membership identification of state-space LPV systems

C. Novara

Dipartimento di Automatica e Informatica
Politecnico di Torino, 10129 Torino, Italy
carlo.novara@polito.it

A Set Membership identification method for state-space Linear Parameter Varying (LPV) systems is proposed in this chapter. The method, based on the assumption of unknown but bounded noise and on a mild regularity assumption on the system to be identified, allows the identification of almost-optimal and optimal state-space LPV models, and the evaluation of tight uncertainty bounds on the elements of the model parameter-varying matrices. A relevant feature of the proposed method is that the obtained model is sparse, i.e., it is described by a linear combination of basis functions, where the vector of linear combination coefficients has only a few non-zero elements. Two simulation examples show the effectiveness of the proposed method: identification of the forced Van der Pol oscillator, and identification and control of a 2-degrees of freedom robot manipulator.

1. Introduction

Linear Parameter Varying (LPV) models are widely used to design control systems for nonlinear systems. The main advantage of LPV models with respect to general nonlinear models is that they allow the application of linear control design techniques, such as gain-scheduling, see e.g. [Packard (1994)], [Rugh and Shamma (2000)], [Hunt *et al.* (2000)], [Lee and Chung (2001)], [Previdi and Carpanzano (2003)], optimal control, see e.g. [Hogberg *et al.* (2003)], and robust control, see e.g. [Matsumura *et al.* (1996)], [He and Yang (2006)], [Wei and del Re (2007)]. Another advantage is that the use of an LPV model structure in place of a general nonlinear structure may reduce the computational complexity of the identification process. Indeed, LPV system identification requires to approximate a function on a domain with lower dimension with respect to nonlinear system identification. Since the computational complexity involved in the approximation of a function grows with the dimension of its domain, the approximation is simpler using an LPV structure, see also Remark 4.1 below.

Several methods for the identification of LPV systems are available in the literature, based on either input-output representations, see e.g. [Campi (1994)], [Bamieh and Giarré (2002)], [Hsu *et al.* (2008)], [Cerone and Regruto (2008)], or state-space representations, see e.g. [Felici *et al.* (2007)], [Lovera and Mercere (2007)], [Lopes dos Santos *et al.* (2008)]. See also [Lovera *et al.* (2011)] and the papers in the related special issue for an overview on LPV system identification. Most of these methods rely on assumptions on the statistical properties of the noise affecting the system to be identified, such as stationarity, uncorrelation, type of distribution. However, in many practical applications, the validity of these assumptions cannot be verified and is certainly lost in the presence of modeling errors. Set Membership (SM) methods have been introduced to overcome this type of problems. Such methods have been mainly developed for Linear Time Invariant (LTI) systems [Milanese and Vicino (1991)], [Milanese and Norton (1996)], [Schweppe (1973)], [Partington (1997)], [Chen and Gu (2000)]. Only a few articles exist on SM identification of nonlinear systems [Milanese and Novara (2004)], and of input-output LPV systems [Cerone and Regruto (2008)].

In this chapter, a method for SM identification of state-space LPV systems is proposed, based on the assumption of unknown but bounded noise and on a mild regularity assumption on the system to be identified. The method allows the identification of almost-optimal and optimal state-space LPV models, and the evaluation of tight uncertainty bounds on the coefficients of the model parameter-varying matrices. A relevant feature of the proposed method is that the obtained model is sparse, i.e., it is described by a linear combination of basis functions, where the vector of linear combination coefficients has only a few non-zero elements. Sparse models are of high practical interest since their evaluation is very fast and can be performed in on-line applications where small sampling periods are used.

Two simulation examples show the effectiveness of the proposed identification method: identification of the forced Van der Pol oscillator, and identification and control of a 2-degrees of freedom robot manipulator.

The chapter is organized as follows: In Section 2, the notation and the basic notions used in the chapter are outlined. In Section 3, the LPV system identification problem is formulated in a SM framework, and the definitions of interpolatory, almost-optimal, and optimal model are provided. In Section 4, an algorithm for the identification of interpolatory models is proposed, and it is shown that these models are optimal under a suitable assumption. In this section, the sparsity properties of the models provided by the algorithm are also discussed. In Section 5, important aspects of the proposed identification method are discussed, such as basis function choice, validation of prior assumptions, and identification parameters choice. In Section 6, the two examples mentioned above are shown.

2. Notation and basic notions

A column vector is denoted by $\theta = [\theta_1 \ \theta_2 \ \cdots \ \theta_N]^T \in \mathbb{R}^{N\times 1}$, a row vector by $\theta^T = [\theta_1 \ \theta_2 \ \cdots \ \theta_N] \in \mathbb{R}^{1\times N}$. The notation $\operatorname{diag}(\theta)$ indicates a matrix with diagonal elements equal to $\theta_1, \ \theta_2, \ ..., \ \theta_N$, and with zero off-diagonal entries.

The ℓ_q norm of a vector θ is defined as

$$\|\theta\|_q \doteq \left(\sum_{i=1}^N |\theta_i|^q\right)^{\frac{1}{q}}, \ q \in [1,\infty),$$
$$\|\theta\|_\infty \doteq \max_{i=1,..,N} |\theta_i| .$$

A discrete-time signal y of length L is denoted by a column vector $y = [y(1)^T \ \cdots \ y(L)^T]^T$, where $y(k) \in \mathbb{R}^{n_y}$. The ℓ_q norm of this signal is defined as

$$\|y\|_q \doteq \left(\sum_{k=1}^L \sum_{i=1}^{n_y} |y_i(k)|^q\right)^{\frac{1}{q}}, \ q \in [1,\infty),$$
$$\|y\|_\infty \doteq \max_{k=1,..,L} \max_{i=1,..,n_y} |y_i(k)| .$$

The L_s norm of a function $f : P \subseteq \mathbb{R}^{n_p} \to \mathbb{R}$ is defined as

$$\|f\|_s \equiv \|f(\cdot)\|_s \doteq \left[\int_P |f(\rho)|^s \, d\rho\right]^{\frac{1}{s}}, \ s \in [1,\infty),$$
$$\|f\|_\infty \equiv \|f(\cdot)\|_\infty \doteq \operatorname{ess\,sup}_{\rho\in P} |f(\rho)| .$$

Consider a generic feasible optimization problem

$$\begin{aligned} \widehat{\theta} &= \arg\min_{\theta} J(\theta) \\ \text{subject to} \quad & g(\theta) \leq 0. \end{aligned}$$

If this problem admits a set of solutions, then $\widehat{\theta}$ indicates one of these solutions. Otherwise, $\widehat{\theta}$ is the unique solution.

3. Set membership identification of state-space LPV systems

Consider an LPV discrete-time system in the following state-space observability form:

$$\begin{aligned} x(k+1) &= A(p(k))\, x(k) + B(p(k))\, u(k) + w(k) \\ y(k) &= Cx(k) + v(k) \end{aligned} \tag{1}$$

$$\begin{aligned} A(p(k)) &= \begin{bmatrix} 0 & 1 & \cdots & 0 \\ \vdots & \vdots & \ddots & \vdots \\ 0 & 0 & \cdots & 1 \\ f_{n_x}(p(k)) & f_{n_x-1}(p(k)) & \cdots & f_1(p(k)) \end{bmatrix} \\ B(p(k)) &= \left[f_{n_x+1}(p(k)) \cdots f_{2n_x}(p(k)) \right]^T \\ C &= \left[1\ 0 \cdots 0 \right], \end{aligned}$$

where $x(k) \in \mathbb{R}^{n_x}$ is the state, $u(k) \in \mathbb{R}$ is the input, $y(k) \in \mathbb{R}$ is the output, $p(k) \in P \subset \mathbb{R}^{n_p}$ is the time-varying parameter vector, P is a compact set, $w(k) \in \mathbb{R}^{n_x}$ is the process noise, $v(k) \in \mathbb{R}$ is the measurement noise, $f_i : \mathbb{R}^{n_p} \to \mathbb{R}$, and $A(p(k))$, $B(p(k))$, and C are matrices/vectors of suitable dimensions. The Single Input Single Output (SISO) case $u(k) \in \mathbb{R}$, $y(k) \in \mathbb{R}$ is considered here for simplicity of notation, the generalization to the Multi Input Multi Output (MIMO) case is treated in Subsection 5.3.

Suppose that system (1) is not known, but a set of data

$$DS = \{u(k), y(k), p(t)\}_{k=1-n_x}^{L}$$

is available. In this chapter, the problem of identifying the LPV system (1) from this data set is considered. Since the LPV system (1) is univocally defined by the vector-valued function $f = [f_1 \cdots f_n]^T$, $n \doteq 2n_x$, $f : \mathbb{R}^{n_p} \to \mathbb{R}^n$, this problem can be formulated as follows.

Problem 4.1. *Identify from the data DS an estimate $\widehat{f}$ of f giving "small" identification error*

$$E\left(f,\widehat{f}\right) \doteq \left\|\xi\left(f,\widehat{f}\right)\right\|_q, \tag{2}$$

where

$$\xi\left(f,\widehat{f}\right) \doteq \left[\xi\left(f_1,\widehat{f_1}\right) \;\cdots\; \xi\left(f_n,\widehat{f_n}\right)\right]^T$$
$$\xi\left(f_i,\widehat{f_i}\right) \doteq \left\|f_i - \widehat{f_i}\right\|_s,$$

$\|\cdot\|_q$ *is a vector ℓ_q norm and* $\|\cdot\|_s$ *is a functional L_s norm evaluated over the set $P \subset \mathbb{R}^{n_p}$.* □

Note that this identification problem is not trivial, due also to the fact that $u(k)$, $y(k)$ and $p(k)$ are measured but the outputs of the functions $f_1, \ldots,$ f_n are not.

The identification error (2) is not known, since the function f is not known. What is known is that $f \in \mathscr{F}_{DS}$, where $\mathscr{F}_{DS}$ is the set of all functions that could have generated the data DS. If no assumptions are made on f, this set is unbounded even in the case of exact measurements, since the mapping generating the data from f is not injective, i.e., infinitely many f measured at $p(k)$, $k = 1, .., L$, correspond to the same input-output values $u(k), y(k)$, $k = 1, .., L$. Then, whatever estimate is chosen, no information on the identification error can be derived, unless some prior assumptions are made on the function f and on the noises $w(k)$ and $v(k)$. The typical approach in the literature is to assume a finitely parametrized functional form for f and statistical models for the noises, [Sjöberg *et al.* (1995)]. In the present chapter, following a SM approach [Milanese and Vicino (1991)], [Milanese and Norton (1996)], [Schweppe (1973)], [Partington (1997)], [Chen and Gu (2000)], [Milanese and Novara (2004)], different and somewhat weaker assumptions are considered. A mild regularity assumption is made on the function f, not requiring any knowledge on its parametric form. The noises are supposed to be unknown but bounded, avoiding statistical assumptions such as stationarity, uncorrelation, type of distribution. The following lemma is required to introduce the SM approach assumptions.

Lemma 1. *The data set* $DS = \{u(k), y(k), p(t)\}_{k=1-n_x}^{L}$ *can be represented as*

$$y = \sum_{i=1}^{n} \phi_i F_i(f) + d \tag{3}$$

where

$$
\begin{aligned}
y &= [y(1) \ \cdots \ y(L)]^T \\
d &= [d(1) \ \cdots \ d(L)]^T \\
d(k) &= \sum_{i=1}^{n_x} w_i(k-i) + v(k) \\
\phi_i &\doteq \text{diag}\left(\left[\varphi_i(1) \cdots \varphi_i(L)\right]\right) \\
\varphi(k) &\doteq [y(k-1) \ \cdots \ y(k-n_x) \ \cdots \ u(k-1) \ \cdots \ u(k-n_x)]^T \\
F_i(f) &\doteq \left[f_i(\eta_i(1)) \cdots f_i(\eta_i(L))\right]^T \\
\eta_i(k) &\doteq \begin{cases} p(k-n_x), & i \leq n_x \\ p(k+n_{x+1}-i), & i > n_x \end{cases}
\end{aligned}
\tag{4}
$$

Proof. For $w(k) = 0$, $v(k) = 0$, the state-space system (1) can be represented in the following input-output form (see [Tóth *et al.* (2007)]):

$$
y(k) = \sum_{i=1}^{n_x} f_i(p(k-n_x))\, y(k-i) + \sum_{i=1}^{n_x} f_{n_x+i}(p(k-i))\, u(k-i).
$$

Analogously, for $u(k) = 0$, the state-space system (1) can be represented in the following input-output form:

$$
y(k) = \sum_{i=1}^{n_x} f_i(p(k-n_x))\, y(k-i) + \sum_{i=1}^{n_x} w_i(k-i) + v(k).
$$

The claim follows from the superposition principle and from the definition of the matrices ϕ_i and vectors $F_i(f)$. □

The following assumption on the noise d in (3), which accounts for both the noises $w(k)$ and $v(k)$ in (1), is made in this chapter.

Assumption on noises. The noise sequence d is bounded:

$$
\|d\|_h \leq \mu. \tag{5}
$$

□

The following assumption is made on the function f. Suppose that, for each $i = 1, \ldots, n$, an approximation f_i^a of f_i has been obtained using any technique. Define the *residue function* Δ_i as

$$
\Delta_i \doteq f_i - f_i^a.
$$

Assumption on f. Each residue function Δ_i, $i = 1, \ldots, n$, is Lipschitz continuous with Lipschitz constant γ_i:

$$\Delta_i \in \mathscr{L}(\gamma_i) \doteq \{g : |g(\rho) - g(\widehat{\rho})| \leq \gamma_i \|\rho - \widehat{\rho}\|_2, \forall \rho, \widehat{\rho} \in P\}.$$

The following assumption on f follows:

$$f \in \mathscr{F}(\gamma, f^a) \doteq \{g : g = f^a + \Delta : \Delta_i \in \mathscr{L}(\gamma_i), \ i = 1, \ldots, n\} \tag{6}$$

where $\gamma = [\gamma_1 \cdots \gamma_n]^T$. □

Note that the assumption (6) is not restrictive. Indeed, it is certainly satisfied for $\gamma_i < \infty$, $i = 1, \ldots, n$, if the functions f_i and f_i^a are Lipschitz continuous. If the f_i's are not continuous and the involved discontinuities are known, these discontinuities can be accounted for by properly choosing the preliminary approximations f_i^a's, so that the residue functions satisfy the Lipschitz continuity assumption. This kind of assumption has been proposed in [Milanese and Novara (2004)], and allows us to analyze the optimality properties of an estimate when no information is available on the parametric form of f.

A necessary and sufficient condition for the validation of the assumptions (5) and (6) will be given in Section 5.2, together with guidelines on how to suitably choose γ and μ. Now, supposing that these assumptions are validated, we have that $f \in FFS$, where FFS is called the Feasible Function Set.

Definition 4.1. The *Feasible Function Set* is

$$FFS \doteq \left\{g : g \in \mathscr{F}(\gamma, f^a), \left\| y - \sum\nolimits_{i=1}^{n} \phi_i F_i(g) \right\|_h \leq \mu \right\}. \quad \square$$

According to this definition, FFS is the set of all functions consistent with prior assumptions and measured data. For a given estimate $\widehat{f}$ of f, a tight bound on the identification error $E\left(f, \widehat{f}\right)$ defined in (2) is given by the following worst-case error.

Definition 4.2. *Worst-case identification error* of an estimate $\widehat{f}$:

$$EW\left(\widehat{f}\right) \doteq \sup_{g \in FFS} E\left(g, \widehat{f}\right). \quad \square$$

An optimal estimate is defined as a function f^{opt} which minimizes the worst-case identification error.

Definition 4.3. An estimate f^{opt} is *optimal* if

$$EW\left(f^{opt}\right) = \inf_{\widehat{f}} EW\left(\widehat{f}\right). \qquad \square$$

However, finding optimal estimates may be hard or not convenient, and sub-optimal solutions are looked for. In particular, estimates called interpolatory are often considered in the literature [Traub *et al.* (1988)], [Milanese and Norton (1996)].

Definition 4.4. An estimate f^I is *interpolatory* if

$$f^I \in FFS. \qquad \square$$

A fundamental property of an interpolatory approximation is that it guarantees a worst-case error degradation of at most 2 [Traub *et al.* (1988)], [Milanese and Norton (1996)]. An approximation with this property is called almost-optimal.

Definition 4.5. An estimate f^{ao} is *almost-optimal* if

$$EW\left(f^{ao}\right) \leq 2 \inf_{\widehat{f}} EW\left(\widehat{f}\right). \qquad \square$$

Remark 4.1. The use of an LPV model structure in place of a general nonlinear structure may reduce the computational complexity of the identification process. Indeed, the identification of a nonlinear system requires to approximate a function on a n_φ-dimensional domain, where $n_\varphi = n_p + 2n_x$, while the identification of an LPV system requires to approximate a function on a n_p-dimensional domain. Since the computational complexity for the approximation of a function grows with the dimension of its domain, and this growth may even be exponential, the approximation can be significantly simpler using an LPV structure. □

4. Interpolatory and optimal estimates

Subsection 4.1 illustrates an algorithm providing an interpolatory (and thus almost-optimal) estimate in the form of a linear combination of given basis functions. Under an additional assumption on the "inner" signals $f_i\left(p(k)\right)$, the interpolatory estimate is optimal. In Subsection 4.2 it is shown that the estimate is sparse, i.e., the vector of linear combination coefficients has only a small number of non-zero elements.

4.1. *Identification algorithm*

Consider a set of basis functions $\beta_i : P \subset \mathbb{R}^{n_p} \to \mathbb{R}$, $i = 1, \ldots, m$, and define the following matrix:

$$\Psi \doteq \left[\phi_1 \boldsymbol{\beta} \cdots \phi_n \boldsymbol{\beta} \right] \in \mathbb{R}^{L \times N} \tag{7}$$

where

$$\boldsymbol{\beta} \doteq \begin{bmatrix} \beta_1 (\eta(1)) & \cdots & \beta_m (\eta(1)) \\ \vdots & \ddots & \vdots \\ \beta_1 (\eta(L)) & \cdots & \beta_m (\eta(L)) \end{bmatrix} \in \mathbb{R}^{L \times m}$$

and $N \doteq nm = 2n_x m$.

Algorithm 1

(1) Solve the following optimization problem:

$$\begin{aligned} \theta^1 &= \arg \min_{\theta \in \mathbb{R}^N} \|\theta\|_1 \\ \text{subject to} &\quad \|y - \Psi\theta\|_h \leq \mu. \end{aligned} \tag{8}$$

(2) Define the set of indices

$$\overline{\text{supp}} \left(\theta^1 \right) \doteq \left\{ i \in \{1, 2, \ldots, N\} : \theta_i^1 = 0 \right\}$$

where θ_i^1 are the components of the vector $\theta^1 = [\theta_1^1 \ \cdots \ \theta_N^1]^T$. Compute the coefficient vector θ^* according to the following optimization problem:

$$\begin{aligned} \theta^* &= \arg \min_{\theta \in \mathbb{R}^N} \|y - \Psi\theta\|_h \\ \text{subject to} &\quad \theta_i = 0, \ i \in \overline{\text{supp}} \left(\theta^1 \right). \end{aligned} \tag{9}$$

□

Consider now the function $f^* = [f_1^* \ \cdots \ f_n^*]^T$ defined by

$$f_i^* (\rho) = \sum_{j=1}^{m} \alpha_{ij}^* \beta_j (\rho) , \tag{10}$$

where $\rho \in P \subset \mathbb{R}^{n_p}$, the coefficients α_{ij}^* are such that

$$\theta^* = \left[\alpha_{11}^* \cdots \alpha_{1m}^* \cdots \alpha_{n1}^* \cdots \alpha_{nm}^* \right]^T \in \mathbb{R}^{N \times 1},$$

and θ^* is the parameter vector provided by Algorithm 1.

The following theorem shows that f^* is an interpolatory (and thus almost-optimal) estimate of the unknown function f. The theorem also provides an explicit expression of the worst-case identification error. Let us define the following function:

$$\overline{\Delta}_i(\rho, \lambda_i) \doteq \min_{k=1,\ldots,L} \left(\lambda_i(k) + \gamma \left\|\rho - \eta_i(k)\right\|_2\right) \tag{11}$$

where $\lambda_i = [\lambda_i(1) \;\cdots\; \lambda_i(L)]^T \in R^{L\times 1}$. Let us also define the following quantities:

$$\begin{aligned} \overline{\xi}(\lambda) &\doteq \left[\overline{\xi}_1(\lambda_1) \cdots \overline{\xi}_n(\lambda_n)\right]^T \\ \overline{\xi}_i(\lambda_i) &\doteq \left\|\overline{\Delta}_i(\cdot, \lambda_i)\right\|_s . \end{aligned} \tag{12}$$

Theorem 1. *Let assumptions (5) and (6) hold with $f^a = f^*$. Then, for any ℓ_q and ℓ_h vector norms, and any L_s functional norm:*
(i) The estimate f^* is interpolatory (and thus almost-optimal).
(ii) The worst-case identification error of f^* is given by

$$EW(f^*) = 2 \max_{\lambda: \left\|y - \sum_{i=1}^{n} \phi_i(F_i(f^*)+\lambda_i)\right\|_h \le \mu} \left\|\overline{\xi}(\lambda)\right\|_q . \tag{13}$$

Proof. (i) In the proof of Theorem 3 in Section 5.2, it is shown that $f^* \in FFS$. This implies that the approximation f^* is interpolatory, see Definition 4.4. Moreover, any interpolatory approximation is almost-optimal, see [Traub *et al.* (1988)], [Milanese and Norton (1996)].

(ii) It has been assumed that $\Delta_i \in \mathscr{L}(\gamma_i)$, i.e., that Δ_i is Lipschitz continuous with Lipschitz constant γ_i. Then,

$$\Delta_i(\rho) \le \Delta_i(\eta_i(k)) + \gamma_i \left\|\rho - \eta_i(k)\right\|_2, \; \forall \rho \in P \tag{14}$$

for $k = 1, \ldots, L$, which yields

$$\Delta_i(\rho) \le \min_{k=1,\ldots,L} \left(\Delta_i(\eta_i(k)) + \gamma_i \left\|\rho - \eta_i(k)\right\|_2\right), \; \forall \rho \in P.$$

Now, suppose that $\lambda_i(k) = \Delta_i(\eta_i(k))$. Then,

$$\begin{aligned} \Delta_i(\rho) &\le \min_{k=1,\ldots,L} \left(\lambda_i(k) + \gamma_i \left\|\rho - \eta_i(k)\right\|_2\right) \\ &\doteq \overline{\Delta}_i(\rho, \lambda_i), \; \forall \rho \in P. \end{aligned}$$

That is, the function $\overline{\Delta}_i(\cdot, \lambda_i)$ is an upper bound of the unknown function $\Delta_i(\cdot)$.

For given $\rho, \widehat{\rho} \in P$, let

$$j \doteq \arg \min_{k=1,\ldots,L} \left(\lambda_i(k) + \gamma_i \left\|\rho - \eta_i(k)\right\|_2\right),$$
$$\widehat{j} \doteq \arg \min_{k=1,\ldots,L} \left(\lambda_i(k) + \gamma_i \left\|\widehat{\rho} - \eta_i(k)\right\|_2\right)$$

Then,

$$\begin{aligned}\overline{\Delta}_i\left(\rho, \lambda_i\right) &= \lambda_i(j) + \gamma_i \left\|\rho - \eta_i(j)\right\|_2 \\ &\leq \lambda_i(\widehat{j}) + \gamma_i \left\|\rho - \eta_i(\widehat{j})\right\|_2, \\ \overline{\Delta}_i\left(\widehat{\rho}, \lambda_i\right) &= \lambda_i(\widehat{j}) + \gamma_i \left\|\widehat{\rho} - \eta_i(\widehat{j})\right\|_2.\end{aligned}$$

This yields

$$\begin{aligned}\overline{\Delta}_i\left(\rho, \lambda_i\right) - \overline{\Delta}_i\left(\widehat{\rho}, \lambda_i\right) &\leq \gamma_i \left\|\rho - \eta_i(\widehat{j})\right\|_2 \\ -\gamma_i \left\|\widehat{\rho} - \eta_i(\widehat{j})\right\|_2 &\leq \gamma_i \left\|\rho - \widehat{\rho}\right\|_2,\end{aligned}$$

which shows that $\overline{\Delta}_i\left(\cdot, \lambda_i\right)$ is Lipschitz continuous with Lipschitz constant γ_i, i.e., that $\overline{\Delta}_i\left(\cdot, \lambda_i\right) \in \mathscr{L}\left(\gamma_i\right)$. Being $\overline{\Delta}_i\left(\cdot, \lambda_i\right)$ an upper bound of $\Delta_i\left(\cdot\right)$, this implies that

$$\sup_{\Delta_i \in \mathscr{L}(\gamma_i)} \Delta_i\left(\rho\right) = \overline{\Delta}_i\left(\rho, \lambda_i\right), \ \forall \rho \in P,$$

i.e., $\overline{\Delta}_i\left(\cdot, \lambda_i\right)$ is the tightest upper bound of the unknown function $\Delta_i\left(\cdot\right)$. Similarly, it can be shown that $-\overline{\Delta}_i\left(\cdot, \lambda_i\right)$ is the tightest lower bound of $\Delta_i\left(\cdot\right)$. It follows that $\left\|f_i - f_i^*\right\|_s$ is tightly bounded as

$$\left\|f_i - f_i^*\right\|_s = \left\|\Delta_i\right\|_s \leq 2 \left\|\overline{\Delta}_i\left(\cdot, \lambda_i\right)\right\|_s \equiv \overline{\xi}_i\left(\lambda_i\right). \tag{15}$$

However, the sequence $\lambda_i(k)$ is not known since it depends on Δ_i, which is not known. What is known from assumption (5) is that

$$\left\|y - \sum_{i=1}^{n} \phi_i F_i\left(f\right)\right\|_h \leq \mu.$$

I.e.,

$$\left\|y - \sum_{i=1}^{n} \phi_i \left(F_i\left(f^*\right) + \lambda_i\right)\right\|_h \leq \mu.$$

This claim follows from (15) and from the definition of worst-case identification error. □

The computation of the worst-case identification error (13) is usually hard, and can be performed only for small dimensions n and n_p. The main difficulty stems from the fact that the internal signals $f_i(\eta_i(k))$ are not measured, but only the external signals $y(k)$ and $u(k)$ are. This lack of information implies the presence of "max" in (13), and this maximum can be easily evaluated only for small n and n_p. In order to de-convolve the individual signals $f_i(\eta_i(k))$ and to allow the computation of EW for large dimensions, an additional assumption is required.

Let λ^* be the solution of the optimization problem in (13). From (11), we have that the function $\overline{\Delta}_i(\rho, \lambda_i^*)$ is bounded for $\rho = \eta_i(k)$, $k = 1, \ldots, L$, $i = 1, \ldots, n$, since λ^* is subject to the constraint in (13). This implies that the sequence $\Delta_i(\eta_i(k))$, $k = 1, \ldots, L$, is also bounded, since $\overline{\Delta}_i(\rho, \lambda_i^*)$ is an upper bound of $\Delta_i(\rho)$ for any $\rho \in P$ (see the proof of Theorem 1). Let us then assume that a bound ε_i on $\Delta_i(\eta_i(k))$ is known for all $k = 1, \ldots, L$, $i = 1, \ldots, n$.

Assumption on δ_i. The sequence $\Delta_i(\eta_i(k))$, $k = 1, \ldots, L$, is bounded:

$$|\Delta_i(\eta_i(k))| \le \varepsilon_i \tag{16}$$

for $i = 1, \ldots, n$. □

Note that this assumption is reasonable, since the bounds $\varepsilon_1 \cdots \varepsilon_n$ can be properly estimated (together with the Lipschitz constants $\gamma_1 \cdots \gamma_n$), see Section 5.2. As shown by the following theorem, under this additional assumption, the estimate f^* is optimal. The theorem also gives an expression of the worst-case error computable for any dimensions n and n_p (see [Milanese and Novara (2007)]), and provides tight point-wise bounds on $f_i(\rho)$ for any $\rho \in P$.

Theorem 2. *Let assumptions (5) and (16) hold, and assumption (6) hold with $f^a = f^*$. Then, for any ℓ_q and ℓ_h vector norms, and any functional norm L_s :*
(i) The estimate f^* is optimal.
(ii) The worst-case identification error of f^* is given by

$$EW(f^*) = \left\| \overline{\xi}(\varepsilon) \right\|_q .$$

(iii) For any $\rho \in P$ and $i = 1, \ldots, n$, $f_i(\rho)$ is tightly bounded as

$$f_i^*(\rho) - \overline{\Delta}_i(\rho, \varepsilon_i) \le f_i(\rho) \le f_i^*(\rho) + \overline{\Delta}_i(\rho, \varepsilon_i) . \tag{17}$$

Proof. (iii) From (14) and (16), we have that

$$\Delta_i(\rho) \le \Delta_i(\eta_i(k)) + \gamma_i \|\rho - \eta_i(k)\|_2 \\ \le \varepsilon_i + \gamma_i \|\rho - \eta_i(k)\|_2, \ \forall \rho \in P,$$

for $k = 1, \ldots, L$. This yields

$$\Delta_i(\rho) \le \min_{k=1,\ldots,L} (\varepsilon_i + \gamma_i \|\rho - \eta_i(k)\|_2) \equiv \overline{\Delta}_i(\rho, \varepsilon_i), \ \forall \rho \in P,$$

which shows that the function $\overline{\Delta}_i(\cdot, \varepsilon_i)$ is an upper bound of the unknown function $\Delta_i(\cdot)$. Following the same argumentation of the proof of Theorem 1, it can be seen that $\overline{\Delta}_i(\cdot, \varepsilon_i)$ is the tightest upper bound of $\Delta_i(\cdot)$. Similarly, it can be seen that $-\overline{\Delta}_i(\cdot, \varepsilon_i)$ is the tightest lower bound of $\Delta_i(\cdot)$. The claim follows from the fact that $f(\rho) = f^*(\rho) + \Delta(\rho)$, $\forall \rho \in P$.

(ii) As shown by statement (iii), $\overline{f}_i(\rho) \doteq f_i^*(\rho) - \overline{\Delta}_i(\rho, \varepsilon_i)$ and $\underline{f}_i(\rho) \doteq f_i^*(\rho) + \overline{\Delta}_i(\rho, \varepsilon_i)$ are tight lower and upper bounds of $f_i(\rho), \forall \rho \in P$. Moreover, $f_i^*(\rho)$ is the mean of these two bounds. It follows that

$$\sup_{g \in FFS} |g_i(\rho) - f_i^*(\rho)| = \max\left\{\overline{f}_i(\rho) - f_i^*(\rho), f_i^*(\rho) - \underline{f}_i(\rho)\right\} \\ = \overline{\Delta}_i(\rho, \varepsilon_i), \forall \rho \in P.$$

The claim follows from the definition of worst-case identification error.

(i) Since $f_i^*(\rho)$ is the mean of the two tight bounds, we have also that

$$\sup_{g \in FFS} |g_i(\rho) - f_i^*(\rho)| = \min_{\widehat{f}} \sup_{g \in FFS} \left| g_i(\rho) - \widehat{f}(\rho) \right|, \forall \rho \in P.$$

The claim follows from the fact that this hold for any $i = 1, \ldots, n$. □

Remark 4.2. The point-wise bounds (17) provide an interval estimate of the unknown value $f_i(\rho)$. Interval estimates are important in system and control applications. Indeed, these estimates can be used e.g. for robust control design, prediction interval evaluation and fault detection. □

4.2. *Sparsity properties*

A *sparse* function is a linear combination of many basis functions, where the vector of linear combination coefficients is sparse, i.e., it has only a few non-zero elements. The *sparsity* of a vector θ is typically measured by the ℓ_0 quasi-norm, defined as the number of its non-zero elements:

$$\|\theta\|_0 \doteq \mathrm{card}(\mathrm{supp}(\theta)), \tag{18}$$

where card $(\cdot)$ is the set cardinality, and supp (θ) is the support of θ, defined as the set of indices at which θ is not null:

$$\operatorname{supp}(\theta) \doteq \{i \in \{1,2,\ldots,N\} : \theta_i \neq 0\}.$$

Sparse identification can thus be performed by looking for a coefficient vector of the basis function linear combination with a small ℓ_0 quasi-norm, a task which can be accomplished by solving the following optimization problem:

$$\begin{aligned} \theta^0 &= \arg\min_{\theta \in \mathbb{R}^N} \|\theta\|_0 \\ \text{subject to} & \quad \|y - \Psi\theta\|_h \leq \mu. \end{aligned} \tag{19}$$

However, the ℓ_0 quasi-norm is a non-convex function and its minimization is in general an NP-hard problem. Two main approaches are commonly adopted to deal with this issue: convex relaxation and greedy algorithms, [Tropp (2004)], [Fuchs (2005)], [Tropp (2006)], [Donoho *et al.* (2006)]. In convex relaxation, a suitable convex function, e.g. the ℓ_1 norm, is minimized instead of the ℓ_0 quasi-norm, [Fuchs (2005)], [Tropp (2006)], [Donoho *et al.* (2006)]. Indeed, the ℓ_1 norm is the convex envelope of the ℓ_0 quasi-norm, and its minimization yields a sparse vector [Fuchs (2005)], [Tropp (2006)], [Donoho *et al.* (2006)]. In greedy algorithms, the sparse solution is obtained iteratively [Tropp (2004)].

Algorithm 1 presented in Section 4 is essentially an improved ℓ_1 algorithm: In Step 1, an optimization problem similar to (19) is solved, where the ℓ_0 quasi-norm is replaced by the ℓ_1 norm. The vector θ^1 derived in step 1 is thus sparse and guarantees a bounded prediction error $\left\|y - \Psi\theta^1\right\|_h$. However, θ^1 is not ensured to give the minimum prediction error among all vectors with the same support. Then, a vector θ^* is obtained in step 2, having the same support of θ^1 and giving a minimum prediction error.

Algorithm 1, thus, provides an estimate θ^* of θ^0 where θ^0 is a maximally sparse coefficient vector, solution of the non-convex optimization problem (19). In [Novara (2011)], a condition is given, under which θ^* has the same support as θ^0, and is thus a maximally sparse vector as well.

Remark 4.3. The form (10), where α_i^* is sparse, is of high practical interest since the evaluation of $f_i^*(\rho)$ is very "fast" and can be performed in on-line applications where small sampling periods are used. □

5. Important aspects of the identification process

In this section, several aspects of the proposed identification procedure are discussed, such as basis function choice, validation of prior assumptions, identification parameter choice, and extension to the MIMO case.

5.1. *Basis function choice*

One important step required before using any parametric identification algorithm is the choice of the basis functions β_i. In several cases of practical interest, it is known *a priori* that the basis functions β_i belong to some large set of functions, see e.g. the example presented in Section 6.1 and the example shown in [Novara (2011)], related to identification of vehicle lateral dynamics. The sparse identification algorithm 1 can be applied in these cases in order to select within this large set only the basis functions which are important for providing an accurate description of the system under investigation.

In other cases, the basis functions are not known *a priori* and their choice can be carried out considering the numerous options available in the literature (e.g. gaussian, sigmoidal, wavelet, polynomial, trigonometric), see [Sjöberg *et al.* (1995)]. Widely used are the gaussian basis functions, defined as

$$\beta_i(\rho) = \exp\left(-\left\|Q\left(\rho - \rho_i(k)\right)\right\|^2\right) \tag{20}$$

where $Q \in \mathbb{R}^{n_p \times n_p}$ is a diagonal matrix where the kth element is proportional to the width of the function along the dimension k. An optimal choice of the diagonal elements of Q can be performed by means of Lemma 2 in [Milanese and Novara (2004)]. These functions are universal approximators, see e.g. [White (1991)], and will be used in the examples presented in Sections 6.1 and 6.2. Polynomial basis functions are also largely used. In Section 6.1, gaussian and polynomial basis functions will be compared in an example where the system to be identified is actually of a polynomial form.

An inappropriately chosen family of basis functions can force the retention of many basis functions by the sparse identification algorithm, and this leads to an increased model complexity and, in some cases, to a large identification error. The approach which can be adopted in order to address this important issue is to use different types of basis functions, e.g. gaussian, sigmoidal, polynomial, gaussian with different wight matrices Q, gaussian with different centers, etc. The larger the set of basis functions, the more succinct is the set of basis function selected by the sparse identification algorithm.

5.2. *Validation of prior assumptions and identification parameter choice*

As typical in any identification theory, the problem of checking the validity of prior assumptions arises. It is usual to introduce the concept of prior assumption validation as consistency with the available data: the prior assumptions are considered validated if at least an estimate consistent with these assumptions and the data exists, i.e., if the Feasible Function Set FFS is not empty, [Milanese and Norton (1996)], [Chen and Gu (2000)], [Milanese and Novara (2004)].

Definition 4.6. Prior assumptions are validated if $FFS \neq \varnothing$. □

The following theorem gives a necessary and sufficient condition for the validation of prior assumptions.

Theorem 3. *$FFS \neq \varnothing$ if and only if the optimization problem (8) is feasible.*

Proof. If the optimization problem (8) is feasible, then the function f^* defined in (10) exists. Now consider that $f^* = f^* + \Delta$, with $\Delta = 0$. Obviously, if $\Delta = 0$, then $\Delta \in \mathscr{L}(\gamma)$, for any $\gamma \geq 0$. Moreover,

$$\left\| y - \sum_{i=1}^{n} \phi_i F_i\left(f^*\right) \right\|_h = \|y - \Psi\theta\|_h .$$

Thus, from (8), we have that

$$\left\| y - \sum_{i=1}^{n} \phi_i F_i\left(f^*\right) \right\|_h = \|y - \Psi\theta\|_h \leq \mu$$

It follows that $f^* \in FFS$, showing that $FFS \neq \varnothing$.

Vice-versa: The fact that $f^a = f^*$ and the non-emptyness of FFS directly imply the existence of the function f^*. But this function is computed by solving the optimization problem (8). This implies the feasibility of this optimization problem. □

The above validation Theorem 3 gives indications on the choice of the bound on noise μ and of the basis functions β_i: When μ is (approximately) known from some prior information available on the noises acting on the system, the feasibility of the optimization problem (8) implies that the basis functions have been correctly chosen, since they are able to explain the measured data. On the contrary, the non-feasibility of the optimization problem (8) indicates that a different choice of the basis functions is needed. When no prior information on the noises is available, μ can be taken slightly larger than the minimum value for which the optimization problem (8) is feasible. Clearly, this value depends on the basis functions, which, thus, should be chosen in order to allow a "small" value of μ.

Once the basis functions and μ have been chosen, Algorithm 1 can be applied to the measured data, and the interpolatory estimate f^* can be obtained according to (10).

The parameters $\gamma = [\gamma_1 \ \cdots \ \gamma_n]^T$ and $\varepsilon = [\varepsilon_1 \ \cdots \ \varepsilon_n]^T$ are not required to identify f^*. They are only needed to evaluate the bounds in (17). For each function f_i^*, $i = 1, \ldots, n$, the parameters γ_i and ε_i can be chosen following the procedure presented in [Milanese and Novara (2004)].

5.3. *Extension to MIMO systems*

Consider a discrete-time LPV system with state $x(k) \in \mathbb{R}^{n_x}$, input $u(k) \in \mathbb{R}^{n_u}$, output $y(k) \in \mathbb{R}^{n_y}$, and time-varying parameter vector $p(k) \in P \subset \mathbb{R}^{n_p}$. Then, this system can be described in the following state-space observability form:

$$\begin{aligned} x(k+1) &= A(p(k))\,x(k) + B(p(k))\,u(k) + w(k) \\ y(k) &= Cx(k) + v(k) \end{aligned} \tag{21}$$

$$A(p(k)) = \begin{bmatrix} \mathbf{0} & I & \cdots & \mathbf{0} \\ \vdots & \vdots & \ddots & \vdots \\ \mathbf{0} & \mathbf{0} & \cdots & I \\ f_\eta(p(k)) & f_{\eta-1}(p(k)) & \cdots & f_1(p(k)) \end{bmatrix}$$

$$\begin{aligned} B(p(k)) &= \left[\, f_{\eta+1}(p(k)) \ \cdots \ f_{2\eta}(p(k)) \,\right]^T \\ C &= \left[\, I \ \mathbf{0} \cdots \mathbf{0} \,\right], \end{aligned}$$

where $\eta : \eta n_y \geq n_x$, I is the $n_y \times n_y$ identity matrix, $\mathbf{0}$ is the $n_y \times n_y$ zero matrix, and

$$f_i : \mathbb{R}^{n_p} \to \begin{cases} \mathbb{R}^{n_y \times n_y}, \ 1 \leq i \leq \eta, \\ \mathbb{R}^{n_y \times n_u}, \ \eta < i \leq 2\eta. \end{cases}$$

Note that the SISO system (1) is in minimal form, see [Tóth *et al.* (2007)], while the MIMO system (21) is in minimal form only if $\eta n_y = n_x$.

Algorithm 1 can be applied to identify an interpolatory (or optimal) model of the MIMO system (21) without significant modifications with respect to the SISO case: In the MIMO case, the regressor $\varphi(k)$ in (4) has to be defined as

$$\varphi(k) \doteq [y(k-1)^T \ \cdots \ y(k-\eta)^T \ \cdots \ u(k-1)^T \ \cdots \ u(k-\eta)^T]^T,$$

where $y(k) \in \mathbb{R}^{n_y}$ and $u(k) \in \mathbb{R}^{n_u}$. The matrix Ψ in (7) has to be defined as

$$\Psi \doteq \left[\phi_1 \boldsymbol{\beta} \cdots \phi_{\eta(n_y+n_u)} \boldsymbol{\beta}\right],$$

where $\phi_i \doteq \mathrm{diag}([\varphi_i(1) \ \cdots \ \varphi_i(L)])$. In Algorithm 1, $\|y - \Psi\theta\|_h$ has to be replaced by $\|y_j - \Psi\theta\|_h$, where $y_j = [y_j(1) \ \cdots \ y_j(L)]^T$ and $y_j(k)$ is the jth component of $y(k)$. For each $j \in \{1, \ldots, n_y\}$, Algorithm 1 provides an interpolatory (or optimal) estimate of the jth row of the functions f_i, $i = 1, \ldots, 2\eta$. An interpolatory (or optimal) estimate of the function $f = [f_1 \cdots f_{2\eta}]^T$, which univocally defines the system (21), can thus be obtained by applying Algorithm 1 for $j = 1, \ldots, n_y$.

6. Examples

In this section, two simulation examples illustrate the proposed identification method. The first regards the identification of the forced Van der Pol oscillator, the second is about the identification and control of a 2-degrees of freedom robot manipulator.

6.1. *Example 1: Identification of the forced Van der Pol oscillator*

The forced van der Pol oscillator is an oscillator with nonlinear damping (see e.g. [Ott (1993)]) described by the following differential equation:

$$\ddot{y}(t) - \sigma\left(1 - p(t)^2\right)\dot{y}(t) + y(t) = u(t). \tag{22}$$

Choosing $z(t) = [y(t)\ \dot{y}(t)]^T$ as the state vector, this system can be written in the following LPV state-space form:

$$\begin{aligned} \dot{z}(t) &= A^c(p(t))z(t) + B^c u(t) \\ y(t) &= Cz(t), \end{aligned} \tag{23}$$

where $p(t) = y(t)$ and

$$A^c(p(t)) = \begin{bmatrix} 0 & 1 \\ -1 & \sigma\left(1 - p(t)^2\right) \end{bmatrix}, \quad B^c = \begin{bmatrix} 0 \\ 1 \end{bmatrix}, \quad C = \begin{bmatrix} 1\ 0 \end{bmatrix}.$$

The continuous-time state-space equations (23) have been discretized by means of the forward Euler method, yielding the following discrete-time state-space LPV system:

$$\begin{aligned} z(k+1) &= A^d(p(k))z(k) + B^d u(k) \\ y(k) &= Cz(k), \end{aligned} \tag{24}$$

where $z(k) = z(T_s k)$, $u(k) = u(T_s k)$, $y(k) = y(T_s k)$, $p(k) = p(T_s k)$, $A^d(p(k)) = I + T_s A^c(p(T_s k))$, $B^d = T_s B^c$, $k = 1, 2, \ldots$, and T_s is the sampling time. Then, the discrete-time LPV system (24) has been implemented in the observability form (1):

$$\begin{aligned} x(k+1) &= A(p(k))\,x(k) + B(p(k))\,u(k) \\ y(k) &= Cx(k) + v(k), \end{aligned} \tag{25}$$

where

$$A(p(k)) = \begin{bmatrix} 0 & 1 \\ f_2(p(k)) & f_1(p(k)) \end{bmatrix}, \quad B(p(k)) = \begin{bmatrix} f_3(p(k)) \\ f_4(p(k)) \end{bmatrix} \tag{26}$$

and a measurement noise $v(k)$ has been added. This observability form has been obtained by means of the transformation

$$\begin{aligned} A(p(k)) &= OA^d(p(k))\,O^{-1} \\ B &= OB^d, \end{aligned}$$

where O is the observability matrix

$$O \doteq \begin{bmatrix} C \\ CA^d(p(k)) \end{bmatrix}$$

which, in this particular case, turns out to be constant in time. Note that this is a Lyapunov transformation, which thus preserves the system trajectories, see [Tóth *et al.* (2007)].

A set $DS = \{u(k), y(k)\}_{k=1-n_x}^{L}$ of $L = 3000$ data has been generated by simulation of the system (25), with a sampling time $T_s = 0.03$ s and a parameter value $\sigma = 5$. In this simulation, the following input signal has been used:

$$u(k) = \begin{cases} 0, \text{ if } 500 < k < 1000 \text{ or } 2000 < k < 2500, \\ 20\sin(0.07k) + 20\sin(0.1k+2), \text{ otherwise,} \end{cases}$$

while $v(k)$ has been simulated as a white gaussian noise with zero mean and standard deviation 0.06 (corresponding to a noise-to-signal standard deviation ratio of about 3%). Note that, for these input signal and parameter values, the system displays a chaotic behavior. The input signal and the corresponding output signal (corrupted by noise) are shown in Figure 1, while the corresponding trajectory in the z plane is shown in Figure 2.

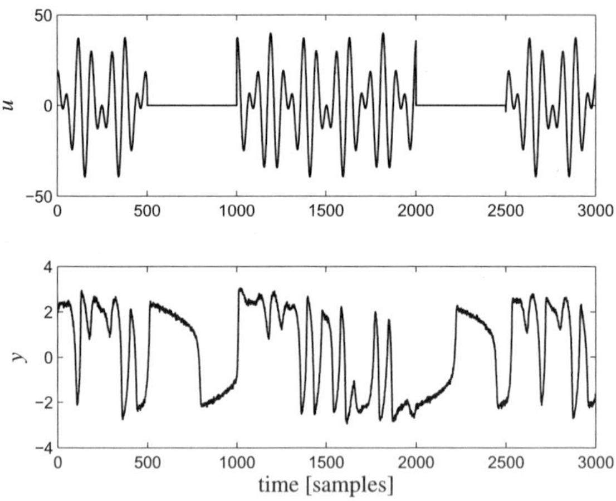

Fig. 1. Van der Pol oscillator. Input and output signals.

Two discrete-time LPV models have been identified from these data by means of Algorithm 1, using $h = 2$ in (8) and (9), and $\mu = 0.02\sqrt{L}$ in (8).

Polynomial basis function model

This model has been obtained by exploiting the *a priori* information that the "true" system (25) is of a polynomial form. Thus, the following polynomial basis functions have been used:

$$\beta_i(\rho) = (\rho - \rho_l)^j,$$

where $j, l = 0, \ldots, 4$, $\rho_l \in \{0, -2, -1, 1, 2\}$, and $i = 5l + j + 1$, giving a total number of $m = 25$ basis functions. Among these, the sparse Algorithm

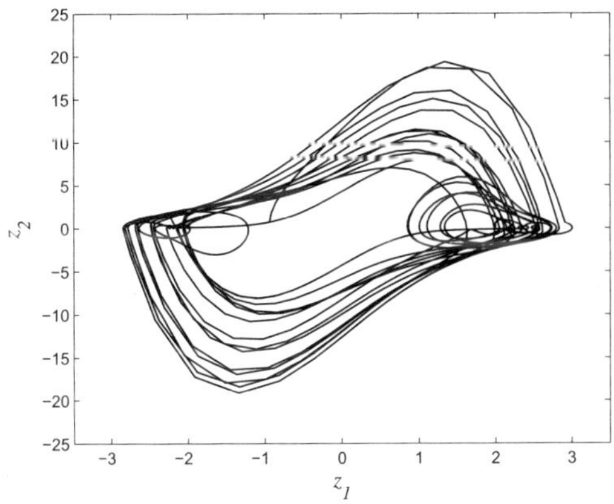

Fig. 2. Van der Pol oscillator. State trajectory.

1 selected 2 basis functions, which resulted to be the "true" ones, i.e., $\beta_1(\rho) = 1$ and $\beta_3(\rho) = \rho^2$. Note that the parameter vector θ^* provided by the algorithm satisfies the maximum sparsity condition of [Novara (2011)].

The comparison between the "true" functions f_i in (26) and the estimates f_i^*, $i = 1, \ldots, 4$, provided by Algorithm 1 is shown in Figure 3. It can be noted that the "true" functions and their respective estimates practically coincide.

Gaussian basis function model

This model has been obtained without exploiting any *a priori* information on the "true" system functional form. $m = L = 3000$ Gaussian basis functions (20) have been considered, where the diagonal elements of the matrix Q have been optimally chosen by means of Lemma 2 in [Milanese and Novara (2004)]. Among these 3000 basis functions, the sparse identification Algorithm 1 selected 8 basis functions.

The comparison between the "true" functions f_i in (26) and the estimates f_i^*, $i = 1, \ldots, 4$, provided by Algorithm 1 is shown in Figure 4. Notably, the estimates are very close to the "true" functions. In Figure 4, the guaranteed bounds (17) are also shown. These bounds provide a description of the uncertainty associated with the identification process and can be used, for example, for robust control design.

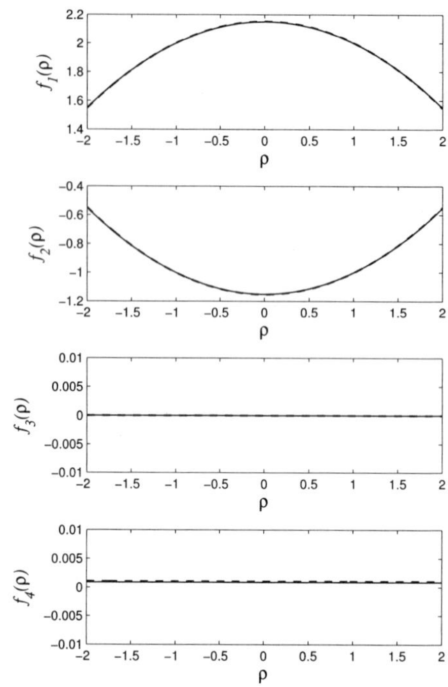

Fig. 3. Van der Pol oscillator. Polynomial basis function model. Continuous line: "true" function f_i. Dashed line: estimate f_i^*.

6.2. *Example 2: Identification and control of a 2-DOF robot manipulator*

The 2-DOF (2-degrees of freedom) robot manipulator depicted in Figure 5 has been considered in this example. In this Figure, ζ_1 and ζ_2 are the angular positions of the two segments of the robot arm, τ_1 and τ_2 are the control torques acting on these segments, l_1 and l_2 are the segment lenghts, and M_1 and M_2 are the segment masses. This robot manipulator can be described by the following continuous-time state-space LPV model:

$$\begin{aligned} \dot{z}(t) &= A^c(p(t))z(t) + B^c(p(t))u(t) \\ y(t) &= Cz(t), \end{aligned} \tag{27}$$

where $z(t) = [\zeta_1(t)\ \zeta_2(t)\ \dot{\zeta}_1(t)\ \dot{\zeta}_2 v]^T$, $u(t) = [\tau_1(t)\ \tau_2(t)]^T$, $y(t) = [\zeta_1(t)\ \zeta_2(t)]^T$, $p(t) = z(t)$, $A^c(p(t)) \in \mathbb{R}^{4\times 4}$, $B^c(p(t)) \in \mathbb{R}^{4\times 2}$, $C = [I\ \mathbf{0}]$, I is the 2×2 identity matrix, and $\mathbf{0}$ is the 2×2 zero matrix. The expression of

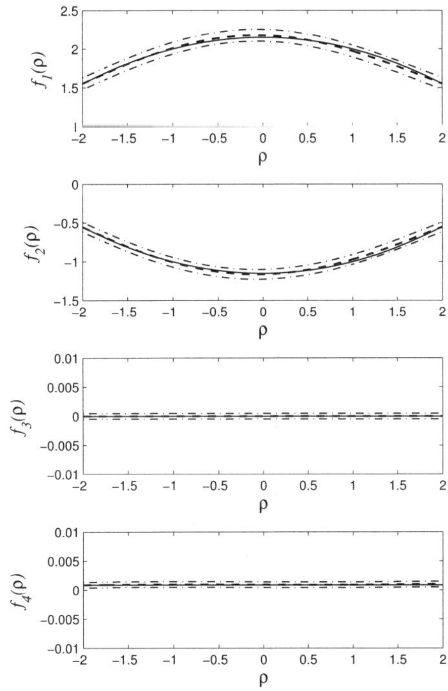

Fig. 4. Van der Pol oscillator. Gaussian basis function model. Continuous line: "true" function f_i. Dashed line: estimate f_i^*. Dot-dashed line: lower and upper bound of f_i.

matrices $A^c(p(t))$ and $B^c(p(t))$ can be found in [Kwiatkowski and Werner (2005)].

A set $DS = \{u(k), y(k)\}_{k=1-n_x}^{L}$ of $L = 5000$ data has been generated by simulation of the system (27), where $l_1 = 0.8$ m, $l_2 = 0.7$ m, $M_1 = 2.5$ Kg, $M_2 = 2$ Kg. The data have been collected with a sampling time $T_s = 0.02$ s, using the following input signals:

$$u_j(k) = \begin{cases} -20z_j(k), \text{ if } |z_j(k)| \geq 1.75 \text{ rad} \\ 0, \quad \text{if } l < k \leq l + 500, \ l = 500, 1500, 2500, 3500, \\ \qquad \text{and } |z_j(k)| < 1.75 \\ U \sin(\omega_{j1}k) + U \sin(\omega_{j2}k), \quad \text{otherwise,} \end{cases} \tag{28}$$

where $U = 100$ Nm, $\omega_{11} = 0.07$ rad/s, $\omega_{12} = 0.8$ rad/s $\omega_{21} = 0.08$ rad/s $\omega_{11} = 0.9$ rad/s, and $j = 1, 2$. The feedback input on the first line of (28) has been applied in order to limit the working range of $z_1(k)$ and $z_2(k)$ to the interval $[-\pi, \pi]$ rad (the gain -20 and the threshold 1.75 rad

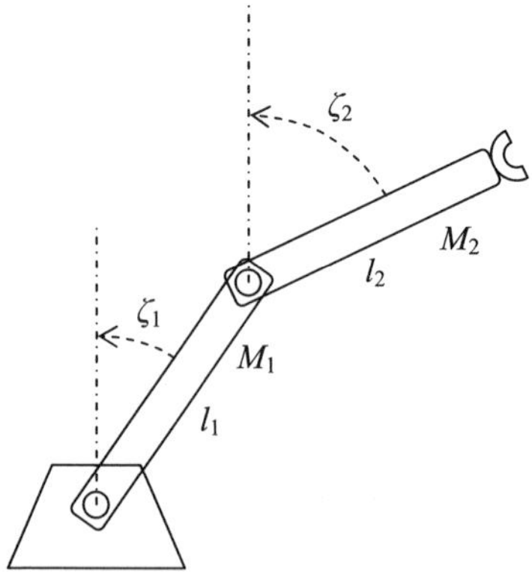

Fig. 5. Robot Manipulator.

have been chosen through several preliminary simulations). Measurement noises $v_j(k)$, $j = 1, 2$, have been added to $y_j(k)$, $j = 1, 2$, simulated as uniform noises with amplitude 0.01 rad. Note that angular position sensors in modern robots guarantee errors smaller than 10^{-5} rad.

A discrete-time LPV model has been identified from the data set DS by means of Algorithm 1, considering the extension to MIMO systems discussed in Subsection 5.3. For both outputs $y_1(k)$ and $y_2(k)$, $m = L = 5000$ Gaussian basis functions (20) have been considered (where the diagonal elements of the matrix Q have been optimally chosen by means of Lemma 2 in [Milanese and Novara (2004)]), the norm $h = 2$ has been used in (8) and (9), and the bound $\mu = 0.01\sqrt{L}$ has been posed in (8). Among these 5000 basis functions, the sparse identification Algorithm 1 selected 8 basis functions for the output $y_1(k)$ and 9 basis functions for the output $y_2(k)$. The identified model is given by

$$\begin{aligned} \widehat{x}(k+1) &= A^*(\widehat{p}(k))\,\widehat{x}(k) + B^*(\widehat{p}(k))\,u(k) \\ \widehat{y}(k) &= C\widehat{x}(k) \end{aligned}$$

$$\begin{aligned} A^*(\widehat{p}(k)) &= \begin{bmatrix} \mathbf{0} & I \\ f_2^*(\widehat{p}(k)) & f_1^*(\widehat{p}(k)) \end{bmatrix} \\ B^*(\widehat{p}(k)) &= \begin{bmatrix} f_3^*(\widehat{p}(k)) & f_4^*(\widehat{p}(k)) \end{bmatrix}^T \\ C &= \begin{bmatrix} I & \mathbf{0} \end{bmatrix} \end{aligned}$$

where $\widehat{x}(k) = \widehat{p}(k) = [\widehat{y}(k)^T \quad \widehat{y}(k+1)^T]^T$, I is the 2×2 identity matrix, $\mathbf{0}$ is the 2×2 zero matrix, and f_i^* are the estimates of the unknown functions

$f_i : \mathbb{R}^4 \to \mathbb{R}^{2\times 2}$, $i = 1, \ldots, 4$, defining the observability representation of the system (27) discretized by means of some technique.

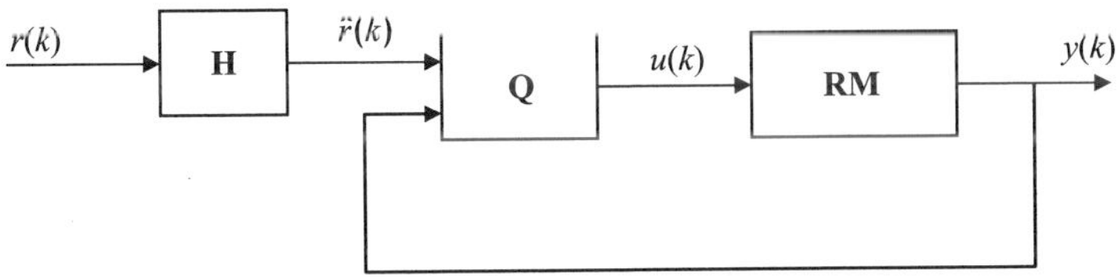

Fig. 6. Robot Manipulator. Control system.

This model has been used for the design of the control system shown in Figure 6.

• **RM** is the robot manipulator system described by the continuous-time state-space equations (27).

• **H** is a second-order filter with cut-off frequency 10 rad/s, introduced in order to avoid abrupt variations and too large amplitudes of the signals involved in the control system.

• **Q** is a gain-scheduling controller defined by

$$u(k) = -K(k)\left(\begin{bmatrix} y(k-1) \\ y(k) \end{bmatrix} - \widehat{r}(k)\right) + M(k)\widehat{r}(k) \tag{29}$$

where $\widehat{r}(k)$ is the filtered reference signal, $K(k)$ is a matrix such that $A^*\left(p\left(k-1\right)\right) - B^*\left(p\left(k-1\right)\right)K(k)$ has eigenvalues $\{0.7, 0.71, 0.72, 0.73\}$, $p\left(k-1\right) = [y(k-1)^T \; y(k)^T]^T$, and $M(k)$ is a matrix such that the controlled system has a unitary steady-state gain. Since $f_3^*\left(p\left(k\right)\right) = \mathbf{0}$, $\forall k$, this matrix has been computed as $M(k) = [\mathbf{0}$ $\mathrm{inv}(f_4^*\left(p\left(k-1\right)\right))]\left(I - A^*\left(p\left(k-1\right)\right)\right)$, where $\mathrm{inv}(\cdot)$ denotes the matrix inverse.

A simulation has been performed to test the designed control system in the task of reference tracking. A reference signal of length 5000 samples (corresponding to 100 s) has been used in this simulation, defined as a sequence of step signals with amplitudes in the interval $[-\pi, \pi]$. In Figure 7, the output of the controlled system is compared with the reference for the first 20 s of this simulation. It can be observed that the control system is quite effective, showing fast tracking and good steady-state precision. Obviously,

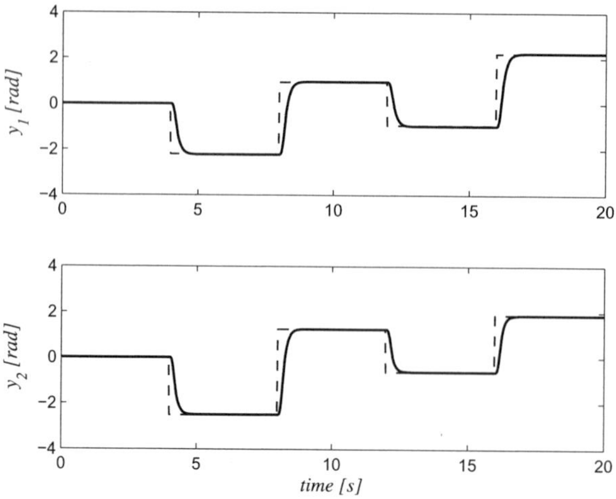

Fig. 7. Robot Manipulator. Reference tracking. Dashed line: reference signal. Continuous line: control system output.

the precision could be improved with a more sophisticated control system. However, the aim of this example is not to design a high-performance control system, but to show that the model identified by the proposed method can be effectively used for control design.

7. Conclusion

LPV models are widely used to design control systems for nonlinear systems. The main advantage of LPV models with respect to general nonlinear models is that they permit the application of linear control design techniques, such as gain-scheduling, optimal control, and robust control. Another advantage is that the use of an LPV model structure in place of a general nonlinear structure may reduce the computational complexity of the identification process. In this chapter, a SM identification method for state-space LPV systems has been proposed. The method allows the identification of almost-optimal and optimal state-space LPV models, and the evaluation of tight uncertainty bounds on the coefficients of the model parameter-varying matrices. A relevant feature of the method is that the obtained model is sparse, and thus allows a very effective implementation in on-line applications where small sampling periods are used. Two simulation examples have been presented to show the effectiveness of the proposed identification method, related to identification of the forced Van der Pol oscillator, and to identification and control of a 2-degrees of freedom robot manipulator.

References

Bamieh, B. and Giarré, L. (2002). Identification of linear parameter varying models, *International Journal of Robust and Nonlinear Control* **12**, 9, pp. 841–853.

Campi, M. (1994). Exponentially weighted least squares identification of time-varying systems with white disturbances, *IEEE Transactions on Signal Processing* **42**, 11, pp. 2906–2914.

Cerone, V. and Regruto, D. (2008). Set-membership identification of LPV models with uncertain measurements of the time-varying parameter, in *Proceedings of the Conference on Decision and Control* (Cancún, Mexico).

Chen, J. and Gu, G. (2000). *Control-oriented system Identification: An H_∞ approach* (John Wiley & Sons, New York City, NY, USA).

Donoho, D., Elad, M. and Temlyakov, V. (2006). Stable recovery of sparse overcomplete representations in the presence of noise, *IEEE Transactions on Information Theory* **52**, 1, pp. 6 – 18.

Felici, F., van Wingerden, J. W. and Verhaegen, M. (2007). Subspace identification of MIMO LPV systems using a periodic scheduling sequence, *Automatica* **43**, 10, pp. 1684–1697.

Fuchs, J. (2005). Recovery of exact sparse representations in the presence of bounded noise, *IEEE Transactions on Information Theory* **51**, 10, pp. 3601–3608.

He, B. and Yang, M. (2006). Robust LPV control of diesel auxiliary power unit for series hybrid electric vehicles, *IEEE Transactions on Power Electronics* **21**, 3, pp. 791–798.

Hogberg, M., Bewley, T. and Henningson, D. (2003). Relaminarization of $re_\tau =$ 100 turbulence using gain scheduling and linear state-feedback control, *Physics of fluids* **15**, 11, pp. 3572–3575.

Hsu, K., Vincent, T. and Poolla, K. (2008). Nonparametric methods for the identification of linear parameter varying systems, in *Proceedings of the IEEE International Conference on Computer-Aided Control Systems* (San Antonio, TX, USA).

Hunt, K., Johansen, T., Kalkkuhl, J., Fritz, H. and Gottsche, T. (2000). Speed control design for an experimental vehicle using a generalized gain scheduling approach, *IEEE Transactions on Control Systems Technology* **8**, 3, pp. 381–395.

Kwiatkowski, A. and Werner, H. (2005). LPV control of a 2-DOF robot using parameter reduction, in *Proceedings of the IEEE Conference on Decision and Control and European Control Conference* (Seville, Spain).

Lee, C. and Chung, M. S. M. (2001). A design of gain-scheduled control for a linear parameter varying system: an application to flight control, *International Journal of Control* **9**, 1, pp. 11–21.

Lopes dos Santos, P., Ramos, J. A. and Martins de Carvalho, J. L. (2008). Identification of LPV systems using successive approximations, in *Proceedings of the IEEE Conference on Decision and Control* (Cancún, Mexico).

Lovera, M. and Mercere, G. (2007). Identification for gain-scheduling: a balanced subspace approach, in *Proceedings of the American Control Conference* (New York City, NY, USA).

Lovera, M., Novara, C., Lopes dos Santos, P. and Rivera, D. (2011). Guest editorial special issue on applied LPV modeling and identification, *IEEE Transactions on Control Systems Technology* **19**, 1, pp. 1–4.

Matsumura, F., Namerikawa, T., Hagiwara, K. and Fujita, M. (1996). Application of gain scheduled h infin; robust controllers to a magnetic bearing, *IEEE Transactions on Control Systems Technology* **4**, 5, pp. 484–493.

Milanese, M. and Norton, J. (1996). *Bounding Approaches to System Identification* (Springer-Verlag, Berlin, Germany).

Milanese, M. and Novara, C. (2004). Set membership identification of nonlinear systems, *Automatica* **40**, pp. 957–975.

Milanese, M. and Novara, C. (2007). Computation of local radius of information in SM-IBC identification of nonlinear systems, *Journal of Complexity* **23**, pp. 937–951.

Milanese, M. and Vicino, A. (1991). Optimal algorithms estimation theory for dynamic systems with set membership uncertainty: an overview, *Automatica* **27**, pp. 997–1009.

Novara, C. (2011). Sparse identification of nonlinear functions and parametric set membership optimality analysis, in *Proceedings of the American Control Conference* (San Francisco, CA, USA).

Ott, E. (1993). *Chaos in Dynamical Systems* (Cambridge University Press, New York City, NY, USA).

Packard, A. (1994). Gain scheduling via linear fractional transformations, *Systems and Control Letters* **22**, 2, pp. 79–92.

Partington, J. R. (1997). *Interpolation, Identification and Sampling*, Vol. 17 (Clarendon Press-Oxford, New York City, NY, USA).

Previdi, F. and Carpanzano, E. (2003). Design of a gain scheduling controller for knee-joint angle control by using functional electrical stimulation, *IEEE Transactions on Control Systems Technology* **11**, 3, pp. 310–324.

Rugh, W. and Shamma, J. (2000). Research on gain scheduling, *Automatica* **36**, pp. 1401–1425.

Schweppe, F. (1973). *Uncertain dynamic systems* (Prentice-Hall, Englewood Cliffs, NJ, USA).

Sjöberg, J., Zhang, Q., Ljung, L., Benveniste, A., Deylon, B., Glorennec, P. Y., Hjalmarsson, H. and Juditsky, A. (1995). Nonlinear black-box modeling in system identification: a unified overview, *Automatica* **31**, pp. 1691–1723.

Tóth, R., Felici, F., Heuberger, P. S. C. and den Hof, P. M. J. V. (2007). Discrete time LPV I/O and state space representations, differences of behavior and pitfalls of interpolation, in *Proceedings of the European Control Conference* (Kos, Greece).

Traub, J. F., Wasilkowski, G. W. and Woźniakowski, H. (1988). *Information-based complexity* (Academic Press, Inc., New York City, NY, USA).

Tropp, J. (2004). Greed is good: algorithmic results for sparse approximation, *IEEE Transactions on Information Theory* **50**, 10, pp. 2231–2242.

Tropp, J. (2006). Just relax: convex programming methods for identifying sparse signals in noise, *IEEE Transactions on Information Theory* **52**, 3, pp. 1030–1051.

Wei, X. and del Re, L. (2007). Gain scheduled H_∞ control for air path systems of diesel engines using LPV techniques, *IEEE Transactions on Control Systems Technology* **15**, 3, pp. 406–415.

White, H. (1991). Universal approximation using radial-basis-function networks, *Neural Computation* **3**, pp. 246–257.

Chapter 5

Identification of input-output LPV models

V. Laurain[†], R. Tóth[‡], M. Gilson[†] & H. Garnier[†]

[†]*Centre de Recherche en Automatique de Nancy*
Nancy-Université, CNRS
BP 239, F-54506 Vandoeuvre-lès-Nancy Cedex, France
vincent.laurain@cran.uhp-nancy.fr
marion.gilson@cran.uhp-nancy.fr
hugues.garnier@cran.uhp-nancy.fr

[‡]*Delft Center for Systems and Control*
Delft University of Technology
Mekelweg 2, 2628 CD Delft, The Netherlands
r.toth@tudelft.nl

This chapter presents an overview of the available methods for identifying input-output LPV models both in discrete time and continuous time with the main focus on noise modeling issues. First, a least-squares approach and an instrumental variable method are presented for dealing with LPV-ARX models. Then, a refined instrumental variable approach is discussed to address more sophisticated noise models like Box-Jenkins in the LPV context. This latter approach is also introduced in continuous time and efficient solutions are proposed for both the problem of time-derivative approximation and the issue of continuous-time modeling of the noise.

1. Introduction

The identification of LPV systems can be addressed with different representations based model structures: *Input-Output* (IO) [Bamieh and Giarré (2002); Abbas and Werner (2009); Butcher *et al.* (2008); Wei and Del Re (2006)], State-Space [Lovera and Mercère (2007); van Wingerden and Verhaegen (2009); Lopes dos Santos *et al.* (2007)] or Orthogonal Basis Functions based models [Tóth *et al.* (2009a)] (see [Tóth (2010)] for an overview of the existing methods). This Chapter focuses on the IO model representation which presents several benefits. Firstly, the recent *Prediction Error Minimization* (PEM) framework extension [Tóth (2010)] makes possible the stochastic analysis of identification methods. Moreover, this representation is well suited for the introduction of different noise models [Laurain *et al.* (2010)] depending on the application considered. Finally, the tremendous need for structure learning/selection in practical applications starts finding answers for both parametric and non-parametric models [Tóth *et al.* (2009b); Tóth *et al.* (2011b)].

Most of the methods developed for IO-LPV models are derived in *discrete-time* (DT) under a linear regression form with static dependence on the scheduling variable p. Moreover, the most widely assumed structure is the *Auto Regressive with eXogeneous Input* (ARX) [Bamieh and Giarré (2002); Wei and Del Re (2006)] model. The ARX noise assumption combined to a linear parametrization of the scheduling dependencies is mainly motivated by the possible direct extension of *Linear Time Invariant* (LTI) PEM framework using linear regression approaches. Though, in the LPV context, not only the ARX assumption means equivalent dynamic properties between the process and the noise, but also identical nonlinear dependencies on the scheduling variable. In other words, the statement that the ARX assumption is unrealistic in the LTI framework is even stronger in the LPV case. However, the extension of the LTI PEM framework to the LPV case for other types of noise models appears to be non-trivial, mostly due to the fact that multiplication with q is not commutative over the p-dependent coefficients. Consequently, it has been only recently understood how a PEM framework can be established for the estimation of general LPV models [Tóth (2010)]. Even if some recursive *Least Squares* (LS) and *Instrumental Variable* (IV) methods have been introduced [Giarré *et al.* (2003); Butcher *et al.* (2008)], it was shown in [Laurain *et al.* (2010)] that despite the apparent similarity to the LTI identification problem, the methods available could not lead to the PEM in the LPV case. Furthermore, it has been detailed how linear regression based methods can still be successfully applied to more realistic noise conditions via the reformulation of the LPV model. Another concern is that even though the identification methods are mainly derived in DT, most physical processes are expressed naturally in *Continuous-Time* (CT) differential equation form. Moreover, LPV controllers are commonly synthesized in CT as stability and performance requirements of the closed-loop behavior can be more conveniently expressed in CT. Therefore, the current design tools focus on CT-LPV controller synthesis requiring accurate and low order CT models of the system. Unfortunately, the DT identification methods cannot cope with this need as the transformation from DT-LPV models to CT-LPV models is in a very immature state. The direct identification of CT-LPV models is consequently an attractive approach but presents some intrinsic issues mainly linked to the sampled nature of the data and the mathematical difficulty to handle CT stochastic processes [Garnier and Wang (2008)]. Based on the recent advances of both CT-LTI and DT-LPV identification frameworks, an IV based method for direct CT-LPV identification has been proposed in [Laurain *et al.* (2011)].

Consequently, by retracing an historical background of the IO-LPV identification framework, this Chapter aims at addressing solutions leading to the PEM in the LPV framework for different noise assumptions, both in DT and CT settings. Section 2 details the structure of DT IO-LPV data-generating systems along with the widely used polynomial model associated. Section 3 focuses on the solutions offered for the estimation of LPV-ARX models. The *Output Error* (OE) and *Box-Jenkins* (BJ) structures are introduced in Section 4, where the importance of the noise assumption as well as the influence of the identification method chosen are illustrated on a representative example. Section 5 addresses the problematic of direct identification of CT systems, presenting some intrinsic difficulties generally involved in CT model identification from sampled data. Section 6 proposes some direct CT identification solution and the advantages over the DT identification scheme are illustrated using a representative example derived from a well-known CT benchmark [Rao and Garnier (2004)]. Finally, the different contributions are summarized and some conclusions are brought in Section 7.

2. Discrete-time LPV polynomials models

DT models suit well the sampled nature of the data and consequently, most models used in the literature are expressed in a DT setting. Moreover, they often assume a static dependence on the scheduling variable p (the system depends on instantaneous values of p): the relevance of this static assumption is not discussed here as the presented methods can easily include delayed version of p in the model for dynamic dependence. This section gives a definition of the DT LPV data-generating systems using polynomial input-output representations based model structures.

2.1. *LPV data generating system*

In DT, the LPV data-generating system $\mathscr{S}_{\mathrm{o}}$ is commonly formulated in the identification literature using the following polynomial IO form:

$$\mathscr{S}_{\mathrm{o}} \begin{cases} A_{\mathrm{o}}(p_k, q^{-1})\chi_{\mathrm{o}}(k) = B_{\mathrm{o}}(p_k, q^{-1})u(k) \\ y(k) = \chi(k) + v_{\mathrm{o}}(k) \end{cases} \tag{1}$$

where p_k is the value of the scheduling parameter p at sample time k, χ_{o} is the noise-free output, u is the input, v_{o} is the additive noise with bounded spectral density, y is the noisy output of the system and q is the time-shift operator, *i.e.* $q^{-i}u(k) = u(k-i)$. The study of these systems is here restricted to the *Single-Input-Single-Ouput* (SISO) case for clarity's sake, but the solutions proposed are all extensions of LTI identification methods which have been also successfully applied to the *Multi-Input-Multi-Ouput* (MIMO) case. Therefore, the signals u and y are considered to be bounded and quasi-stationnary signals on $\mathbb{R}$. $A_{\mathrm{o}}(p_k, q^{-1})$ and $B_{\mathrm{o}}(p_k, q^{-1})$ are polynomials in q^{-1} of degree n_{a} and n_{b} respectively:

$$A_{\mathrm{o}}(p_k, q^{-1}) = 1 + \sum_{i=1}^{n_{\mathrm{a}}} a_i^{\mathrm{o}}(p_k)q^{-i}, \quad \text{and} \quad B_{\mathrm{o}}(p_k, q^{-1}) = \sum_{j=0}^{n_{\mathrm{b}}} b_j^{\mathrm{o}}(p_k)q^{-j}, \tag{2}$$

where the coefficients a_i and b_j are real meromorphic functions ($f : \mathbb{R}^n \to \mathbb{R}$ is a real meromorphic function if $f = g/h$ with g, h analytic and $h \neq 0$) with static dependence on p. It is assumed that these coefficients are bounded on $\mathbb{P}$, thus the solutions of $\mathscr{S}_{\mathrm{o}}$ are well-defined and the process part is completely characterized by the coefficient functions $\{a_i^{\mathrm{o}}\}_{i=1}^{n_{\mathrm{a}}}$ and $\{b_j^{\mathrm{o}}\}_{j=0}^{n_{\mathrm{b}}}$. Like in any identification problem, it is assumed that a data sequence $\mathcal{D}_N = \{y(k), u(k), p(k)\}_{k=1}^{N}$ generated by $\mathscr{S}_{\mathrm{o}}$ is available. The data-generating system in Eq. (1) is then identified by defining a model set and some information criterion characterizing the quality of the given model based on $\mathcal{D}_N$. As a next step, we explore how such model structures and criteria can be chosen to efficiently solve the identification problem.

2.2. *Polynomial LPV model*

2.2.1. *Process model*

The studied process is fully characterized by the knowledge of functions $\{a_i^{\rm o}\}_{i=1}^{n_{\rm a}}$ and $\{b_j^{\rm o}\}_{j=0}^{n_{\rm b}}$, but in practical cases these functions are *a priori* unknown nonlinear functions. Even if some recent work gives some preliminary results oriented towards a model structure selection [Tóth *et al.* (2009b); Tóth *et al.* (2011b)], the usual solution to overcome this problem is to parameterize these functions using a sum of *a priori* known basis functions. Consequently, this Chapter focuses on the usual assumption used in the literature where the process model denoted by $\mathscr{G}_\rho$ is well defined in an IO-LPV representation form:

$$\mathscr{G}_\rho : \left(A(p_k, q^{-1}, \rho), B(p_k, q^{-1}, \rho)\right) = (\mathscr{A}_\rho, \mathscr{B}_\rho) \tag{3}$$

where the p-dependent polynomials A and B are parameterized as

$$\mathscr{A}_\rho \begin{cases} A(p_k, q^{-1}, \rho) = 1 + \sum_{i=1}^{n_{\rm a}} a_i(p_k) q^{-i}, \\ a_i(p_k) = \rho_{i,0} + \sum_{l=1}^{s_i} \rho_{i,l} \psi_{i,l}(p_k), \quad i = 1, \dots, n_{\rm a} \end{cases}$$

$$\mathscr{B}_\rho \begin{cases} B(p_k, q^{-1}, \rho) = \sum_{j=0}^{n_{\rm b}} b_j(p_k) q^{-j}, \\ b_j(p_k) = \rho_{i,0} + \sum_{l=1}^{s_i} \rho_{i,l} \psi_{i,l}(p_k), \quad i = j + n_{\rm a} + 1, \quad j = 0, \dots, n_{\rm b}. \end{cases}$$

In this parametrization, $\{\psi_{i,l}\}_{i=1,l=1}^{n_{\rm a}+n_{\rm b}+1,s_i}$ are *a priori* chosen meromorphic functions of p, with static dependence on p, allowing the identifiability of the model (linearly independent functions on $\mathbb{P}$ for example). The associated model parameters ρ are stacked columnwise:

$$\rho = \left[\rho_{1,0} \cdots \rho_{n_{\rm a},0}, \rho_{1,1} \cdots \rho_{n_{\rm a},s_{n_{\rm a}}}, \rho_{n_{\rm a}+1,0} \cdots \rho_{n_{\rm g},s_{n_{\rm g}}}\right]^\top \in \mathbb{R}^{n_\rho}, \tag{4}$$

with $n_{\rm g} = n_{\rm a} + n_{\rm b} + 1$ and $n_\rho = \sum_{i=1}^{n_{\rm g}} s_i + 1$. We introduce also $\mathcal{G} = \{\mathscr{G}_\rho \mid \rho \in \mathbb{R}^{n_\rho}\}$, as the collection of all process models in the form of Eq. (3).

2.2.2. *Noise model*

Another term which needs to be modeled is the process noise v_o. In the PEM framework, the noise model implicitly defines the error criterion which assesses the quality of a model when minimized. In most practical applications, information about the noise structure or properties is unavailable. It is a critical issue considering that the quality of the estimated model will highly depend on the noise assumption. Consequently, it is important to introduce a noise structure able to model a large set of behaviors. A rather general description in the IO-LPV context is a noise model denoted by $\mathscr{H}_\eta$ and defined by the following representation:

$$\mathscr{H}_\eta : D(p_k, q^{-1}, \eta)v(k) = C(p_k, q^{-1}, \eta)e(k), \tag{5}$$

where $e(k)$ is assumed to be a white noise process with a normal distribution, and the p-dependent polynomials are defined as

$$\begin{aligned} C(p_k, q^{-1}, \eta) &= 1 + \sum_{i=1}^{n_\mathrm{a}} c_i(p_k)q^{-i}, i = 1, \ldots, n_c \\ D(p_k, q^{-1}, \eta) &= 1 + \sum_{j=1}^{n_\mathrm{b}} d_j(p_k)q^{-j}, j = 1, \ldots, n_d. \end{aligned} \tag{6}$$

The focus has recently been given to simpler structures in the literature and the next sections present estimation methods relying on different simplifications of this noise model.

3. Estimating LPV-ARX models in DT

The first model structure which appeared in the IO-LPV identification framework is the LPV-ARX model [Bamieh and Giarré (2002); Butcher *et al.* (2008)]. This very specific assumption on the noise is motivated by a substantial simplification of the model which renders possible the direct application of the LTI PEM framework using regression based methods.

3.1. *LPV-ARX models*

This model is characterized by the fact that the noise model is assumed to have the same dynamics and nonlinearities as the process model. More specifically, the noise model is denoted by $\mathscr{H}_\rho$ and defined by the following

LPV-IO representation:

$$\mathscr{H}_\rho : A(p_k, q^{-1}, \rho)v(k) = e(k), \tag{7}$$

where $e(k)$ is assumed to be a white noise process. Under this noise assumption and using the process model given in Eq. (3), the full LPV-ARX model denoted $\mathscr{M}_\rho$ can be written in the linear regression form:

$$\mathscr{M}_\rho : y(k) = \varphi^\top(k)\rho + e(k) \tag{8}$$

with ρ as defined in Eq. (4), and

$$\begin{aligned}\varphi(k) = \big[-y(k-1) \quad \dots \quad & - y(k-n_\mathrm{a}), -y(k-1)\psi_{1,1}(p_k) \quad \dots \\ & - y(k-n_\mathrm{a})\psi_{n_\mathrm{a},s_{n_\mathrm{a}}}(p_k), u(k) \quad \dots \quad u(k-n_\mathrm{b}), \\ & u(k)\psi_{n_\mathrm{a}+1,1}(p_k) \quad \dots \quad u(k-n_\mathrm{b})\psi_{n_\mathrm{g},s_{n_\mathrm{g}}}(p_k)\big].\end{aligned} \tag{9}$$

It must be pointed out that under this polynomial LPV-ARX model assumption, the time-varying nature of the system is transfered to the n_ρ signals composing the regressor $\varphi(k)$. Moreover, the parameter vector ρ is time-invariant and the error equation $e(k)$ is a white noise with a normal distribution: consequently, Eq. (8) defines an LTI model.

3.2. *The LS solution*

By acknowledging that the LPV-ARX model leads to a MIMO LTI formulation of the LPV problem, the LTI PEM framework can be directly used: the LS solution dedicated to IO-LPV models was introduced by Giarré [Bamieh and Giarré (2002)] and is described here. Let $\mathcal{M} = \{\mathscr{M}_\rho \mid \rho \in \mathbb{R}^{n_\rho}\}$ be the collection of all models in the form of Eq. (8). If the studied system is well described by the ARX structure ($\mathrm{S_o} \in \mathcal{M}$), then $\mathscr{S}_\mathrm{o}$ can also be written as a linear regression:

$$y(k) = \rho_\mathrm{o}^\top \varphi(k) + e_\mathrm{o}(k), \tag{10}$$

where $\rho_\mathrm{o} \in \mathbb{R}^{n_\rho}$ is the *true* parameter vector ($\mathscr{M}_{\rho_\mathrm{o}} = \mathscr{S}_\mathrm{o}$) and $\varphi(k) \in \mathbb{R}^{n_\rho}$ is the regressor vector (9). In order to estimate ρ_o, the quality of the model fit is formulated in terms of a cost function $\mathscr{V}(\rho, \mathcal{D}_N)$. The minimization of $\mathscr{V}(\rho, \mathcal{D}_N)$ corresponds to the estimation of ρ. In case of the so-called LS estimation, the cost function is the *squared equation error*:

$$\mathscr{V}(\rho, \mathcal{D}_N) = \frac{1}{N}\sum_{k=1}^{N} e^2(k) = \frac{1}{N}\|e(k)\|_{\ell_2}^2, \tag{11}$$

where e is given by Eq. (8). $\mathcal{M}$ represents the set of models in which we are searching for the "best" $\mathscr{M}_\rho$ that describes $\mathscr{S}_o$ given a dataset $\mathcal{D}_N = \{y(k), u(k), p(k)\}_{k=1}^N$ generated by $\mathscr{S}_o$. As Eq. (11) is quadratic in ρ and under this LTI representation, the LS solution is analytic and given by:

$$\hat{\rho}_{\mathrm{LS}} = \left[\sum_{k=1}^{N} \varphi(k)\varphi^\top(k)\right]^{-1} \sum_{k=1}^{N} \varphi(k)y(k). \tag{12}$$

If the data generating the system is actually in the model set defined, Eq. (12) is a statistically optimal estimator (minimum variance and unbiased). Nonetheless, it must be pointed out that this model assumption is very limited in practice. Even though it might be a fair assumption to consider that the process is well parameterized using Eq. (3), the probability that $v_o(t_k)$ is correctly described by Eq. (7) is very low. Indeed, in most cases, there is no reasonable explanation to justify why the noise v_o and the process part of $\mathscr{S}_o$ should contain the same dynamics and nonlinearities. In terms of estimation, it means that using the LS method in practical applications will most often lead to biased estimates. Consequently, some methods have been developed in order to cope with the error induced by this invalid assumption on the noise. A solution introduced in [Butcher *et al.* (2008)] and relying on the IV method is described in the next section.

3.3. *Non-white noise: An IV solution*

In the LTI context, the PEM methods have evolved in order to deal with general and realistic types of noise [Ljung (1999)]. Their efficiency and convergence have been proven for generic types of models such as: ARX, OE or BJ for example. The most common assumption is the inclusion of the system studied in the model set defined, both for the process and the noise. Nonetheless, the physical insight of the noise process is very unrealistic in practice and some methods have especially been designed to consistently estimate models when the noise assumption is incorrect: a set of these approaches are the IVs methods [Young and Jakeman (1980); Söderström and Stoica (1983)].

3.3.1. *Instrumental variable method in the LTI case*

It is well-known from the LTI theory that when considering a regression form $y(k) = \varphi(k)\rho + \tilde{v}(k)$, the LS estimate from Eq. (12) is also:

$$\hat{\rho}_{\mathrm{LS}} = \rho_{\mathrm{o}} + \left[\sum_{k=1}^{N} \varphi(k)\varphi^{\top}(k)\right]^{1} \sum_{k=1}^{N} \varphi(k)\tilde{v}(k). \tag{13}$$

This equation induces that $\hat{\rho}_{LS}$ is a consistent estimate of ρ_{o} (unbiased for finite data) if both the following conditions are respected [a]:

CLS1 $\bar{\mathbb{E}}\{\varphi(k)\varphi^{\top}(k)\}$ is full column rank.

CLS2 $\bar{\mathbb{E}}\{\varphi(k)\tilde{v}(k)\} = 0$.

While CLS1 is usually respected if the input is exciting enough, CLS2 only holds when $\tilde{v}(k)$ is a white noise. Unfortunately, this assumption is only verified for the ARX case [Ljung (1999)]. Hence, as previously pointed out, using the LS estimate is not the most pertinent choice in most practical applications. The original aim of IV methods is to cope with the fact that in most cases, $\tilde{v}(k)$ is a colored process. The idea is to introduce an instrument ζ which is used to produce a consistent estimate independently on the noise model taken. The IV estimate is given as:

$$\hat{\rho}_{\mathrm{IV}} = \left[\sum_{k=1}^{N} \zeta(k)\varphi^{\top}(k)\right]^{-1} \sum_{k=1}^{N} \zeta(k)y(k), \tag{14}$$

and using this definition, it can be seen that

$$\hat{\rho}_{\mathrm{IV}} = \rho_{\mathrm{o}} + \left[\sum_{k=1}^{N} \zeta(k)\varphi^{\top}(k)\right]^{-1} \sum_{k=1}^{N} \zeta(k)\tilde{v}(k). \tag{15}$$

Therefore, and similarly to the LS solution $\hat{\rho}_{\mathrm{IV}}$ is a consistent estimate of ρ_{o} if

C1 $\bar{\mathbb{E}}\{\zeta(k)\varphi^{\top}(k)\}$ is full column rank.

C2 $\bar{\mathbb{E}}\{\zeta(k)\tilde{v}(k)\} = 0$.

There is a considerable amount of freedom in the choice of an instrument respecting simultaneously C1 and C2. Nonetheless, even if any of these instruments leads to a consistent estimate, the amount of correlation between the instrument and the regressor will have a serious impact on the variance of the estimated parameters.

---

[a] The notation $\bar{\mathbb{E}}\{.\} = \lim_{N\to\infty} \frac{1}{N} \sum_{t=1}^{N} \mathbb{E}\{.\}$ is adopted from the prediction error framework of [Ljung (1999)].

3.3.2. *The optimal instrument for LPV-ARX models*

In the LTI context, the choice of the instrument has been widely discussed and most advanced IV methods offer similar performance as extended LS methods or other PEM methods (see [Rao and Garnier (2004); Ljung (2009)]) while providing consistent results even for an imperfect noise structure. It can be proven that under the ARX model assumption, the variance of the IV estimate is minimal if the instrument is chosen as the noise-free version of the regressor. In other words, when directly applying the IV theory to the LPV-ARX model from Eq. (8) (the LPV-ARX model is an LTI model), the optimal IV estimate is given by:

$$\hat{\rho}_{\mathrm{IV}}^{\mathrm{opt}} = \left[\sum_{k=1}^{N} \zeta_{\mathrm{opt}}(k)\varphi^{\top}(k)\right]^{-1} \sum_{k=1}^{N} \zeta_{\mathrm{opt}}(k)y(k), \tag{16}$$

where the optimal instrument is defined as [Söderström and Stoica (1983)]:

$$\begin{aligned}\zeta_{\mathrm{opt}}(k) = \big[-\chi_{\mathrm{o}}(k-1) \quad \ldots \quad -\chi_{\mathrm{o}}(k-n_{\mathrm{a}}), -\chi_{\mathrm{o}}(k-1)\psi_{1,1}(p_k) \quad \ldots \\ -\chi_{\mathrm{o}}(k-n_{\mathrm{a}})\psi_{n_{\mathrm{a}},s_{n_{\mathrm{a}}}}(p_k), u(k) \quad \ldots \quad u(k-n_{\mathrm{b}}), \\ u(k)\psi_{n_{\mathrm{a}}+1,1}(p_k) \quad \ldots \quad u(k-n_{\mathrm{b}})\psi_{n_{\mathrm{g}},s_{n_{\mathrm{g}}}}(p_k)\big]\end{aligned} \tag{17}$$

It can be noticed that in the LPV-ARX assumption both the IV solution from Eq. (16) and the LS solution from Eq. (12) are statistically optimal estimators. Nonetheless, the deterministic output terms χ_{o} contained in ζ_{opt} are *a priori* unknown in practice. Consequently, some estimates of χ_{o} commonly compose the instrument, requiring some iterative process and implying some difficulties for the convergence analysis. It could consequently be argued that the LS estimate is simpler and hence, better suited to solve the identification problem: this statement is true as long as the true system respects the ARX structure. Nonetheless, this is even more unlikely in the LPV context than in the LTI context: the ARX assumption implies that not only the dynamics of the noise and process are the same but that the measuring devices are affected by the external scheduling variable in the exact same nonlinear way as the system studied. In the probable situation where the noise assumption is violated, the IV method is a consistent estimate while the LS is inevitably biased.

3.3.3. *The LPV-IV4 method*

As the knowledge of the deterministic output χ_{o} is unavailable, an algorithm is needed to approximate the deterministic output terms and to build the optimal instrument ζ_{opt}. Therefore, the use of an IV method similar to the IV4 method ([Ljung (1999)]) is proposed in [Butcher *et al.* (2008)], where the instrument is built using the simulated data generated from an estimated auxiliary ARX model. This method is detailed as:

Algorithm 1 (LPV-IV4 method).

Step 1 *Estimate an ARX model by the LS method (minimizing Eq. (11)) using the extended regressor* (9).

Step 2 *Generate an estimate* $\hat{\chi}(k)$ *of* $\chi(k)$ *based on the resulting ARX model of the previous step. Build an instrument based on* $\hat{\chi}(k)$ *and then estimate* ρ *using the IV method.*

In [Butcher *et al.* (2008)], it was shown that according to the theory, in case $\mathscr{S}_o$ corresponds to an LPV-OE model ($v_o = e_o$), Algorithm 1 leads to an unbiased estimate. Moreover, like in the LTI case, the noise modeling error results in a bias for the LS estimates while only the variance is affected when using this IV method. Nevertheless, the bigger the difference between the true noise process and the noise model assumed is, the higher the resulting variance in the IV estimates will be. Depending on the size of the dataset, the increasing variance of the IV estimates can lead to less relevant results than the LS method (for which the variance is known to remain low). Consequently, it is highly important to assume a noise model as realistic as possible in the first place. In the LTI case, many IV methods are dedicated to more general noise models such as OE of BJ [Young (2008)]. The next section describes some available methods for LPV-OE and LPV-BJ which were introduced in [Laurain *et al.* (2010)].

4. Addressing estimation with general noise models

In order to assess more efficiently the quality of a given model, more realistic noise models have to be taken into consideration. Hence, this section introduces noise structures which have been widely used in the LTI framework and which have not been dealt with in the LPV case until recently.

4.1. *LPV-BJ model*

The general LPV-BJ model structure from Eq. (5) has never been dealt with in the literature so far. Nevertheless, as a preliminary step towards this p-dependent noise, it can be assumed that the rational spectral density is not dependent on p. In the case of high-tech applications where the major source of noise is due to measurement error (e.g. wafer-scanners), this assumption often holds true and is in general far more realistic than the LPV-ARX assumption.

4.1.1. *Noise model*

In the defined LPV-BJ structure, the noise model denoted by $\mathscr{H}$ is defined as a DT *autoregressive moving average* (ARMA) model

$$\mathscr{H}_\eta : (H(q,\eta)), \tag{18}$$

where H is a monic rational function given in the form of

$$H(q,\eta) = \frac{C(q^{-1},\eta)}{D(q^{-1},\eta)} = \frac{1 + c_1 q^{-1} + \ldots + c_{n_\mathrm{c}} q^{-n_\mathrm{c}}}{1 + d_1 q^{-1} + \ldots + d_{n_\mathrm{d}} q^{-n_\mathrm{d}}} \tag{19}$$

and all roots of $z^{n_\mathrm{d}} D(z^{-1})$ and $z^{n_\mathrm{c}} C(z^{-1})$ are inside the unit disc. It can be noticed that in case $C(q^{-1}) = D(q^{-1}) = 1$, Eq. (19) defines an OE model. The associated model parameters η are stacked columnwise in the parameter vector,

$$\eta = \begin{bmatrix} c_1 & \ldots & c_{n_\mathrm{c}} & d_1 & \ldots & d_{n_\mathrm{d}} \end{bmatrix}^\top \in \mathbb{R}^{n_\eta}, \tag{20}$$

where $n_\eta = n_\mathrm{c} + n_\mathrm{d}$. Additionally, denote $\mathcal{H} = \{\mathscr{H}_\eta \mid \eta \in \mathbb{R}^{n_\eta}\}$, the collection of all noise models in the form of Eq. (18).

4.1.2. *Whole model*

With respect to a given process and noise part $(\mathscr{G}_\rho, \mathscr{H}_\eta)$, the parameters can be collected as $\theta = [\rho^\top \ \eta^\top]$ and the signal relations of the LPV-BJ model $\mathscr{M}_\theta$, are defined as:

$$\mathscr{M}_\theta \begin{cases} A(p_k, q^{-1}, \rho)\chi(k) = B(p_k, q^{-1}, \rho)u(k) \\ v(k) = \dfrac{C(q^{-1},\eta)}{D(q^{-1},\eta)} e(k) \\ y(k) = \chi(k) + v(k) \end{cases} \tag{21}$$

Based on this model structure, the model set, denoted as $\mathcal{M}$, with process ($\mathscr{G}_\rho$) and noise ($\mathscr{H}_\eta$) models parameterized independently, takes the form

$$\mathcal{M} = \left\{ (\mathscr{G}_\rho, \mathscr{H}_\eta) \mid [\rho^\top, \eta^\top]^\top = \theta \in \mathbb{R}^{n_\rho + n_\eta} \right\}. \tag{22}$$

4.1.3. *The LPV-BJ model issue*

If the system belongs to the model set defined in Eq. (22), then $y(k)$ can be written in the linear regression form:

$$y(k) = \varphi^\top(k)\rho + \tilde{v}(k) \tag{23}$$

with ρ and $\varphi(k)$ as defined in Eq. (4) and Eq. (9), respectively. Under the LPV-BJ assumption, it can be seen from Eq. (21) that

$$\tilde{v}(k) = A(p_k, q^{-1}, \rho)\frac{C(q^{-1}, \eta)}{D(q^{-1}, \eta)}e(k) = Q(p_k, q^{-1}, \rho, \eta)e(k). \tag{24}$$

Hence, using Eq. (12) to estimate ρ corresponds to the minimization of the LS criterion on $\tilde{v}(k)$, which is clearly different from the PEM (the minimization of the LS criterion on $e(k)$) [Laurain *et al.* (2010)]. It must be pointed out that despite the apparent similarity between Eq. (8) and Eq. (23), the estimation problem of the LPV-BJ model cannot be considered as an LTI problem like it is under the LPV-ARX assumption. This statement is motivated by the presence of the filter Q in Eq. (24) which is obviously p-dependent. In an LTI context, Q is p-independent and most estimation methods use the inverse filtering of Q to solve the estimation problem in an optimal way. Nonetheless, in this LPV case, it was shown in [Laurain *et al.* (2010)] that the non-commutativity of LPV filters as well as the impossibility to compute the exact inverse of an LPV filter in practice renders the PEM impossible under the regression form (23). A possible solution to overcome this problem is to reformulate the model from Eq. (21) in order to use directly the LTI theory. This reformulation is detailed in the next section.

4.1.4. *Reformulation of the model equations*

In order to introduce a method which provides a solution to the identification problem of LPV-BJ models, rewrite the signal relations of Eq. (21)

as

$$
\mathcal{M}_\theta \begin{cases} \underbrace{\chi(k) + \sum_{i=1}^{n_\mathrm{a}} a_{i,0}\chi(k-i)}_{F(q^{-1},\rho)\chi(k)} + \sum_{i=1}^{n_\mathrm{a}}\sum_{l=1}^{s_i} a_{i,l}\underbrace{\psi_{i,l}(p_k)\chi(k-i)}_{\chi_{i,l}(k)} \\ \qquad\qquad = \sum_{j=0}^{n_\mathrm{b}}\sum_{l=0}^{s_{\tilde{j}}} b_{j,l}\underbrace{\psi_{\tilde{j},l}(p_k)u(k-j)}_{u_{j,l}(k)} \\ v(k) = \dfrac{C(q^{-1},\eta)}{D(q^{-1},\eta)}e(k) \\ y(k) = \chi(k) + v(k), \end{cases} \tag{25}
$$

where $F(q^{-1},\rho) = 1 + \sum_{i=1}^{n_\mathrm{a}} a_{i,0}q^{-i}$, $\tilde{j} = j + n_\mathrm{a} + 1$ and $\psi_{\tilde{j},0}(k) = 1$. Note that in this way, the LPV-BJ model is rewritten as a Multiple Input Single Output (MISO) system with $\sum_{i=1}^{n_\mathrm{g}} s_i + 1$ inputs $\{\chi_{i,l}\}_{i=1,l=1}^{n_\mathrm{a},s_i}$ and $\{u_{j,l}\}_{j=0,l=0}^{n_\mathrm{b},s_{\tilde{j}}}$ as represented in Fig. 1. Given the fact that the polynomial operator commutes in this representation ($F(q^{-1},\rho)$ does not depend on p_k), Eq. (25) can be rewritten as

$$
y(k) = -\sum_{i=1}^{n_\mathrm{a}}\sum_{l=1}^{s_i} \frac{a_{i,l}}{F(q^{-1},\rho)}\chi_{i,l}(k) + \sum_{j=0}^{n_\mathrm{b}}\sum_{l=0}^{s_{\tilde{j}}} \frac{b_{j,l}}{F(q^{-1},\rho)}u_{j,l}(k) + H(q,\eta)e(k), \tag{26}
$$

which is an LTI representation. Consequently, it becomes possible to apply the LTI theory in order to solve the estimation problem.

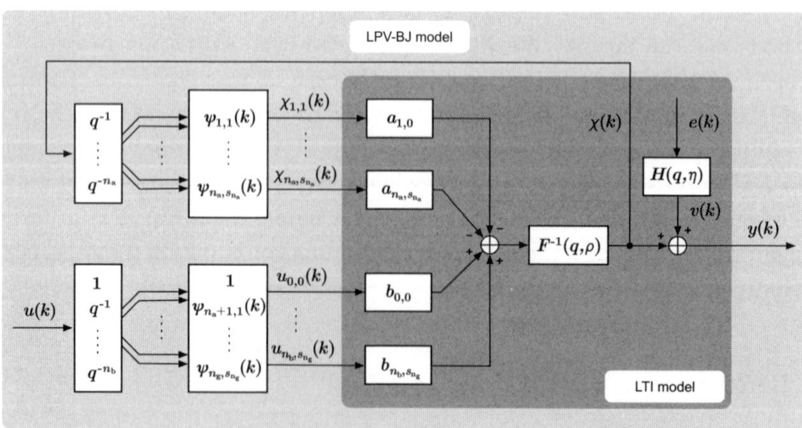

Fig. 1. MISO LTI interpretation of the LPV-BJ model.

4.2. A refined instrumental approach

Based on the MISO-LTI formulation Eq. (26), it becomes possible in theory to achieve optimal PEM using linear regression. The only method available for LPV-BJ models in the literature so far is an extended version of the *Refined IV* (RIV) approach of the LTI identification framework [Laurain *et al.* (2010)] which provides an efficient way of identifying LPV-BJ models. It derives directly from the extended IV scheme developed for LTI-BJ models which is described in the next section.

4.2.1. *Instrumental variable for LTI-BJ models*

Based on Eq. (26), it is possible to introduce some LS-based methods to solve the identification problem. However, it is important to realize that in experimental conditions, even if a realistic noise assumption will lead to more accurate estimates, the chances that the true noise process is not perfectly described are high. In this case, similarly to the LPV-ARX case, an LS based method will produce biased results. Therefore, the need to cope with noise modeling errors remains. As a refinement of the IV scheme presented in Section 3.3, IV methods have been developed to cope with more general structures such as the BJ case, which can be expressed in the linear regression form $y(k) = \varphi^\top(k)\rho + Q(q)e(k)$ ($Q(q)$ is an LTI transfer function, with $Q^{-1}(q)$ stable and $e(k)$ a white noise). Based on this form, the extended-IV estimate can be given as [Söderström and Stoica (1983)]:

$$\hat{\rho}_{\mathrm{XIV}}(N) = \arg\min_{\rho\in\mathbb{R}^{n_\rho}} \left\| \left[\frac{1}{N}\sum_{k=1}^{N} L(q)\zeta(k)L(q)\varphi^\top(k)\right]\rho - \left[\frac{1}{N}\sum_{t=1}^{N} L(q)\zeta(k)L(q)y(k)\right]\right\|_W^2, \quad (27)$$

where $\zeta(k)$ is the instrument, $\|x\|_W^2 = x^T W x$, with W a positive definite weighting matrix and $L(q)$ is a stable prefilter. The conditions for consistency now read:

CX1 $\bar{\mathbb{E}}\{L(q)\zeta(k)L(q)\varphi^\top(k)\}$ is full column rank.

CX2 $\bar{\mathbb{E}}\{L(q)\zeta(k)L(q)\tilde{v}(k)\} = 0$.

Again, there is a considerable amount of freedom in the choice of the instruments. Nonetheless, if the studied system belongs to the model set defined, *i.e.* if there exists ρ_o and Q_o (Q_o^{-1} being stable) such that

$$\mathscr{S}_o : y(k) = \varphi^\top(k)\rho_o + Q_o(q)e_o(k), \tag{28}$$

with $e_o(k)$ a white noise, it has been shown in [Söderström and Stoica (1983)] and [Young (1984)] that the minimum variance estimator can be achieved in the BJ case if:

CX3 $W = I$.

CX4 ζ is chosen as the noise-free version of the extended regressor φ.

CX5 $L(q)$ is chosen as $Q_o^{-1}(q)$.

In case of noise modeling error, the extended IV method is consistent and the variance of the estimates should be significantly decreased with respect to the IV4 method: even if the noise process is not in the noise model set defined, it is more likely to be better described by the BJ model than by the ARX model.

4.2.2. *The optimal instrument for LPV-BJ systems*

Using Eq. (26), $y(k)$ can be written in the regression form:

$$y(k) = \varphi^\top(k)\rho + \tilde{v}(k), \tag{29}$$

where

$$\begin{aligned}\varphi(k) = \Big[& -y(k-1) \ldots -y(k-n_\text{a}) \; -\chi_{1,1}(k) \ldots \\ & -\chi_{n_\text{a},s_{n_\text{a}}}(k) \; u_{0,0}(k) \ldots u_{n_\text{b},s_{n_\text{g}}}(K)\Big]^\top, \\ \tilde{v}(k) &= F(q^{-1},\rho)v(k), \\ &= F(q^{-1},\rho)\frac{C(q^{-1},\eta)}{D(q^{-1},\eta)}e(k) = Q(q,\rho,\eta)e(k). \end{aligned} \tag{30}$$

It is important to notice that Eq. (29) and Eq. (23) are not equivalent as the extended regressor in Eq. (29) contains the noise-free output terms $\{\chi_{i,l}\}$. Therefore, by momentary assuming that $\{\chi_{i,l}(k)\}_{i=1,l=0}^{n_\text{a},s_i}$ are known *a priori* and that the data-generating system $\mathscr{S}_o$ is in the model set defined ($v_o(k) = C_o(q^{-1},\eta)/D_o(q^{-1},\eta)e_o(k)$ and $e_o(k)$ is a white noise), then the conditions for optimal estimates expressed in [CX3-CX5] lead to the choice of optimal instrument given in this LPV-BJ case as [Laurain *et al.* (2010)]:

$$\zeta_{\text{opt}}(k) = \big[-\chi_{\text{o}}(k-1) \quad \dots \quad -\chi_{\text{o}}(k-n_{\text{a}}) \quad -\chi^{\text{o}}_{1,1}(k) \\ -\chi^{\text{o}}_{n_{\text{a}},s_{n_{\text{a}}}}(k) \quad u_{0,0}(k) \quad \dots \quad u_{n_{\text{b}},s_{n_{\text{g}}}}(k)\big]^\top \tag{31}$$

while the optimal filter is given as:

$$L_{\text{opt}}(q) = \frac{D_{\text{o}}(q^{-1})}{C_{\text{o}}(q^{-1})F_{\text{o}}(q^{-1})}. \tag{32}$$

4.2.3. *Iterative LPV-RIV algorithm for BJ models*

In a practical situation, the optimal instrument (31) and filter (32) are unknown *a priori*. Therefore, the RIV estimation normally involves an iterative (or relaxation) algorithm in which, at each iteration, an 'auxiliary model' is used to generate the instrument (which guarantees CX2), as well as the associated filters. This auxiliary model is based on the parameter estimates obtained at the previous iteration. Consequently, if convergence occurs, CX4 and CX5 are fulfilled. Based on the previous considerations, the iterative scheme of the RIV algorithm dedicated to the LPV case is given as:

Algorithm 2 (LPV-RIV).

Step 1 *Generate an initial estimate of the process model parameter* $\hat{\rho}^{(0)}$ *(e.g. using the LS method). Set* $C(q^{-1},\hat{\eta}^{(0)}) = D(q^{-1},\hat{\eta}^{(0)}) = 1$. *Set* $\tau = 0$.

Step 2 *Compute an estimate of* $\chi(k)$ *via*

$$A(p_k, q^{-1}, \hat{\rho}^{(\tau)})\hat{\chi}(k) = B(p_k, q^{-1}, \hat{\rho}^{(\tau)})u(k),$$

where $\hat{\rho}^{(\tau)}$ *is the estimated obtained at the previous iteration. Based on* $\mathscr{M}_{\hat{\theta}^{(\tau)}}$, *deduce* $\{\hat{\chi}_{i,l}(k)\}_{i=1,l=0}^{n_{\text{a}},s_i}$ *as given in Eq. (25).*

Step 3 *Compute the filter:*

$$L(q^{-1},\hat{\theta}^{(\tau)}) = \frac{D(q^{-1},\hat{\eta}^{(\tau)})}{C(q^{-1},\hat{\eta}^{(\tau)})F(q^{-1},\hat{\rho}^{(\tau)})}$$

and the associated filtered signals $\{u^{\text{f}}_{j,l}(k)\}_{j=0,l=0}^{n_{\text{b}},s_j}$, $y_{\text{f}}(k)$ *and* $\{\chi^{\text{f}}_{i,l}(k)\}_{i=1,l=0}^{n_{\text{a}},s_i}$.

Step 4 *Build the filtered estimated regressor* $\hat{\varphi}_{\rm f}(k)$ *and in terms of CX4 the filtered instrument* $\hat{\zeta}_{\rm f}(k)$ *as:*

$$\hat{\varphi}_{\rm f}(k) = \big[-y_{\rm f}(k-1) \ldots -y_{\rm f}(k-n_{\rm a}) \; -\hat{\chi}^{\rm f}_{1,1}(k) \\ \ldots -\hat{\chi}^{\rm f}_{n_{\rm a},s_{\rm na}}(k) \; u^{\rm f}_{0,0}(k) \ldots u^{\rm f}_{n_{\rm b},s_{\rm ng}}(k)\big]^\top$$

$$\hat{\zeta}_{\rm f}(k) = \big[-\hat{\chi}_{\rm f}(k-1) \ldots -\hat{\chi}_{\rm f}(k-n_{\rm a}) \; -\hat{\chi}^{\rm f}_{1,1}(k) \\ \ldots -\hat{\chi}^{\rm f}_{n_{\rm a},s_{\rm na}}(k) \; u^{\rm f}_{0,0}(k) \ldots u^{\rm f}_{n_{\rm b},s_{\rm ng}}(k)\big]^\top$$

Step 5 *The IV solution is obtained as*

$$\hat{\rho}^{(\tau+1)}(N) = \left[\sum_{k=1}^{N} \hat{\zeta}_{\rm f}(k)\hat{\varphi}_{\rm f}^\top(k)\right]^{-1} \sum_{k=1}^{N} \hat{\zeta}_{\rm f}(k) y_{\rm f}(k).$$

The resulting $\hat{\rho}^{(\tau+1)}(N)$ *is the IV estimate of the process model associated parameter vector at iteration* $\tau+1$ *based on the prefiltered input/output data.*

Step 6 *An estimate of the noise signal* v *is obtained as*

$$\hat{v}(k) = y(k) - \hat{\chi}(k, \hat{\rho}^{(\tau)}). \tag{33}$$

Based on $\hat{v}$, *the estimation of the noise model parameter vector* $\hat{\eta}^{(\tau+1)}$ *follows, using in this case the ARMA estimation algorithm of the MATLAB identification toolbox (an IV approach can also be used for this purpose, see [Young (2008)]).*

Step 7 *If* $\theta^{(\tau+1)}$ *has converged or the maximum number of iterations is reached, then stop, else increase* τ *by* 1 *and go to Step 2.*

4.2.4. *LPV-SRIV algorithm for OE models*

Based on a similar concept, the so-called *simplified* LPV-RIV (LPV-SRIV) method, can also be developed for the estimation of LPV-OE models. This method is based on a model structure Eq. (21) with $C(q^{-1},\eta) = D(q^{-1},\eta) = 1$ and consequently, Step 6 of Algorithm 2 can be skipped. In practical cases, it is a fair assumption to consider that the noise model assumed is incorrect for both LPV-OE and LPV-BJ models. In this case, the LPV-SRIV algorithm might perform as well as the LPV-RIV algorithm: the BJ assumption might be more realistic, but this is compensated by the reduced number of parameters to be estimated under the OE assumption.

4.3. *Examples*

This section presents the performance of the presented algorithm on a representative set of examples.

4.3.1. *Data-generating system*

The system taken into consideration is inspired by the example in [Butcher *et al.* (2008)] and is mathematically described as

$$\mathscr{S}_{\mathrm{o}}\begin{cases} A_{\mathrm{o}}(q,p_k) = 1 + a_1^{\mathrm{o}}(p_k)q^{-1} + a_2^{\mathrm{o}}(p_k)q^{-2} \\ B_{\mathrm{o}}(q,p_k) = b_0^{\mathrm{o}}(p_k)q^{-1} + b_1^{\mathrm{o}}(p_k)q^{-2} \\ v_{\mathrm{o}}(k) = \dfrac{1}{1-q^{-1}+0.2q^{-2}}e_{\mathrm{o}}(k), \end{cases} \tag{34}$$

where $v(k) = H_{\mathrm{o}}(q)e(k)$ and

$$a_1^{\mathrm{o}}(p_k) = 1 - 0.5p_k - 0.1p_k^2,\ a_2^{\mathrm{o}}(p_k) = 0.5 - 0.7p_k - 0.1p_k^2, \tag{35a}$$

$$b_0^{\mathrm{o}}(p_k) = 0.5 - 0.4p_k + 0.01p_k^2,\ b_1^{\mathrm{o}}(p_k) = 0.2 - 0.3p_k - 0.02p_k^2. \tag{35b}$$

4.3.2. *Model assumptions*

In order to identify the presented data generating system by using the different methods described, three different model structures are defined: LPV-ARX, LPV-OE and LPV-BJ. In the example presented, all the methods assume a correct process model by using the representation:

$$\mathscr{G}_\rho\begin{cases} A(p_k,q^{-1},\rho) = 1 + a_1(p_k)q^{-1} + a_2(p_k)q^{-2} \\ B(p_k,q^{-1},\rho) = b_0(p_k)q^{-1} + b_1(p_k)q^{-2}, \end{cases}$$

where

$$a_1(p_k) = a_{1,0} + a_{1,1}p_k + a_{1,2}p_k^2,\quad a_2(p_k) = a_{2,0} + a_{2,1}p_k + a_{2,2}p_k^2 \tag{36a}$$

$$b_0(p_k) = b_{0,0} + b_{0,1}p_k + b_{0,2}p_k^2,\quad b_1(p_k) = b_{1,0} + b_{1,1}p_k + b_{1,2}p_k^2 \tag{36b}$$

Both the LS and LPV-IV4 methods [Bamieh and Giarré (2002); Butcher *et al.* (2008)] presented in Section 3 assume an LPV-ARX model structure and therefore the following noise model:

$$\mathscr{H}_\rho^{\mathrm{LS,IV4}}\left\{A(p_k,q^{-1},\rho)v(t_k) = e(t_k).\right.$$

The LPV Simplified RIV (LPV-SRIV) assumes a more realistic type of noise using the OE hypothesis:

$$\mathscr{H}_\rho^{\mathrm{LPV-SRIV}} \left\{ v(t_k) = e(t_k). \right.$$

Finally, the LPV *Refined Instrumental Variable* method (LPV-RIV) represents the situation where both the process and noise are correcty modeled using the following LPV-BJ representation:

$$\mathscr{H}_\eta^{\mathrm{LPV-RIV}} \left\{ v(t_k) = \frac{1}{1 + d_1 q^{-1} + d_2 q^{-2}} e(t_k). \right.$$

The robustness of the proposed and existing algorithms are investigated with respect to different *signal-to-noise ratios* SNR $= 10 \log \frac{P_{\chi_\mathrm{o}}}{P_{e_\mathrm{o}}}$, where P_{χ_o} and P_{e_o} are the average power of signals χ_o and e_o, respectively.

4.3.3. *Results*

In the upcoming examples, the scheduling signal p is considered as a periodic function of time: $p_k = 0.5 \sin(0.35\pi k) + 0.5$. The input $u(k)$ is taken as a white noise with a uniform distribution $\mathscr{U}(-1, 1)$ and with length $N = 4000$ to generate data sets $\mathcal{D}_N$ of $\mathscr{S}_\mathrm{o}$. To provide representative results, a Monte Carlo simulation is realized based on $N_{\mathrm{run}} = 100$ random realization, with the Gaussian white noise e_o being selected randomly for each realization at different noise levels: 15dB, 10dB, 5dB and 0dB. With respect to the considered methods, Table 1 shows the norm of the bias (BN) and standard deviation (SN) of the estimated parameter vector with

$$\begin{cases} \mathrm{BN} = ||\rho_o - \bar{\mathbb{E}}(\hat{\rho})||_{\ell_2}, \\ \mathrm{SN} = ||\bar{\mathbb{E}}(\hat{\rho} - \bar{\mathbb{E}}(\hat{\rho}))||_{\ell_2}, \end{cases} \tag{37}$$

where $\bar{\mathbb{E}}$ is the mean operator over the Monte Carlo simulation. The table also displays the mean number of iterations (Nit) the algorithms needed to converge to the estimated parameter vector.

It can be seen from Table 1 that the IV methods are unbiased according to the theoretical results. It might not appear clearly for the LPV-IV4 method when using SNR under 10dB but considering the variances induced, the bias is only due to the relatively low number of simulation runs. The LPV-SRIV method offers satisfying results, considering that the noise model assumption is also not correct. The better behavior of the LPV-SRIV method with respect to the LPV-IV4 method is justified by two facts. On the one hand, the LPV-IV4 method suffers from the ARX assumption which

Table 1. Estimator bias and variance norm at different SNR.

Method		15dB	10dB	5dB	0dB
LPV-LS	BN	2.9107	3.2897	3.0007	2.8050
	SN	0.0074	0.0151	0.0215	0.0326
LPV-IV4	BN	0.1961	1.8265	6.9337	10.85
	SN	1.3353	179.42	590.78	11782
	BN	0.0072	0.0426	0.1775	0.2988
LPV-SRIV	SN	0.0149	0.0537	0.4425	0.4781
	Nit	22	22	25	30
	BN	0.0068	0.0184	0.0408	0.1649
LPV -RIV	SN	0.0063	0.0219	0.0696	0.2214
	Nit	31	30	30	32

is not as realistic as the OE assumption for the considered BJ system. On the other hand, the LPV-SRIV method profits from an iterative process which leads to a close estimate of the optimal instrument and filter. The consequences are a variance in close range to the LS method for the LPV-SRIV algorithm while the LPV-IV4 method can be considered as irrelevant for a SNR under 10dB. Finally, and for SNR down to 5dB, the LPV-RIV method produces variance in the estimated parameters which is very close to the LPV-LS method, not mentioning that the bias has been completely suppressed. This latter result shows on this example, that the LPV-RIV algorithm dramatically improves the accuracy of the estimates.

4.4. *LPV noise system example*

The assumption of a p-independent noise model assumption is a restriction in comparison to the general case presented in Section 2.2.1. In practice, the true noise might actually be of a p-dependent nature, in which case none of the presented algorithms is suited for statistically optimal estimation. Any IV-based method is unbiased as long as the instrument is not correlated to noise and the noise modeling error will mainly have impact on the variance. Furthermore, the variance of the estimators will depend on how close the noise model and the true noise process are. The example chosen is derived from the system (34) in which the noise model is turned into an LPV model and the studied system is consequently expressed by:

$$\mathscr{S}_o \begin{cases} A_o(q, p_k) = 1 + a_1^o(p_k)q^{-1} + a_2^o(p_k)q^{-2}, \\ B_o(q, p_k) = b_0^o(p_k)q^{-1} + b_1^o(p_k)q^{-2}, \\ \left(1 + d_1^o(p_k)q^{-1} + d_2^o(p_k)q^{-2}\right) v_o(k) = e_o(k), \end{cases} \tag{38}$$

where a_i^o and b_j^o are chosen as in Eq. (36a-b) and

$$d_1^o(p_k) = -1 + 0.2p_k - 0.1p_k^2, \quad d_2^o(p_k) = 0.2 - 0.1p_k + 0.05p_k^2. \tag{39}$$

The excitation conditions are the same as previously, and the results of a Monte Carlo simulation are exposed in Table 2 with a new realization of e_o for each run. The compared methods and their respective associated models are the same as the ones considered in Section 4.3.2. The Monte Carlo simulation results are based on $N_{\rm run} = 100$ using a SNR = 10dB and a number of samples $N = 4000$. Again, the bias and standard deviation norms of the resulting estimates are analyzed and given in Table 2. From this example, it can be seen that the methods LPV-LS, LPV-IV4, LPV-SRIV give similar performance as in the previous example. This can be easily explained considering that in both examples, the noise assumption is false for these methods. Consequently, the bias of the LPV-LS method is identical, the LPV-IV4 method displays a very large parameter estimate variance, while the LPV-SRIV technique keeps a low variance and removes the bias. The variance for the LPV-RIV parameter estimates is larger than in the p-independent noise case: it can be explained by the wrong noise model assumption in this example. Nonetheless, the standard deviation remains lower than the one obtained with LPV-SRIV. In practice, for more general type of noise (e.g. if it cannot be expressed in a stochastic way), it might be better to use LPV-SRIV as the number of parameters to be estimated is lower.

Table 2. Results for a LPV noise model.

	LPV-LS	LPV-IV4	LPV-SRIV	LPV-RIV
BN	3.4943	4.2293	0.0335	0.0257
SN	0.0148	1377	0.0466	0.0318

5. Direct estimation of continuous-time LPV systems

Commonly LPV controllers are synthesized in CT as stability and performance requirements of the closed loop behavior can be more conveniently expressed in CT, like in a mixed-sensitivity setting [Zhou and Doyle (1998)].

Furthermore, developing CT-LPV models based on first principle laws is a very costly and time consuming process, often resulting in a high order model unsuitable for control design. On the other hand, identification methods are almost exclusively expressed in DT, this being mainly motivated by an easier estimation of the parameter-varying dynamics. Unfortunately, transformation of DT-LPV models to CT-LPV models is more complicated than in the LTI case and despite recent advances in LPV discretization theory (see [Tóth *et al.* (2010)]), the theory of CT realization of DT models is still in an immature state. Thus, it is often difficult in practice to obtain an adequate CT realization from an identified DT model. In order to illustrate the underlying problems, consider the following simple CT-LPV model

$$\frac{\mathrm{d}}{\mathrm{d}t}y(t) + p(t)y(t) = bu(t), \tag{40}$$

where p, y and u are the scheduling, output and input signals of the system respectively and $b \in \mathbb{R}$ is a constant parameter. When approximating the derivative in DT by, for example, using the backward Euler approximation: $\frac{\mathrm{d}}{\mathrm{d}t}y(k) = \frac{y(k)-y(k-1)}{T_\mathrm{s}}$ with $T_\mathrm{s} > 0$ the sampling period, Eq. (40) transforms into:

$$y(k) = \frac{1}{1+T_\mathrm{s}p(k)}y(k-1) + \frac{bT_\mathrm{s}}{1+T_\mathrm{s}p(k)}u(k). \tag{41}$$

This discretized model has now two p-dependent coefficients to be estimated instead of the one single constant parameter in Eq. (40). Moreover, the dependence of the coefficients on p is not linear anymore but rational with a singularity whenever $p(k) = -\frac{1}{T_\mathrm{s}}$. An alternative way to approximate derivatives in DT is to apply a forward Euler approximation: $\frac{\mathrm{d}}{\mathrm{d}t}y(k) = \frac{y(k+1)-y(k)}{T_\mathrm{s}}$, which turns the CT model into:

$$y(k) = (1 - T_\mathrm{s}p(k-1))y(k-1) + bT_\mathrm{s}u(k-1). \tag{42}$$

This discretized model has only one p-dependent coefficient and the linearity of the dependence is preserved, however the model equation depends now on $p(k-1)$ instead of $p(k)$. This so-called dynamic dependence (dependence of the model coefficients on time-shifted versions of p) is a common result of model transformations in the LPV case and rises problems in LPV system identification and control alike (see [Tóth *et al.* (2008)]). Furthermore, it is well-known in numerical analysis that the forward Euler approximation is more sensitive for the choice of T_s in terms of numerical stability than the backward Euler approximation [Atkinson (1989)]. This means that Eq. (42) requires much faster sampling rate than Eq. (41) to give a stable approximation of the system and it is more sensitive to parameter uncertainties which rises problems if Eq. (42) is used for estimation. Consequently, it can be seen that even for a very simple CT-LPV model, estimation of a DT model with the purpose of obtaining afterwards a CT realization is a tedious task with many underlying problems for which there are no general theoretical solutions available: there is a large gap between the possibilities offered in terms of CT-LPV system modeling and the need of the control framework. A possible solution to the CT system identification is to use a direct approach in which the parameters of the CT model are directly identified [Garnier and Wang (2008)]. However, in practice, CT systems can only be identified based on sampled measured data records which results in several intrinsic issues for which some solutions are presented in this section.

5.1. *System description*

Consider the data generating CT-LPV system with static dependence on p described by the following equations

$$\mathscr{S}_{\mathrm{o}} \left\{ \begin{array}{l} A_{\mathrm{o}}(p_t, d)\chi_{\mathrm{o}}(t) = B_{\mathrm{o}}(p_t, d)u(t) \\ y(t) = \chi_{\mathrm{o}}(t) + v_{\mathrm{o}}(t), \end{array} \right. \tag{43}$$

where d denotes the differentiation operator with respect to time, i.e., $d = \frac{\mathrm{d}}{\mathrm{d}t}$, $p : \mathbb{R} \rightarrow \mathbb{P}$ is the scheduling variable with $p_t = p(t)$, χ_{o} is the noise-free output, v_{o} is a quasi stationary noise process with bounded spectral density and uncorrelated to p. A_{o} and B_{o} are polynomials in d with coefficients a_i^{o} and b_i^{o} that are meromorphic functions of p

$$A_{\mathrm{o}}(p_t, d) = d^{n_{\mathrm{a}}} + \sum_{i=1}^{n_{\mathrm{a}}} a_i^{\mathrm{o}}(p_t) d^{n_{\mathrm{a}}-i}, \quad \text{and} \quad B_{\mathrm{o}}(p_t, d) = \sum_{j=0}^{n_{\mathrm{b}}} b_j^{\mathrm{o}}(p_t) d^{n_{\mathrm{b}}-j}. \tag{44}$$

In terms of identification we can assume that sampled measurements of (y, p, u) are available using a sampling period $T_{\mathrm{s}} > 0$. Hence, we will denote the DT samples of these signals as $u(k) = u(kT_{\mathrm{s}})$, where $k \in \mathbb{Z}$.

5.2. *Handling the time derivatives*

In comparison with the DT counterpart, direct CT model identification raises several technical issues. The first is related to implementation: the differential equation model is not a linear combination of the sampled process input and output signals but contains time-derivatives of the signals.

A suitable model fitting the system description assumes that these time-derivatives are available, and therefore, the parameter estimation procedure can be directly applied on them. However, these input and output time-derivatives are not available as measured data in most practical cases. A standard approach used in CT model identification is to introduce a low-pass stable filter $\mathscr{F}$, i.e., define

$$x_f(t) = f(d)x(t), \tag{45}$$

where the subscript f is used to denote the prefiltered forms of the associated variable x. The filtered time-derivatives can then be obtained by applying an array of filters in the form $f(d)p^i$:

$$x_f^{(i)}(t) = (f(d)d^i x)(t), \tag{46}$$

where $x^{(i)}(k)$ denotes the value of the time-derivative of signal $x(t)$. Various types of CT filters have been devised to deal with the need to reconstruct the time-derivatives [Garnier *et al.* (2003)]. Nonetheless, a critical issue is that these filters depend on design parameters, that need to be adequately chosen with respect to the underlying system (which is actually unknown in practice) to achieve reasonable performance of the estimation.

5.3. *Hybrid models*

Another technical issue is that direct CT model estimation, even in the LTI case, implies dealing with CT random noise sources, requiring a complicated mathematical machinery and introducing many problematic issues regarding estimation [Pintelon *et al.* (2000)]. Therefore, the basic idea to solve the noisy CT modeling problem is to assume that the CT noise process can be written at the sampling instances as a DT noise process filtered by a DT transfer function [Johansson (1994)]. This avoids the rather difficult mathematical problem of treating sampled CT random process and their equivalent in terms of a filtered piece-wise constant CT noise source (see [Pintelon *et al.* (2000); Johansson (1994)]). In terms of Eq. (43), $y(k)$ can be written as

$$y(k) = \chi_{\mathrm{o}}(k) + v_{\mathrm{o}}(k), \tag{47}$$

which corresponds to a so called hybrid Box-Jenkins system concept already used in CT identification of LTI systems (see [Pintelon *et al.* (2000); Johansson (1994); Laurain *et al.* (2008)]).

5.4. *Hybrid LPV-BJ polynomial models*

This section aims at presenting some possible tools in order to handle the estimation of CT LPV models. In this case, the p-dependent polynomials A and B are now polynomials in d:

$$A(p_t, d, \rho) = d^{n_\mathrm{a}} + \sum_{i=1}^{n_\mathrm{a}} a_i(p_t) d^{n_\mathrm{a}-i}, \quad \text{and } B(p_t, d, \rho) = \sum_{j=0}^{n_\mathrm{b}} b_j(p_t) d^{n_\mathrm{b}-j}.$$

Their parametrization is defined using the set of functions $\psi_{i,l}$ as given in Section 4.1 and the coefficients are gathered into ρ in a very similar way as in the DT case. Moreover, in order to avoid the complexity of CT noise processes, the noise process is expressed in DT using the exact same model as in Eq. (19) and the parameters are stored in the vector η. Consequently, the full hybrid LPV polynomial model is expressed as:

$$\mathscr{M}_\theta \begin{cases} \underbrace{\chi(t) + \sum_{i=1}^{n_\mathrm{a}} a_{i,0} \chi^{(n_\mathrm{a}-i)}(t)}_{F(d,\rho)\chi(t)} + \sum_{i=1}^{n_\mathrm{a}} \sum_{l=1}^{s_i} a_{i,l} \underbrace{\psi_{i,l}(p(t)) \chi^{(n_\mathrm{a}-i)}(t)}_{\chi_{i,l}(t)} \\ \qquad = \sum_{j=0}^{n_\mathrm{b}} \sum_{l=0}^{s_{\tilde{j}}} b_{j,l} \underbrace{\psi_{\tilde{j},l}(p(t)) u^{(n_\mathrm{b}-j)}(t)}_{u_{j,l}(t)} \\ v(k) = \dfrac{C(q^{-1}, \eta)}{D(q^{-1}, \eta)} e(k) \\ y(k) = \chi(k) + v(k) \end{cases} \tag{48}$$

where $\tilde{j} = j + n_\mathrm{a} + 1$ $F(d, \rho) = 1 + \sum_{i=1}^{n_\mathrm{a}} a_{i,0} d^{n_\mathrm{a}-i}$ and $\psi(i, 0)(t) = 1$. In this way, just like as in the DT case, the hybrid LPV-BJ model is rewritten as a MISO system with $\sum_{i=1}^{n_\mathrm{a}+n_\mathrm{b}+1} s_i + 1$ inputs $\{\chi_{i,l}\}_{i=1,l=1}^{n_\mathrm{a},s_i}$ and $\{u_{j,l}\}_{j=0,l=0}^{n_\mathrm{b},s_{\tilde{j}}}$:

$$\chi_{i,l}(t) = \psi_{i,l}(p(t)) \chi^{(n_\mathrm{a}-i)}(t) \qquad \{i, l\} \in \{1 \ldots n_\mathrm{a}, 1 \ldots s_i\}, \tag{49}$$

$$u_{j,l}(t) = \psi_{\tilde{j},l}(p(t)) u^{(n_\mathrm{b}-j)}(t) \qquad \{j, l\} \in \{1 \ldots n_\mathrm{b}, 1 \ldots s_{\tilde{j}}\}. \tag{50}$$

The hybrid representation of the model in Eq. (48) is a combined operation of CT and DT filtering. In order to clearly define the coexistence of DT and CT filtering, a detailed investigation and discussion about the assumptions and the structure of the model are needed. As previously pointed out, the practically feasible situation such that only sampled measurements of the CT signals (y, p, u) are available is considered here. In order to apply a CT filter on sampled data one can either interpolate the samples to obtain a CT signal and apply the CT filter on this reconstructed signal or use a numerical approximation, i.e., DT approximation of the considered system. This is a common problem for simulation of CT systems. For simulation purposes, DT approximation of the system can efficiently be dealt with by using powerful numerical algorithms available [Atkinson (1989)]. Note that to derive an accurate DT approximation of the system, it is often sufficient in terms of the classical discretization theory to assume that the sampled free CT signals of the system are restricted to be constant in the sampling period [Goodwin *et al.* (2000)], which has also been shown in case of LPV systems with static dependence [Tóth (2010)]. This provides the hypothesis, also used in [Pintelon *et al.* (2000); Johansson (1994)], that if CT (p, u) are piecewise constant between two samples, then the trajectory of y is completely determined by its observations at the sample period $T_s k$. Therefore, under these inter-sampling conditions, the following operation is well-defined [Garnier and Wang (2008)]:

$$(F(d)y)(t_k) = F(d)y(t_k), \tag{51}$$

Under this assumption, and considering that a CT filter can only be applied to sampled data through numerical approximation, the usual filter properties such as commutativity holds between a DT filter and the numerical approximation of a CT filter. Nevertheless, it is important to notice that the numerical approximation method used for the evaluation of a CT filter does not have any impact on the coefficients to be estimated which remain, the coefficients of the parsimonious CT model. Consequently, $\mathscr{M}_\theta$ can be rewritten as

$$y(k) = -\left(\sum_{i=1}^{n_a}\sum_{l=1}^{s_i}\frac{a_{i,l}}{F(d,\rho)}\chi_{i,l}\right)(k) + \left(\sum_{j=0}^{n_b}\sum_{l=0}^{s_{\bar{j}}}\frac{b_{j,l}}{F(d,\rho)}u_{k,j}\right)(k) + H(q,\eta)e(k). \tag{52}$$

which is an LTI representation.

Based on this model structure, the model set, denoted as $\mathcal{M}$, with process takes the form

$$\mathcal{M} = \left\{\mathscr{M}_\theta \mid \left[\rho^\top, \eta^\top\right]^\top = \theta \in \mathbb{R}^{n_\rho + n_\eta}\right\}. \tag{53}$$

6. Instrumental variable approach in continuous-time

Based on the MISO-LTI formulation Eq. (52), the LTI theory can be extended to achieve optimal PEM using linear regression. Nonetheless, the inherent problem of the time-derivative estimation still holds and a solution based on the RIV method for CT models (RIVC) proposed in [Laurain *et al.* (2011)] is detailed in this section.

6.1. *The optimal instrument for hybrid LPV-BJ systems*

Using the LTI model Eq. (48), $y(k)$ can be written in this CT case under the regression form:

$$y^{(n_a)}(k) = \varphi^\top(k)\rho + \tilde{v}(k), \tag{54}$$

where

$$\varphi(k) = [-y^{(n_a-1)}(k) \dots -y(k) - \chi_{1,1}(k) \dots -\chi_{n_a, s_{n_a}}(k) u_{0,0}(k) \dots u_{n_b, s_{n_g}}(k)]^\top$$

$$\tilde{v}(k) = F(d,\rho)v(k) = F(d,\rho)\frac{C(q^{-1},\eta)}{D(q^{-1},\eta)}e(k).$$

The definition of the optimal filter and instrument can be defined in a very similar way as in the DT case [Laurain *et al.* (2011)]. The optimal instrument is given as:

$$\zeta_{\text{opt}}(k) = \Big[-\chi_{\text{o}}^{(n_a-1)}(k-1) \quad \dots \quad -\chi_{\text{o}}(k) - \chi_{1,1}^{\text{o}}(k) \quad \dots \quad -\chi_{n_a,s_{n_a}}^{\text{o}}(k) \quad u_{0,0}(k) \quad \dots \quad u_{n_b,s_{n_g}}(k)\Big]^\top \tag{55}$$

while the optimal filter is given as [Young *et al.* (2007)]

$$y_{\text{f}}(k) = \frac{D_{\text{o}}(q^{-1})}{C_{\text{o}}(q^{-1})}\left[\left(\frac{1}{F_{\text{o}}(d)}y\right)(k)\right]. \tag{56}$$

As pointed out in Section 5.2, it is important to note that the regressor and instrument contain time-derivatives of y and u which, in the assumed framework considering sampled data, can only be approximated and therefore need some low pass filtering. However, it can be seen from Eq. (27), that the IV optimization problem does not require the knowledge of the instrument and regressor, but of their filtered version. It can be moreover seen from the filter expression (56) that $F_{\text{o}}(d)$ (or its estimated version $F(d,\rho)$ in practice) achieves this stable low-pass filtering directly. Therefore, it is a particular strength of the presented reformulation that the estimated filter $F(d,\rho)$ is not only used for the minimization of the prediction error but it also provides an efficient solution to the signal time-derivatives approximation problem.

6.1.1. The LPV-RIVC algorithm for BJ models

Just as in the DT case, the signals $\{\chi_{i,l}(k)\}_{i=1,l=0}^{n_a,s_i}$ and the knowledge of the different filters involved in Eq. (56) are unknown *a priori*. Therefore, similarly to the DT case, an iterative procedure is proposed to refine these unknowns and to aim at the PEM. The iterative RIV algorithm for hybrid LPV-BJ models reads as follows.

Algorithm 3 (LPV-RIVC).

Step 1 *Generate an initial estimate $\hat{\rho}^{(0)}$ using the SRIVC algorithm (by forcing an LTI structure). Set $C(d,\hat{\eta}^{(0)}) = D(d,\hat{\eta}^{(0)}) = 1$. Set $\tau = 0$.*

Step 2 *Compute an estimate of $\chi(k)$ via numerical approximation of*

$$A(p_t, d, \hat{\rho}^{(\tau)})\hat{\chi}(t) = B(p_t, d, \hat{\rho}^{(\tau)})u(t).$$

Step 3 *Compute the estimated CT filter using:*

$$L_c(d,\hat{\rho}^{(\tau)}) = \frac{1}{F(d,\hat{\rho}^{(\tau)})} \tag{57}$$

where $F(d,\hat{\rho}^{(\tau)})$ is as given in Eq. (48).

Step 4 *Use the CT filter $L_c(d,\hat{\rho}^{(\tau)})$ as well as $\hat{\chi}(k)$ in order to generate the estimates of the derivatives which are needed:*

$$\{L_c(d,\hat{\rho}^{(\tau)})\hat{\chi}^{(i)}(k)\}_{i=0}^{n_a-1}, \qquad \{L_c(d,\hat{\rho}^{(\tau)})\hat{y}^{(i)}(k)\}_{i=0}^{n_a-1}$$

$$\{L_c(d,\hat{\rho}^{(\tau)})u_{j,l}(k)\}_{j=0,l=0}^{n_b,s_{\bar{j}}}, \qquad \{L_c(d,\hat{\rho}^{(\tau)})\hat{\chi}_{i,l}(k)\}_{i=1,l=0}^{n_a,s_i}$$

Step 5 *Compute the estimated DT filter:*

$$L_d(q,\hat{\eta}^{(\tau)}) = \frac{D(q^{-1},\hat{\eta}^{(\tau)})}{C(q^{-1},\hat{\eta}^{(\tau)})}.$$

Step 6 *The filtered signals $\{u^f_{j,l}(k)\}_{j=0,l=0}^{n_b,s_{\bar{j}}}$, $y_f(k)$ and $\{\chi^f_{i,l}(k)\}_{i=1,l=0}^{n_a,s_i}$ are computed using the DT filter on the estimated derivatives obtained from Step 4.*

Step 7 *Build the filtered regressor $\hat{\varphi}_f(k)$ and, in terms of C4, the filtered instrument $\hat{\zeta}_f(k)$ as:*

$$\hat{\varphi}_f(k) = \Big[-y_f^{(n_a-1)}(k) \ldots -y_f(k)$$
$$-\hat{\chi}^f_{1,1}(k) \ldots -\hat{\chi}^f_{n_a,s_{n_a}}(k)\; u^f_{0,0}(k) \ldots u^f_{n_b,s_{n_g}}(k)\Big]^\top$$

$$\hat{\zeta}_{\mathrm{f}}(k) = \Big[-\hat{\chi}_{\mathrm{f}}^{(n_{\mathrm{a}}-1)}(k-1) \ldots -\hat{\chi}_{\mathrm{f}}(k) \\ -\hat{\chi}_{1,1}^{\mathrm{f}}(k) \ldots -\hat{\chi}_{n_{\mathrm{a}},s_{n_{\mathrm{a}}}}^{\mathrm{f}}(k)\; u_{0,0}^{\mathrm{f}}(k) \ldots u_{n_{\mathrm{b}},s_{n_{\mathrm{g}}}}^{\mathrm{f}}(k) \Big]^{\top}.$$

Step 8 *The IV solution is obtained as*

$$\hat{\rho}^{(\tau+1)}(N) = \left[\sum_{k=1}^{N} \hat{\zeta}_{\mathrm{f}}(k) \hat{\varphi}_{\mathrm{f}}^{\top}(k) \right]^{-1} \sum_{k=1}^{N} \hat{\zeta}_{\mathrm{f}}(k) y_{\mathrm{f}}^{(n_{\mathrm{a}})}(k).$$

The resulting $\hat{\rho}^{(\tau+1)}(N)$ is the IV estimate of the process model associated parameter vector at iteration $\tau+1$ based on the prefiltered input/output data.

Step 9 *An estimate of the noise signal v is obtained as*

$$\hat{v}(k) = y(k) - \hat{\chi}(k, \hat{\rho}^{(\tau)}). \tag{58}$$

Based on $\hat{v}$, the estimation of the noise model parameter vector $\hat{\eta}^{(\tau+1)}$ follows, using in this case the ARMA estimation algorithm of the MATLAB identification toolbox (an IV approach can also be used for this purpose, see [Young (2008)]).

Step 10 *If $\theta^{(\tau+1)}$ has converged or the maximum number of iterations is reached, then stop, else increase τ by 1 and go to Step 2.*

6.1.2. *The LPV-SRIVC algorithm for OE models*

Based on a similar concept and like in the DT framework, the so-called *simplified* LPV-RIVC (LPV-SRIVC) method, can also be developed for the estimation of CT-LPV-OE models. Therefore LPV-SRIVC uses the same algorithm as LPV-RIVC except for Step 5, 6 and 10 of Algorithm 3 which are skipped. It can be noticed, as well, that CT-LPV-OE models do not involve any DT filtering and, consequently, their structure is fully CT unlike the hybrid BJ LPV model.

6.2. *Examples*

As a next step, the performance of the described algorithms are presented on a representative simulation example. The system taken into consideration is inspired by a benchmark example proposed by Rao and Garnier in

[Rao and Garnier (2004)]. Since then, it has been widely used to demonstrate the performance of direct CT identification methods [Ljung (2003); Campi *et al.* (2008); Laurain *et al.* (2008); Rao and Garnier (2004)]. In order to create a CT-LPV system on which the strength of direct CT identification can be demonstrated, a "moving pole" is considered. A particular feature of LPV systems is that they have an LTI representation for every constant trajectory of p. Such an LTI representation describes the so-called frozen behavior of the system and can be expressed in a transfer function form. In terms of the frozen concept, the "moving pole" means that a particular pole of these frozen transfer functions of $\mathscr{S}_o$ is a function of p. This phenomenom often occurs in mechatronic applications such as for instance, wafer scanners [Wassink (2002)]. In our case, the Rao-Garnier benchmark inspired "moving pole" LPV system is a fourth order system with non-minimum phase frozen dynamics and a p-dependent complex pole pair. It is defined as follows:

$$\mathscr{S}_o \begin{cases} A_o(d,p) = \omega_2^2(p)\omega_1^2 + \left(2\zeta_2\omega_2(p)\omega_1^2 + 2\zeta_1\omega_2^2(p)\omega_1\right) d + \\ \quad \left(\omega_1^2 + \omega_2^2(p) + 4\zeta_2\zeta_1\omega_2(p)\omega_1\right) d^2 + \left(2\zeta_2\omega_2(p) + 2\zeta_1\omega_1\right) d^3 + d^4 \\ B_o(d,p) = -K\omega_2^2(p)\omega_1^2 d + \omega_2^2(p)\omega_1^2 \\ H_o(q) = \dfrac{1}{1 - q^{-1} + 0.2q^{-2}}, \end{cases} \tag{59}$$

where $K = 4$ [sec], $\omega_1 = 20$ [rad/s], $\zeta_1 = 0.1$, $\zeta_2 = 0.5$. The slow frozen mode ω_2 is p-dependant and chosen as: $\omega_2 = 2{+}0.5p$. Notice that the frozen behavior (p is fixed to a constant trajectory) of $\mathscr{S}_o$ for $p = 0$ corresponds exactly to the Rao-Garnier benchmark defined as

$$G_{\mathrm{RG}}(d) = \frac{-Kd + 1}{\left(\frac{d^2}{\omega_1^2} + 2\zeta_1\frac{d}{\omega_1} + 1\right)\left(\frac{d^2}{\omega_2^2(0)} + 2\zeta_2\frac{d}{\omega_2(0)} + 1\right)}. \tag{60}$$

$\mathscr{S}_o$ takes therefore the following form

$$\mathscr{S}_o \begin{cases} A_o(d,p) = d^4 + a_1^o(p)d^3 + a_2^o(p)d^2 + a_3^o(p)d + a_4^o(p) \\ B_o(d,p) = b_0^o(p)d + b_1^o(p) \\ H_o(q) = \dfrac{1}{1 - q^{-1} + 0.2q^{-2}}, \end{cases} \tag{61}$$

where

$$a_1^o(p) = 5 + 0.25p, \qquad a_2^o(p) = 408 + 3p + 0.25p^2, \tag{62a}$$

$$a_3^o(p) = 416 + 108p + p^2, \qquad a_4^o(p) = 1600 + 800p + 100p^2, \tag{62b}$$

$$b_0^o(p) = -6400 - 3200p - 400p^2, \quad b_1^o(p) = 1600 + 800p + 100p^2. \tag{62c}$$

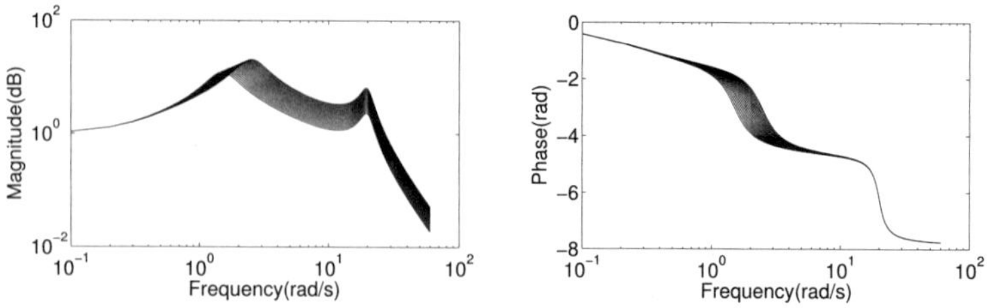

Fig. 2. Bode plot of the frozen behaviors of the true LPV system for 20 values of the scheduling variable p (between -1 to 1).

The Bode plot of 20 frozen behaviors of $\mathscr{S}_o$ is depicted in Fig. 2 for 20 fixed values of the scheduling variable p equally distributed from -1 to 1 where the consequence of the moving low frequency mode can be clearly observed. To obtain data records for identification purposes, the input signal u is chosen as a uniformly distributed sequence $\mathscr{U}(-1,1)$, with a scheduling trajectory $p(t) = \sin(\pi t)$. Furthermore, the sampling period is chosen as $T_s = 1$ms, and the simulation time is $T_{\max} = 10$s which, considering the currently available acquisition possibilities, is a fair assumption. In the sequel, the LPV-RIVC algorithm is studied for the identification of the data generating system $\mathscr{S}_o$. The proposed LPV-RIVC method is applicable to hybrid LPV-BJ model and assumes the following model structure:

$$\mathscr{M}_{\text{LPV-RIVC}} \begin{cases} A(d,p) = d^4 + a_1(p)d^3 + a_2(p)d^2 + a_3(p)d + a_4(p) \\ B(d,p) = b_0(p)d + b_1(p) \\ H(q) = \dfrac{1}{1 + d_1 q^{-1} + d_2 q^{-2}}, \end{cases}$$

where

$$a_1(p) = a_{1,0} + a_{1,1}p, \qquad a_2(p) = a_{2,0} + a_{2,1}p + a_{2,2}p^2, \tag{63a}$$

$$a_3(p) = a_{3,0} + a_{3,1}p + a_{3,2}p^2, \qquad a_4(p) = a_{4,0} + a_{4,1}p + a_{4,2}p^2, \tag{63b}$$

$$b_0(p) = b_{0,0} + b_{0,1}p + b_{0,2}p^2, \qquad b_1(p) = b_{1,0} + b_{1,1}p + b_{1,2}p^2. \tag{63c}$$

In terms of identification, the model $\mathscr{M}_{\text{LPV-RIVC}}$ corresponds to the case $\mathscr{S}_o \in \mathcal{M}$ and 19 parameters are to be estimated. The results of a Monte Carlo simulation are presented, using $N_{\text{run}} = 200$ random realizations with a SNR of 20dB. In order to assess the performance of the presented method, the Bode plot of each estimated model ($N_{\text{MC}} = 200$) at 20 frozen values of p from -1 to 1 are depicted in Fig. 3. It can be clearly observed from the Bode plot that there is a lack of excitation in the low frequencies area. To analyze the distribution of responses at low frequencies, the densities of curves

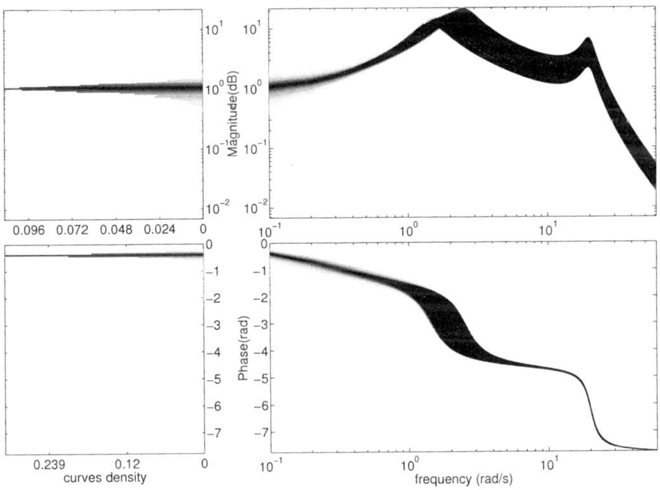

Fig. 3. Bode plots of the estimated models.

at frequency $\omega = 0.1$ rad/s are displayed on the left hand-side of Fig. 3 for both magnitude and phase diagrams. Furthermore, this distribution density is used as color coding for drawing the frequency responses in the Bode plot: the darker a given line, the higher the number of estimated models having the same response at $\omega = 0.1$ rad/s. It becomes clear that the estimated models are normally distributed and centered on the true model. Finally, the most interesting advantage of the direct CT estimation in LPV framework is shown in Fig. 4 where the poles of the 200 estimated models at 20 fixed values of p between -1 and 1 are plotted along with the true model poles. This figure also displays in the top part the density of the pole real parts. Using the same idea as in Fig. 3, the intensity of each displayed pole is related to the number of estimated poles which have the same real part. Using this representation, it can be clearly seen that the pole distribution around the "fixed pole" (around -2 ± 20i) has a Gaussian distribution while there is a sparse repartition around the "moving pole" (around $(-0.5 \pm 0.16) \pm (2 + 0.5p)$i). Furthermore, the imaginary part of all estimated poles is close to the true imaginary part of the "fixed pole". This means that by using a parsimonious CT model, the estimated models perfectly capture the time-varying nature of $\mathscr{S}_{\rm o}$ as well as the transient behavior from one operating condition to the other, which is a relevant achievement considering the complexity of the studied system and the level of additive noise. It must be noticed that trying to identify a DT model

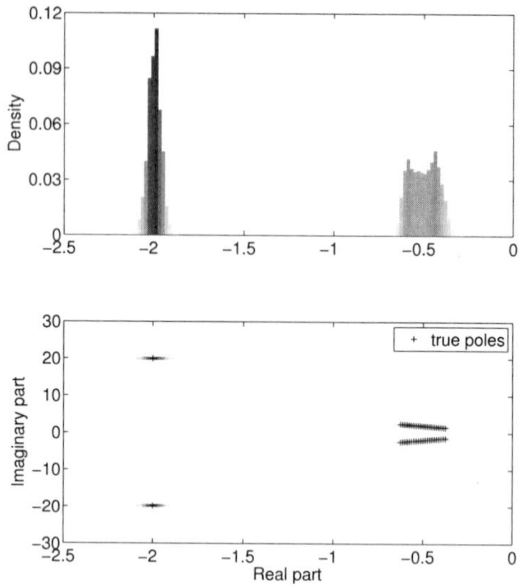

Fig. 4. Poles of the estimated models and of the true system.

for such a system is quite a tedious task. As pointed out in Section 5, the discretization of the system Eq. (59) would result in i) the augmentation of the number of parameters to be estimated, ii) the non-linear-in-the-parameter dependence on p iii) some dynamic dependence on p (appearance of $p(k-1), p(k-2)\dots$ terms). Moreover, the time-varying property of the LPV system (in this case, a "moving pole") would not anymore explicitly appear in the DT model coefficients and might not be clearly identified.

7. Conclusion

Input-output LPV models are widely used in real world applications but the dedicated identification methods have not focused on realistic noise conditions until recently. This chapter aimed at presenting some recently developed methods dedicated to the identification of IO models under different noise assumptions. It was pointed out that the most common assumption of an LPV-ARX model is attractive in the sense that it allows the direct extension of well-known LTI methods (LS and IV4). Unfortunately, this noise assumption is very restrictive and is very unlikely to be verified in practice. In order to cope with more general types of noise, a reformulation of the LPV model into a MISO LTI representation has been proposed along with some RIV based methods to solve the estimation problem. The importance of the noise model assumption and of the associated identification methods have been demonstrated through a bias and variance analysis on representative examples. Whereas most methods are developed in DT settings, very few solutions have been proposed in the literature to deal with the direct identification of CT-LPV models. Therefore, there is a large gap between the control needs and the identification solutions proposed. The inherent difficulties of direct CT identification have been listed and some solutions have been proposed relying on a hybrid representation of CT systems. An RIVC based method has been introduced, based on the same model reformulation as the one presented in the DT framework. In the CT framework however, the iterative filtering used for aiming at the PEM allows also an accurate time-derivative approximation. The method was analyzed on a fourth order moving pole system with colored noise and the advantages of direct approaches such as the possibility of capturing the true dynamical behavior of the system have been addressed. It can be finally pointed out that most of the methods presented here can be extended to the closed-loop case [Boonto and Werner (2008); Abbas and Werner (2009); Butcher *et al.* (2008); Tóth *et al.* (2011a)] for which the added difficulty lies in the noise corrupted input.

References

Abbas, H. and Werner, H. (2009). An instrumental variable technique for open-loop and closed-loop identification of input-output LPV models, in *Proceedings of the European Control Conference* (Budapest, Hungary).

Atkinson, K. E. (1989). *An introduction to numerical analysis* (John Wiley & Sons, Singapore).

Bamieh, B. and Giarré, L. (2002). Identification of linear parameter varying models, *International Journal of Robust and Nonlinear Control* **12**, 9, pp. 841–853.

Boonto, S. and Werner, H. (2008). Closed-loop system identification of LPV input-output models - application to an arm-driven pendulum, in *Proceedings of the IEEE Conference on Decision and Control* (Cancún, Mexico).

Butcher, M., Karimi, A. and Longchamp, R. (2008). On the consistency of certain identification methods for linear parameter varying systems, in *Proceedings of the IFAC World Congress* (Seoul, Korea).

Campi, M., Sugie, T. and Sakai, F. (2008). An iterative identification method for linear continuous-time systems, *IEEE Transactions on Automatic Control* **53**, 7, pp. 1661–1669.

Garnier, H., Mensler, M. and Richard, A. (2003). Continuous-time model identification from sampled data: implementation issues and performance evaluation, *International Journal of Control* **76**, 13, pp. 1337–1357.

Garnier, H. and Wang, L. e. (2008). *Identification of continuous-time models from sampled data* (Springer-Verlag, London, UK).

Giarré, L., Bauso, D., Falugi, P. and Bamieh, B. (2003). Identification for gain scheduling control: an application to rotating stall and surge control problem, *Control Engineering Practice* **14**, 4, pp. 351–361.

Goodwin, G., Graebe, S. and Salgado, M. (2000). *Control system design* (Prentice-Hall, Englewood Cliffs, NJ, USA).

Johansson, R. (1994). Identification of continuous–time models, *IEEE Transactions on Signal Processing* **42**, 4, pp. 887–897.

Laurain, V., Gilson, M., Garnier, H. and Young, P. C. (2008). Refined instrumental variable methods for identification of Hammerstein continuous time Box-Jenkins models, in *Proceedings of the IEEE Conference on Decision and Control* (Cancún, Mexico).

Laurain, V., Gilson, M., Tóth, R. and Garnier, H. (2010). Refined instrumental variable methods for identification of LPV output-error and Box-Jenkins models, *Automatica* **46**, 6, pp. 959–967.

Laurain, V., Tóth, R., Gilson, M. and Garnier, H. (2011). Direct identification of continuous-time LPV input/ouput models, *IET Control Theory & Applications* **5**, 7, pp. 878–888.

Ljung, L. (1999). *System identification: theory for the user - Second edition* (Prentice-Hall, Upper Saddle River, NJ, USA).

Ljung, L. (2003). Initialisation aspects for subspace and output-error identification methods, in *Proceedings of the European Control Conference* (Cambridge, U.K.).

Ljung, L. (2009). Experiments with identification of continuous time models, in *Proceedings of the IFAC Symposium on System Identification* (Saint-Malo, France).

Lopes dos Santos, P., Ramos, J. A. and Martins de Carvalho, J. L. (2007). Identification of linear parameter varying systems using an iterative deterministic-stochastic subspace approach, in *Proceedings of the European Control Conference.*

Lovera, M. and Mercère, G. (2007). Identification for gain-scheduling: a balanced subspace approach, in *Proceedings of the American Control Conference* (New York City, NY, USA).

Pintelon, R., Schoukens, J. and Rolain, Y. (2000). Box–Jenkins continuous–time modeling, *Automatica* **36**, 7, pp. 983–991.

Rao, G. P. and Garnier, H. (2004). Identification of continuous-time systems: direct or indirect ? *Systems Science* **30**, 3, pp. 25–50.

Söderström, T. and Stoica, P. (1983). *Instrumental variable methods for system identification* (Springer-Verlag, New York, NY, USA).

Tóth, R. (2010). *Modeling and identification of linear parameter-varying systems*, *Lecture Notes in Control and Information Sciences*, Vol. 403 (Springer-Verlag, Berlin/Heidelberg, Germany).

Tóth, R., Heuberger, P. S. C. and Van den Hof, P. M. J. (2008). Flexible model structures for LPV identification with static scheduling dependency, in *Proceedings of the IEEE Conference on Decision and Control* (Cancún, Mexico).

Tóth, R., Heuberger, P. S. C. and Van den Hof, P. M. J. (2009a). Asymptotically optimal orthonormal basis functions for LPV system identification, *Automatica* **45**, 6, pp. 1359–1370.

Tóth, R., Heuberger, P. S. C. and Van den Hof, P. M. J. (2010). On the discretization of LPV state-space representations, *IET Control Theory & Applications* **4**, 4, pp. 2082–2096.

Tóth, R., Laurain, V., Gilson, M. and Garnier, H. (2011a). On the closed loop identification of LPV models using instrumental variables, in *Proceedings of the the IFAC World Congress* (Milan, Italy).

Tóth, R., Laurain, V., Zheng, W. and Poolla, K. J. (2011b). Model structure learning: a support vector machine approach for LPV linear-regression models, in *submitted to the 50th IEEE Conference on Decision and Control* (Orlando, FL, USA).

Tóth, R., Lyzell, C., Enqvist, M., Heuberger, P. S. C. and Van den Hof, P. M. J. (2009b). Order and structural dependence selection of LPV-ARX models using a nonnegative garrote approach, in *Proceedings of the IEEE Conference on Decision and Control* (Shanghai, China).

van Wingerden, J. W. and Verhaegen, M. (2009). Subspace identification of bilinear and LPV systems for open- and closed-loop data, *Automatica* **45**, 2, pp. 372–381.

Wassink, M. G. (2002). *LPV control for wafer stages*, Msc thesis, CTB595-02-2097, Philips CFT, Eindhoven, The Netherlands.

Wei, X. and Del Re, L. (2006). On persistent excitation for parameter estimation of quasi-LPV systems and its application in modeling of diesel engine torque, in *Proceedings of the IFAC Symposium on System Identification* (Newcastle, Australia).

Young, P. (2008). The refined instrumental variable method: Unified estimation of discrete and continuous-time transfer function models, *Journal Européen des Systèmes Automatisés* **42**, pp. 149–179.

Young, P., Garnier, H. and Gilson, M. (2007). *Identification of continuous-time models from sampled data*, chap. Refined instrumental variable identification of continous-time hybrid Box-Jenkins models (Springer-Verlag, London, UK), pp. 91–131.

Young, P. C. (1984). *Recursive estimation and time-series analysis* (Springer-Verlag, Berlin, Germany).

Young, P. C. and Jakeman, A. (1980). Refined instrumental variable methods of recursive time-series analysis —part III. extensions, *International Journal of Control* **31**, pp. 441–764.

Zhou, K. and Doyle, J. C. (1998). *Essentials of robust control* (Prentice-Hall, Upper Saddle River, NJ, USA).

Chapter 6

Reducing the dimensions of data matrices involved in LPV subspace identification methods

V. Verdult & M. Verhaegen

Delft Center for Systems and Control
Delft University of Technology
Mekelweg 2, Delft, 2628 CD, The Netherlands
vverdult@caiway.nl
m.verhaegen@tudelft.nl

Subspace identification methods for linear parameter-varying state-space systems aim at the reconstruction of the state sequence based on the available measurements of the inputs, outputs and time-varying parameters. Several methods have been proposed in the literature that perform computations with data matrices of which the number of rows grow exponentially with the order of the system. The data matrices for relatively low-order systems with only a few inputs, outputs and time-varying parameters can easily have more than 100 000 rows. Besides computational issues, these methods require a huge amount of measurements (larger than or equal to the number of rows in the data matrices). This severely limits the practical applicability of the LPV subspace methods. In this chapter a method to reduce the dimensions of the data matrices is discussed. It is based on an efficient RQ factorization in which only the rows that have the largest influence on the output-error of the LPV model are used. Also a kernel method is discussed that avoids computations with large dimensional matrices.

1. Introduction

Subspace identification is a well-accepted method for the identification of multivariable linear time-invariant (LTI) systems (Katayama, 2005; Van Overschee and De Moor, 1996; Verhaegen and Verdult, 2007). In many cases these methods provide a good alternative to the classical nonlinear optimization-based prediction-error methods (Ljung, 1999). Subspace methods do not require a particular parametrization of the system; this makes them numerically attractive and especially suitable for multivariable systems. Subspace methods can also be used to generate an initial starting point for the iterative prediction-error methods. This combination of subspace and prediction-error methods is a powerful tool for determining an LTI system from input and output measurements.

Subspace identification methods for LPV systems aim at estimating the state sequence and use this estimate in a second step to determine the system matrices. Unfortunately, it is only possible to determine an approximation of the state sequence, because the influence of the initial state is unknown. This problem also occurs for LTI subspace methods based on estimating the state sequence, as pointed out by Van Overschee and De Moor (1994) and Picci and Katayama (1996). The approximation error that results from neglecting the initial state in LTI subspace methods, can be made arbitrarily small by taking a sufficiently large block size for the data matrices involved. A similar observation can be made for the LPV case. However, to keep the computational requirements of the LPV subspace methods within reasonable limits, it is often necessary to use a small block size: the number of rows in the data matrices used in the LPV subspace methods grows exponentially with this block size. Different LPV subspace methods use different ways of approximating the state, leading to different approximation errors.

Similar to the LTI case, the LPV system obtained from the subspace identification method can be used as an initial starting point for a nonlinear-optimization based prediction-error method. Such nonlinear-optimization based methods have been described by Lee and Poolla (1996, 1999).

The subspace methods discussed in this chapter are designed for an LPV state-space system with affine parameter dependence in the state equation and a time-invariant output equation:

$$
\begin{aligned}
x(k+1) &= \left(A_0 + \sum_{i=1}^{n_p} p_i(k) A_i\right) x(k) + \left(B_0 + \sum_{i=1}^{n_p} p_i(k) B_i\right) u(k) + w(k), \\
y(k) &= Cx(k) + Du(k) + v(k).
\end{aligned}
$$

where $x(k) \in \mathbb{R}^{n_x}$ is the state, $u(k) \in \mathbb{R}^{n_u}$ is the input, $y(k) \in \mathbb{R}^{n_y}$ is the output, $p(k) \in \mathbb{R}^{n_p}$ is the time-varying parameter, $w(k) \in \mathbb{R}^{n_x}$ is the process noise, $v(k) \in \mathbb{R}^{n_y}$ is the measurement noise, $p_i(k)$ denotes the ith element of the vector $p(k)$, and $A_i \in \mathbb{R}^{n_x \times n_x}$, $B_i \in \mathbb{R}^{n_x \times n_u}$, for $i = 0, 1, 2, \ldots, n_p$, $C \in \mathbb{R}^{n_y \times n_x}$, and $D \in \mathbb{R}^{n_y \times n_u}$. By defining

$$
\begin{aligned}
A &:= [A_0 \; A_1 \; A_2 \cdots A_{n_p}], \\
B &:= [B_0 \; B_1 \; B_2 \cdots B_{n_p}],
\end{aligned}
$$

the LPV system can also be written in the more compact form

$$
x(k+1) = A \begin{bmatrix} x(k) \\ p(k) \otimes x(k) \end{bmatrix} + B \begin{bmatrix} u(k) \\ p(k) \otimes u(k) \end{bmatrix} + w(k), \tag{1}
$$

$$
y(k) = Cx(k) + Du(k) + v(k), \tag{2}
$$

where $\otimes$ denotes the *Kronecker product*, which is defined as follows: let $M \in \mathbb{R}^{p \times q}$ and $N \in \mathbb{R}^{r \times t}$ be two arbitrary matrices, then the Kronecker product of these matrices is a $pr \times qt$ matrix given by

$$
M \otimes N = \begin{bmatrix} m_{1,1}N & m_{1,2}N & \cdots & m_{1,q}N \\ m_{2,1}N & m_{2,2}N & \cdots & m_{2,q}N \\ \vdots & \vdots & \ddots & \vdots \\ m_{p,1}N & m_{p,2}N & \cdots & m_{p,q}N \end{bmatrix},
$$

where $m_{i,j}$ denotes the (i,j)th entry of the matrix M.

The identification problem that is addressed is: given a finite number of measurements of the input $u(k)$, the output $y(k)$ and the time-varying parameters $p(k)$, determine the system matrices A, B, C and D of (1)–(2) up to a similarity transformation. It is assumed that the LPV system is bounded-input, bounded-output stable (Rugh, 1996), that the LTI system constructed from A_0 and C is observable, and that $w(k)$ and $v(k)$ are sequences of independent random variables with zero-mean and bounded second- and fourth-order moments. Further it will be assumed that the data is recorded in open-loop, that is, the random variables $w(k)$ and $v(k)$ are independent from the input $u(k)$.

The LPV system (1)–(2) is closely related to the bilinear system. The bilinear system can be regarded as a special type of LPV system in which the time-varying parameter $p(k)$ equals the input $u(k)$. In fact, the theory for LPV subspace identification presented in this chapter is an extension of the theory for bilinear subspace identification developed by Favoreel (1999) and Chen and Maciejowski (2000b).

Subspace identification methods for the LPV system (1)–(2) can also be derived for an innovation type of model where $v(k) = e(k)$ and $w(k) = Kv(k)$.

The derivation of such methods goes along the same lines as presented in this chapter following the original work presented by Verdult and Verhaegen (2002). Though this paper belongs to the early developments in LPV subspace identification, the key ideas presented are general in nature and can also be used in more recent developments. One example is the use of the kernel approach in LPV subspace identification as reported by van Wingerden and Verhaegen (2009).

A major problem with subspace methods for LPV and bilinear systems is the enormous dimension of the data matrices involved. The number of rows that needs to be included in the data matrices grows exponentially with the order of the system. This limits the applicability of the methods to high-order systems with several inputs and outputs. In this chapter a method to reduce the dimensions of the data matrices is discussed. It is based on an efficient RQ factorization in which only the rows that have the largest influence on the output-error of the LPV model are used. A second implementation method is discussed that avoids computations with large dimensional matrices by using a kernel method.

This chapter is organized as follows: In Section 2 the notation is introduced, and the data equations for the LPV system are derived. Section 3 contains the basic ideas underlying subspace identification for LPV systems. Based on these ideas a number of subspace identification methods can be derived. One such method, the two-block method is described in Section 4. Section 5 describes the practical implementation of the LPV subspace identification methods based on selection of only the dominant rows of the data matrices. Section 6 describes the kernel-based method.

2. Data equations

The *data equations* describe the relation between structured matrices built from the available measurements of the input $u(k)$, output $y(k)$, and time-varying parameters $p(k)$. These structured matrices are called the *data matrices*. The definitions of the data matrices make use of the *Khatri-Rao* product (Khatri and Rao, 1968). This product, denoted by the symbol $\odot$, is a column-wise Kronecker product for two matrices with an equal number of columns. The formal definition is as follows: let $M \in \mathbb{R}^{p \times q}$ and $N \in \mathbb{R}^{r \times q}$ be two arbitrary matrices, then the Khatri-Rao product of these matrices equals:

$$M \odot N = \left[m_1 \otimes n_1 \; m_2 \otimes n_2 \; \dots \; m_q \otimes n_q \right],$$

where m_i and n_i $(i = 1, 2, \dots, q)$ denote the columns of the matrices M and N, respectively.

The following data matrices are defined:

$$P_k := [p(k), p(k+1), \dots, p(k+N-1)] \in \mathbb{R}^{n_p \times N}, \tag{3}$$

$$X_k := [x(k), x(k+1), \dots, x(k+N-1)] \in \mathbb{R}^{n_x \times N}, \tag{4}$$

$$X_{j|j} := \begin{bmatrix} X_j \\ P_j \odot X_j \end{bmatrix} \in \mathbb{R}^{(n_p+1)n_x \times N}, \tag{5}$$

$$X_{k+j|j} := \begin{bmatrix} X_{k+j-1|j} \\ P_{k+j} \odot X_{k+j-1|j} \end{bmatrix} \in \mathbb{R}^{((n_p+1)^{k+1})n_x \times N}, \tag{6}$$

$$U_k := [u(k), u(k+1), \dots, u(k+N-1)] \in \mathbb{R}^{n_u \times N}, \tag{7}$$

$$U_{j|j} := \begin{bmatrix} U_j \\ P_j \odot U_j \end{bmatrix} \in \mathbb{R}^{(n_p+1)n_u \times N}, \tag{8}$$

$$U_{k+j|j} := \begin{bmatrix} U_{k+j} \\ P_{k+j} \odot U_{k+j} \\ U_{k+j-1|j} \\ P_{k+j} \odot U_{k+j-1|j} \end{bmatrix} \in \mathbb{R}^{((n_p+1)^{k+1}-1)n_u(n_p+1)/n_p \times N}, \tag{9}$$

$$W_k := [w(k), w(k+1), \dots, w(k+N-1)] \in \mathbb{R}^{n_x \times N}, \tag{10}$$

$$W_{j|j} := W_j \in \mathbb{R}^{n_x \times N}, \tag{11}$$

$$W_{k+j|j} := \begin{bmatrix} W_{k+j} \\ W_{k+j-1|j} \\ P_{k+j} \odot W_{k+j-1|j} \end{bmatrix} \in \mathbb{R}^{((n_p+1)^{k+1}-1)n_x/n_p \times N}. \tag{12}$$

Note that the matrices $X_{k+j|j}$, $W_{k+j|j}$ and $U_{k+j|j}$ have different structures. The integer j in the subscript $k+j|j$ indicates that the data matrix contains measurements from time instant j and further; the integer k is called the *block size* and determines the number of block rows in the data matrix.

Some additional data matrices are:

$$Y_k := [y(k), y(k+1), \dots, y(k+N-1)] \in \mathbb{R}^{n_y \times N}, \tag{13}$$

$$\mathscr{Y}_{j|j} := Y_j \in \mathbb{R}^{n_y \times N}, \tag{14}$$

$$\mathscr{Y}_{k+j|j} := \begin{bmatrix} Y_{k+j} \\ \mathscr{Y}_{k+j-1|j} \end{bmatrix} \in \mathbb{R}^{(k+1)n_y \times N}, \tag{15}$$

$$V_k := [v(k), v(k+1), \ldots, v(k+N-1)] \in \mathbb{R}^{n_y \times N}, \tag{16}$$

$$\mathscr{V}_{j|j} := V_j \in \mathbb{R}^{n_y \times N}, \tag{17}$$

$$\mathscr{V}_{k+j|j} := \begin{bmatrix} V_{k+j} \\ \mathscr{V}_{k+j-1|j} \end{bmatrix} \in \mathbb{R}^{(k+1)n_y \times N}. \tag{18}$$

Note that the matrix $\mathscr{Y}_{k+j|j}$ is constructed from the same rows as the data matrix corresponding to $y(k)$ in linear subspace identification.

The data matrices are related to each other through the data equations given in the following three lemmas. The proof of these lemmas follows a simple induction argument (Verdult, 2002).

Lemma 1. *The relation between X_j and $X_{k+j|j}$ is*

$$X_{k+j|j} = \begin{bmatrix} X_j \\ P_{k+j|j} \odot X_j \end{bmatrix}, \tag{19}$$

where $P_{j|j} := P_j \in \mathbb{R}^{n_p \times N}$,

$$P_{k+j|j} := \begin{bmatrix} P_{k+j-1|j} \\ P_{k+j}(1,:) \\ P_{k+j}(1,:) \odot P_{k+j-1|j} \\ \vdots \\ P_{k+j}(n_p,:) \\ P_{k+j}(n_p,:) \odot P_{k+j-1|j} \end{bmatrix} \in \mathbb{R}^{((n_p+1)^{k+1}-1) \times N},$$

and $P_k(i,:)$ denotes the ith row of the matrix P_k.

Lemma 2. *For the LPV system (1)–(2) it holds that*

$$X_{k+j} = \Delta_k^x X_{k+j-1|j} + \Delta_k^u U_{k+j-1|j} + \Delta_k^w W_{k+j-1|j}, \tag{20}$$

where

$$\Delta_1^x := [A_0, A_1, \ldots, A_{n_p}],$$
$$\Delta_k^x := [A_0 \Delta_{k-1}^x, A_1 \Delta_{k-1}^x, \ldots, A_{n_p} \Delta_{k-1}^x],$$
$$\Delta_1^u := [B_0, B_1, \ldots, B_{n_p}],$$

$$\Delta_k^u := [\Delta_1^u, A_0\Delta_{k-1}^u, A_1\Delta_{k-1}^u, \dots, A_{n_p}\Delta_{k-1}^u],$$
$$\Delta_1^w := I_n,$$
$$\Delta_k^w := [\Delta_1^w, A_0\Delta_{k-1}^w, A_1\Delta_{k-1}^w, \dots, A_{n_p}\Delta_{k-1}^w].$$

Lemma 3. *For the LPV system (1)–(2) it holds that*

$$\mathscr{Y}_{k+j|j} = H_k^x X_{k+j-1|j} + H_k^u U_{k+j-1|j} + H_k^w W_{k+j-1|j} + G_k^u U_{k+j} + \mathscr{V}_{k+j|j}, \tag{21}$$

where

$$H_1^x := \begin{bmatrix} CA_0 & CA_1 & \cdots & CA_{n_p} \\ C & 0 & \cdots & 0 \end{bmatrix},$$
$$H_k^x := \begin{bmatrix} CA_0\Delta_{k-1}^x & CA_1\Delta_{k-1}^x & \cdots & CA_{n_p}\Delta_{k-1}^x \\ H_{k-1}^x & 0 & \cdots & 0 \end{bmatrix},$$
$$H_1^u := \begin{bmatrix} CB_0 & CB_1 & \cdots & CB_{n_p} \\ D & 0 & \cdots & 0 \end{bmatrix},$$
$$H_k^u := \begin{bmatrix} CB_0 & CB_1 & \cdots & CB_{n_p} & CA_0\Delta_{k-1}^u & CA_1\Delta_{k-1}^u & \cdots & CA_{n_p}\Delta_{k-1}^u \\ G_{k-1}^u & 0 & \cdots & 0 & H_{k-1}^u & 0 & \cdots & 0 \end{bmatrix},$$
$$H_1^w := \begin{bmatrix} C \\ 0 \end{bmatrix},$$
$$H_k^w := \begin{bmatrix} C & CA_0\Delta_{k-1}^w & CA_1\Delta_{k-1}^w & \cdots & CA_{n_p}\Delta_{k-1}^w \\ 0 & H_{k-1}^w & 0 & \cdots & 0 \end{bmatrix},$$
$$G_k^u := \begin{bmatrix} D \\ 0 \end{bmatrix},$$

and where Δ_k^x, Δ_k^u, *and* Δ_k^w *are as in Lemma 2.*

3. Basic ideas behind the methods

The subspace identification methods for LPV systems consist of two steps. First, the state sequence is approximated by manipulating the data equations derived in Section 2. Second, the estimated state sequence is used to set up a linear least-squares problem from which the system matrices of the LPV system follow. There are different ways to approximate the state; this gives rise to a number of different identification methods. In this section, a general framework is created that allows a unified treatment of these different subspace identification methods.

3.1. Generalized data equation

Subspace identification is not based on only one data equation, but on a whole set of data equations starting at different time instants. Since these data equations all have the same structure, it is convenient to work with a generalized expression, the so-called *generalized data equation.*

To derive the generalized data equation, the result of Lemma 1 is combined with Lemma 3. For this purpose the matrix

$$\Gamma_k := H_k^x(:\,,1\colon n) = \begin{bmatrix} CA_0^k \\ \vdots \\ CA_0 \\ C \end{bmatrix} \tag{22}$$

is introduced. Note that for $k \geq n$ this matrix equals the extended observability matrix for a linear system, constructed from the matrices A_0 and C, with its block rows turned upside down. The matrix Γ_k is used to partition the matrix H_k^x defined in Lemma 3 as follows:

$$H_k^x = \begin{bmatrix} \Gamma_k \ \widetilde{H}_k^x \end{bmatrix}, \tag{23}$$

where $\widetilde{H}_k^x$ denotes the remaining columns of H_k^x. Combining Lemma 1 with Lemma 3 transforms Eq. (21) into

$$\begin{aligned}\mathscr{Y}_{k+j|j} = \Gamma_k X_j &+ \widetilde{H}_k^x(P_{k+j-1|j} \odot X_j) + H_k^u U_{k+j-1|j} + G_k^u U_{k+j} \\ &+ H_k^w W_{k+j-1|j} + \mathscr{V}_{k+j|j}.\end{aligned} \tag{24}$$

This equation has the following basic structure:

$$Y = \Gamma X + H(P \odot X) + \Phi_U U + V, \tag{25}$$

where the time indices have been neglected. Equation (25) is called the *generalized data equation*; it consists of four parts:

(1) A linear combination of the rows of the state sequence X. This linear combination is determined by Γ.

(2) A linear combination of the rows of the matrix $P \odot X$. This linear combination is determined by H.

(3) A linear combination of the rows of the input data matrix U. This linear combination is determined by Φ_U.

(4) Noise terms, represented by V.

An important step in LPV subspace identification is the approximation of the state sequence from an estimate of the product ΓX. Provided that Γ and X are of rank n, a singular value decomposition (SVD) of their product can be used to recover the state sequence X up to a similarity transformation: let the SVD of ΓX be

$$\Gamma X = U_n \Sigma_n V_n^T,$$

with $\Sigma_n \in \mathbb{R}^{n_x \times n_x}$ containing the nonzero singular values, then

$$\Gamma = U_n \Sigma_n^{1/2} T,$$
$$X = T^{-1} \Sigma_n^{1/2} V_n^T,$$

with $T \in \mathbb{R}^{n_x \times n_x}$ an arbitrary nonsingular matrix that represents the similarity transformation. The requirement that Γ has full rank is fulfilled if the block size satisfies $k \geq n-1$, and the linear system constructed from A_0 and C is observable. The requirement that X has full rank can be thought of as a persistency of excitation condition, like in linear subspace methods (Verhaegen and Dewilde, 1992; Jansson and Wahlberg, 1998).

Since the state sequence can be recovered from the product ΓX, the goal is to approximate this product. The first step is to determine a matrix Z_X that can be constructed from the available measurements of $u(k)$, $p(k)$ and $y(k)$, such that it has approximately the same row space as the matrix X. An appropriate linear combination of the rows of Z_X then yields $\Phi_X Z_X \approx \Gamma X$. The matrix Z_X can be determined in different ways, by manipulation of the data equations derived in Section 2. The different possibilities are discussed in more detail by Verdult (2002). In Section 4 one possibility is described. In general, the following relation holds:

$$\Gamma X = \Phi_X Z_X + V_X + \Xi_X, \tag{26}$$

where V_X denotes noise terms that appear due to the manipulation of the data equation, and Ξ_X is an error due to neglecting certain terms in the data equations. Equation (26) will be clarified with an example. Based on Eq. (20), a possible approximation would be $\Gamma_k X_{k+j} \approx \Gamma_k \Delta_k^u U_{k+j-1|j}$, that is, the influence of the noise $W_{k+j-1|j}$ and of the past history of the state sequence $X_{k+j-1|j}$ are neglected. With this choice, the matrices appearing on the right-hand side of Eq. (26) can be identified as follows:

$$\Phi_X \leftarrow \Gamma_k \Delta_k^u, \tag{27}$$

$$Z_X \leftarrow U_{k+j-1|j}, \tag{28}$$

$$V_X \leftarrow \Gamma_k \Delta_k^w W_{k+j-1|j}, \tag{29}$$

$$\Xi_X \leftarrow \Gamma_k \Delta_k^x X_{k+j-1|j}. \tag{30}$$

Since the state sequence also appears in the term $P \odot X$ in the generalized data equation (25), the product $H(P \odot X)$ must also be approximated. Let $Z_{P \odot X}$ be a matrix constructed from the available measurements of $u(k)$, $p(k)$, and $y(k)$, such that it has approximately the same row space as the matrix $P \odot X$, then similar to Eq. (26):

$$H(P \odot X) = \Phi_{P \odot X} Z_{P \odot X} + V_{P \odot X} + \Xi_{P \odot X}, \tag{31}$$

where $V_{P \odot X}$ represents noise terms and $\Xi_{P \odot X}$ is an error due to neglecting certain terms.

Using Eq. (26) and Eq. (31), the generalized data equation (25) can be reformulated as

$$Y = \Phi_X Z_X + \Phi_{P \odot X} Z_{P \odot X} + \Phi_U U + \Phi_V Z_V + \Xi_X + \Xi_{P \odot X}, \tag{32}$$

with $\Phi_V Z_V = V + V_X + V_{P \odot X}$. This equation is used in the following subsection to explain how an approximation of the state sequence can be computed. The terms Ξ_X and $\Xi_{P \odot X}$ can be thought of as bias terms.

3.2. *Approximating the state sequence*

From the previous discussion it follows that $\Gamma X \approx \Phi_X Z_X$, thus it seems natural to estimate the state sequence from an SVD of the product $\Phi_X Z_X$. The matrix Z_X in this product is known, because it is constructed from the available measurements. The matrix Φ_X on the other hand is unknown. To explain how this matrix can be determined, the matrices Φ and Ω are introduced:

$$\Phi := \begin{bmatrix} \Phi_X & \Phi_{P \odot X} & \Phi_U \end{bmatrix} \tag{33}$$

$$\Omega := \begin{bmatrix} Z_X \\ Z_{P \odot X} \\ U \end{bmatrix}. \tag{34}$$

Based on Eq. (32), an estimate of Φ can be obtained by solving the linear least-squares problem

$$\min_{\Phi} \|Y - \Phi\Omega\|_F^2, \tag{35}$$

where $\|\cdot\|_F$ denotes the Frobenius norm. Recall that the Frobenius norm of a matrix $A \in \mathbb{R}^{m\times n}$ equals

$$\|A\|_F = \left(\sum_{i=1}^{m}\sum_{j=1}^{n} a_{i,j}^2\right)^{1/2} = \left(\text{trace}(AA^T)\right)^{1/2} = \|\text{vec}(A)\|_2,$$

where $a_{i,j}$ denotes the (i,j)th element of the matrix A.

The least-squares estimate of Φ equals

$$\widehat{\Phi} = Y\Omega^T\left(\Omega\Omega^T\right)^{-1}. \tag{36}$$

From this it is immediately clear that the matrix Ω must have full rank. This full rank condition can be considered as a kind of persistency of excitation requirement. It means that certain restrictions must be imposed on the sequences $u(k)$ and $p(k)$, and the system (1)–(2). The difference between Φ and its least-squares estimate $\widehat{\Phi}$ equals

$$\widehat{\Phi} - \Phi = (\Phi_V Z_V + \Xi_X + \Xi_{P\odot X})\Omega^T\left(\Omega\Omega^T\right)^{-1}.$$

So the estimation error depends on the noise term Z_V and the approximation errors Ξ_X and $\Xi_{P\odot X}$. If Ω is uncorrelated with Z_V in the sense that

$$\lim_{N\to\infty} \frac{1}{N} Z_V\Omega^T = 0, \tag{37}$$

where N is the number of columns in Z_V and Ω, then the estimation error $\widehat{\Phi} - \Phi$ becomes independent of the noise term Z_V for $N \to \infty$.

An estimate of Φ_X, denoted by $\widehat{\Phi}_X$, can now be constructed from the first set of columns of $\widehat{\Phi}$ that correspond to Z_X. Since the goal is to use $\widehat{\Phi}_X Z_X$ as an approximation for ΓX, it is important to analyse the error $\Gamma X - \widehat{\Phi}_X Z_X$. This error can be split into three parts

$$\begin{aligned}\Gamma X - \widehat{\Phi}_X Z_X &= (\Gamma X - \Phi_X Z_X) + (\Phi_X Z_X - \widehat{\Phi}_X Z_X)\\ &= \Xi_X + V_X + (\Phi_X - \widehat{\Phi}_X)Z_X.\end{aligned} \tag{38}$$

The first two terms come from the approximation $\Gamma X \approx \Phi_X Z_X$, the third term is due to the estimation of Φ_X. The following theorem gives some insight in the size of the third term in terms of the Frobenius norm:

Theorem 1. *(Verdult, 2002) Given Eq. (32) and the matrices Φ and Ω by Eq. (33)) and Eq. (34), respectively. Let $\widehat{\Phi}$ be given by Eq. (36), and let the following conditions be satisfied:*

(1) $\lim_{N\to\infty} \frac{1}{N} Z_V \Omega^T = 0,$

(2) $\lim_{N\to\infty} \frac{1}{N} \Omega\Omega^T = M$, *with* $\|M\|_F < \infty$ *and such that* M *has full row rank,*

(3) $\limsup_{N\to\infty} \frac{1}{N} \|\Xi_X + \Xi_{P\odot X}\|_F^2 \leq \xi < \infty,$

(4) $\sup_i \|\Omega(:,i)\|_2^2 = \gamma < \infty$ *with* $\Omega(:,i)$ *the* i*th column of the matrix* Ω.

Then

$$\limsup_{N\to\infty} \frac{1}{N} \left\| (\widehat{\Phi} - \Phi)\Omega \right\|_F^2 \leq r\xi, \tag{39}$$

with r *the total number of rows of the matrix* Ω.

Proof. Combining Eq. (32) with Eq. (36) and using some properties of the Frobenius norm it follows that:

$$\begin{aligned} \|(\widehat{\Phi} - \Phi)\Omega\|_F \leq & \left\| \frac{1}{N} \Phi_V Z_V \Omega^T \right\|_F \left\| \left(\frac{1}{N} \Omega\Omega^T \right)^{-1} \right\|_F \|\Omega\|_F \\ & + \|\Xi_X + \Xi_{P\odot X}\|_F \left\| \frac{1}{N} \Omega^T \left(\frac{1}{N} \Omega\Omega^T \right)^{-1} \Omega \right\|_F . \end{aligned}$$

Evaluate the norms in this expression:

$$\|\Omega\|_F^2 = \sum_{i=1}^{N} \|\Omega(:,i)\|_2^2 \leq N\gamma,$$

$$\left\| \frac{1}{N} \Omega^T \left(\frac{1}{N} \Omega\Omega^T \right)^{-1} \Omega \right\|_F^2 = r.$$

Therefore,

$$\begin{aligned} \frac{1}{\sqrt{N}} \|(\widehat{\Phi} - \Phi)\Omega\|_F \leq & \sqrt{\gamma} \left\| \frac{1}{N} \Phi_V Z_V \Omega^T \right\|_F \left\| \left(\frac{1}{N} \Omega\Omega^T \right)^{-1} \right\|_F \\ & + \frac{\sqrt{r}}{\sqrt{N}} \|\Xi_X + \Xi_{P\odot X}\|_F . \end{aligned}$$

Taking the limit for $N \to \infty$ yields the desired result. □

This theorem shows that with the assumption (37), the error $(\widehat{\Phi}-\Phi)\Omega$ does not depend on the noise Z_V; it only depends on the approximation errors Ξ_X and $\Xi_{P\odot X}$.

If Eq. (37) is not satisfied, it is still possible to remove the noise term Z_V from the error $(\widehat{\Phi}-\Phi)\Omega$, by using an appropriate instrumental variable matrix (Söderström and Stoica, 1983). Methods based on such approach are discussed in Verdult (2002).

From the previous discussion it follows that an estimate $\widehat{\Phi}_X$ can be obtained by Eq. (36). The next step is to approximate the state sequence by computing the SVD of $\widehat{\Phi}_X Z_X$. Let this SVD be given by

$$\widehat{\Phi}_X Z_X = \begin{bmatrix} U_1 \ U_2 \end{bmatrix} \begin{bmatrix} \Sigma_1 & 0 \\ 0 & \Sigma_2 \end{bmatrix} \begin{bmatrix} V_1^T \\ V_2^T \end{bmatrix}, \tag{40}$$

where $\Sigma_1 \in \mathbb{R}^{n_x \times n_x}$ is a diagonal matrix, then an approximation of the state sequence is given by

$$\widehat{X} = \Sigma_1^{1/2} V_1^T.$$

The nonzero singular values Σ_2 result from the noise and the approximation errors Ξ_X and $\Xi_{P\odot X}$. If the noise and the approximation errors are not too large, the singular values contained in Σ_1 will be much larger than the ones in Σ_2. In this case, the order n of the system can be determined from this SVD, by locating a gap in the singular value plot.

The estimate of the state $\widehat{X}$ is only an approximation of the real state sequence, because of Eq. (38). From this equation it follows that the error on the state can be divided into three parts:

- the error due to the difference between Φ_X and $\widehat{\Phi}_X$, caused by Ξ_X and $\Xi_{P\odot X}$;
- the error due to neglecting certain terms in the approximation of ΓX (the term Ξ_X);
- the error due to the noise (the term V_X).

The approximation errors Ξ_X and $\Xi_{P\odot X}$ depend of course on the particular choice of the matrices Z_X and $Z_{P\odot X}$. These matrices can be constructed from the data matrices $U_{k+j|j}$ and $P_{k+j|j}$ defined in Section 2 in different ways. Since the number of rows in these data matrices grows exponentially with the block size, the number of rows in Z_X and $Z_{P\odot X}$ can become very large. A very large number of rows limits the practical applicability of the identification method. Therefore, Z_X and $Z_{P\odot X}$ should be chosen carefully. The particular choice of Z_X and $Z_{P\odot X}$ is always a trade-off between the size of the approximation errors Ξ_X and $\Xi_{P\odot X}$, and the dimensions of the matrices Z_X and $Z_{P\odot X}$.

3.3. *Estimating the system matrices*

Once an estimate of the state sequence has been determined, the system matrices can be estimated by solving a set of linear least-squares problems. The system equations (1)–(2) motivate the following two least-squares problems:

$$\min_{A,B} \|\widehat{X}_{k+1} - A\widehat{X}_{k|k} - BU_{k|k}\|_2^2,$$

$$\min_{C,D} \|Y_k - C\widehat{X}_k - DU_k\|_2^2,$$

where $\widehat{X}_k$ denotes an estimate of the state sequence, and

$$\widehat{X}_{k|k} := \begin{bmatrix} \widehat{X}_k \\ P_k \odot \widehat{X}_k \end{bmatrix}.$$

The least-squares solutions are

$$[\widehat{A}, \widehat{B}] = \widehat{X}_{k+1} \begin{bmatrix} \widehat{X}_{k|k}^T & U_{k|k}^T \end{bmatrix} \left(\begin{bmatrix} \widehat{X}_{k|k} \\ U_{k|k} \end{bmatrix} \begin{bmatrix} \widehat{X}_{k|k}^T & U_{k|k}^T \end{bmatrix} \right)^{-1}, \quad (41)$$

$$[\widehat{C}, \widehat{D}] = Y_k \begin{bmatrix} \widehat{X}_k^T & U_k^T \end{bmatrix} \left(\begin{bmatrix} \widehat{X}_k \\ U_k \end{bmatrix} \begin{bmatrix} \widehat{X}_k^T & U_k^T \end{bmatrix} \right)^{-1}. \quad (42)$$

The system matrices are estimated up to a similarity transformation compared to the real system matrices, because the estimate of the state sequence is obtained by using an SVD as explained in the previous subsection. Obviously, the approximation used in estimating the state sequence will influence the estimates of the system matrices.

4. Two-block identification method

As mentioned in the previous section, it is possible to derive different identification methods by changing the approximation used for the state sequence. In other words, by making different choices for the matrices Z_X

and $Z_{P\odot X}$ in the generalized data equation (32). One such choice is discussed in this section: the two-block method in which two sets of data matrices are used; one set consists of matrices starting at $j = 0$, the other set consists of matrices starting at $j = k$, where k is the block size used for the data matrices in both sets. The first set of matrices is used to approximate the state sequence and the second set is used to identify the system. For $j = k$ it follows from Eq. (24) that the data matrices are related by

$$\begin{aligned}\mathscr{Y}_{2k|k} &= \Gamma_k X_k + \widetilde{H}_k^x (P_{2k-1|k} \odot X_k) \\ &\quad + H_k^u U_{2k-1|k} + G_k^u U_{2k} + H_k^w W_{2k-1|k} + \mathscr{V}_{2k|k}. \qquad (43)\end{aligned}$$

The state sequence X_k in this equation is approximated based on Eq. (20) for $j = 0$:

$$X_k = \Delta_k^x X_{k-1|0} + \Delta_k^u U_{k-1|0} + \Delta_k^w W_{k-1|0}.$$

The terms $\Delta_k^w W_{k-1|0}$ and $\Delta_k^x X_{k-1|0}$ in this equation are neglected, and the approximation becomes

$$X_k \approx \Delta_k^u U_{k-1|0}. \qquad (44)$$

The error due to neglecting $\Delta_k^w W_{k-1|0}$ depends, of course, on the signal-to-noise ratio. The error due to neglecting the term $\Delta_k^x X_{k-1|0}$ is bounded if the LPV system is uniformly exponentially stable (Rugh, 1996):

Lemma 4. *(Verdult, 2002) Consider the LPV system (1)–(2). If the state equation (1) for $u(k) = 0$ is uniformly exponentially stable, and for any j*

$$\sup_{k\geq j} \|u(k)\|_2 < \infty, \qquad \sup_{k\geq j} \|w(k)\|_2 < \infty,$$

then given a finite constant β, there exist a finite constant $\gamma > 0$ and a constant λ, $0 \leq \lambda < 1$, such that

$$\frac{1}{\sqrt{N}} \|\Delta_k^x X_{k+j-1|j}\|_F \leq \gamma \lambda^k,$$

for all k, j, such that $k \geq j$, and such that

$$\sup_k \|p(k)\|_2 \leq \beta,$$

where Δ_k^x is as in Lemma 2 and $\|\cdot\|_F$ denotes the Frobenius norm.

Proof. Lemma 1 and the definition of the Frobenius norm yields:

$$\left|\left|\Delta_k^x X_{k+j-1|j}\right|\right|_F^2 = \sum_{q=j}^{N+j-1} \left|\left|\Delta_k^x \begin{bmatrix} x(q) \\ P_{k+j-1|j}(:, q-j+1) \odot x(q) \end{bmatrix}\right|\right|_2^2,$$

where $P_{k+j-1|j}(:, q-j+1)$ is the $(q-j+1)$th column of $P_{k+j-1|j}$. An induction argument can be used to show that:

$$\prod_{t=0}^{k-1}\left(A_0 + \sum_{i=1}^{n_p} p_i(t+q)A_i\right)x(q) = \Delta_k^x \begin{bmatrix} x(q) \\ P_{k+j-1|j}(:, q-j+1) \odot x(q) \end{bmatrix}. \tag{45}$$

Since the state equation (1) is uniformly exponentially stable, there exist a finite constant $\kappa > 0$ and a constant λ, $0 \leq \lambda < 1$, such that for every $k > 0$ (Rugh, 1996)

$$\left|\left|\prod_{t=0}^{k-1}\left(A_0 + \sum_{i=1}^{n_p} p_i(t+q)A_i\right)\right|\right|_2 \leq \kappa\lambda^k.$$

Thus, the left-hand side of Eq. (45) satisfies

$$\left|\left|\prod_{t=0}^{k-1}\left(A_0 + \sum_{i=1}^{n_p} p_i(t+q)A_i\right)x(q)\right|\right|_2 \leq ||x(q)||_2 \cdot \kappa\lambda^k.$$

Now putting everything together, we get

$$\left|\left|\Delta_k^x X_{k+j-1|j}\right|\right|_F \leq \sqrt{N}\gamma\lambda^k, \qquad \gamma = \kappa \sup_{q \geq j} ||x(q)||_2,$$

The constant γ is finite, because the system (1)–(2) is uniformly bounded-input, bounded-output stable, and because the signals $p(k)$, $u(k)$, and $w(k)$ are bounded. □

This lemma shows that if the LPV system is uniformly asymptotically stable, the perturbation due to neglecting $\Delta_k^x X_{k+j-1|j}$ decreases with increasing block size k. However, it does not give the rate of decrease: it only shows that there exists a λ, but does not provide its value. The rate of decrease can be quantified by using

$$\sigma = \sup_{q \geq j}\left|\left|A_0 + \sum_{i=1}^{n_p} p_i(q)A_i\right|\right|_2, \tag{46}$$

provided that $\sigma < 1$. Similar to Lemma 4 it can be deduced that

$$\frac{1}{\sqrt{N}}||\Delta_k^x X_{k+j-1|j}|| \leq \beta\sigma^k,$$

for some finite constant $\beta > 0$. The idea of using (46) to analyse the error with respect to the block size k has been put forward by Chen and Maciejowski (2000a,b) in the context of subspace identification for bilinear systems. It is however a conservative approach, because there exist uniformly asymptotically stable LPV systems with $\sigma > 1$.

The two-block method is obtained by using the approximation (44) for the terms X_k and $(P_{2k-1|k} \odot X_k)$ in equation (43). The next theorem shows the relation of the two-block method with the generalized data equation (32) and with the procedure outlined in Section 3.2 to approximate the state sequence. Recall that the key step in this procedure is the estimation of the matrix Φ given by Eq. (33).

Theorem 2. *Given the bounded-input, bounded-output stable system (1)–(2) with bounded sequences $p(k)$, $u(k)$, $w(k)$, and $v(k)$, the data equations of Lemmas 1, 2, and 3, and the generalized data equation (32):*

$$Y = \Phi_X Z_X + \Phi_{P\odot X} Z_{P\odot X} + \Phi_U U + \Phi_V Z_V + \Xi_X + \Xi_{P\odot X},$$

with the following substitutions:

$$\begin{aligned} Y &\leftarrow \mathscr{Y}_{2k|k}, \\ Z_X &\leftarrow U_{k-1|0}, \\ Z_{P\odot X} &\leftarrow P_{2k-1|k} \odot U_{k-1|0}, \\ U &\leftarrow \begin{bmatrix} U_{2k-1|k} \\ U_{2k} \end{bmatrix}, \\ Z_V &\leftarrow \begin{bmatrix} W_{2k-1|k} \\ \mathscr{V}_{2k|k} \\ W_{k-1|0} \\ P_{2k-1|k} \odot W_{k-1|0} \end{bmatrix}. \end{aligned}$$

Then

$$\lim_{N\to\infty} \frac{1}{N} Z_V \Omega^T = 0 \quad \textit{with} \quad \Omega := \begin{bmatrix} Z_X \\ Z_{P\odot X} \\ U \end{bmatrix} \tag{47}$$

holds with probability one, and the approximation errors are

$$\begin{aligned} \Xi_X &\leftarrow \Gamma_k \Delta_k^x X_{k-1|0}, \\ \Xi_{P\odot X} &\leftarrow (\widetilde{H}_k^x \otimes \Delta_k^x)(P_{2k-1|k} \odot X_{k-1|0}), \end{aligned}$$

with Γ_k given by (22) and $\widetilde{H}_k^x$ given by (23).

If the matrix Ω *has full row rank, the matrix*

$$\Phi := \left[\Phi_X \; \Phi_{P\odot X} \; \Phi_U\right]$$

can be estimated as

$$\widehat{\Phi} = Y\Omega^T \left(\Omega\Omega^T\right)^{-1},$$

and $(\widehat{\Phi} - \Phi)\Omega$ *does not depend on the noise term* Z_V *with probability one for* $N \to \infty$.

Proof. The proof of (47) follows from the fact that $p(k)$ and $u(k)$ are independent from the noise sequences $v(k)$ and $w(k)$. The expressions for the approximation errors follow by comparing Eq. (43) with the generalized data equation (32). The approximation errors are bounded, because the system is bounded-input, bounded-output stable. The remaining part of the proof follows by applying Theorem 1. □

It is expected that the approximation errors, Ξ_X and $\Xi_{P\odot X}$, decrease with increasing block size k. This statement is supported by two facts. First, the matrices Ξ_X and $\Xi_{P\odot X}$ result from neglecting the term $\Delta_k^x X_{k+j-1|j}$ for some j. Second, Lemma 4 shows that the Frobenius norm of $\Delta_k^x X_{k+j-1|j}$ decreases with increasing block size k. To make a precise statement, the propagation of the error due to neglecting the term $\Delta_k^x X_{k+j-1|j}$ should be computed. However, it is difficult to come up with norm bounds, because the block size varies, and thus the number of rows in the matrices involved varies. Furthermore, it is not sufficient to have a norm bound on Ξ_X and $\Xi_{P\odot X}$, because the bound in Theorem 1 also depends on the number of rows in the data matrix Ω.

The approximation error is not the only concern in choosing the block size. Assuming that the approximation error decreases with increasing block size k, a large block size is desirable. However, a large block size increases the number of rows in the data matrices. Since the number of rows in these matrices grows exponentially with the block size, huge dimensional matrices can result. Therefore, the way that the identification methods are implemented is of paramount importance. Below two implementation methods that avoid computations with huge dimensional matrices are discussed.

Table 1. Examples of the total number of rows in the RQ factorization for the two-block identification method.

	$k = 1$	$k = 2$	$k = 3$	$k = 4$
$n_p = n_u = n_y = 1$	33	130	515	2 052
$n_p = n_u = n_y = 2$	246	2 192	19 690	177 156

5. Implementation by selection of dominant rows

The RQ factorization provides an efficient way to compute an estimate of the matrix Φ that is needed to determine the state sequence. It can be used to implement Eq. (36). It is immediately clear that Eq. (36) is the solution to a least-squares problem by solving the normal equations. It is well-known that a numerically more reliable and computationally less demanding solution can be computed using an RQ factorization (Golub and Van Loan, 1996). Given the RQ factorization

$$\begin{bmatrix} Z_X \\ \begin{bmatrix} Z_{P\odot X} \\ U \end{bmatrix} \\ Y \end{bmatrix} = \begin{bmatrix} R_{11} & 0 & 0 \\ R_{21} & R_{22} & 0 \\ R_{31} & R_{32} & R_{33} \end{bmatrix} \begin{bmatrix} Q_1 \\ Q_2 \\ Q_3 \end{bmatrix}, \tag{48}$$

Eq. (36) becomes

$$\widehat{\Phi} = \begin{bmatrix} R_{31} & R_{32} \end{bmatrix} \begin{bmatrix} R_{11} & 0 \\ R_{21} & R_{22} \end{bmatrix}^{-1}, \tag{49}$$

where it is assumed that N is sufficiently large, such that Ω has full row rank. Equation (49) shows that the least-squares estimate of Φ_X is given by

$$\widehat{\Phi}_X = R_{31}R_{11}^{-1} - R_{32}R_{22}^{-1}R_{21}R_{11}^{-1}. \tag{50}$$

The number of rows in the RQ factorization for the two-block method is given by:

$$(2d_k + d_k^2)\frac{n_u(n_p+1)}{n_p} + n_u + (k+1)n_y.$$

where $d_k := (n_p+1)^{k+1} - 1$. Table 1 illustrates this expression. Clearly, the number of rows in the RQ factorization grows very rapidly with the block size k.

To avoid computations with large dimensional matrices, the idea is to select a subset of rows from the data matrices Z_X, $Z_{P\odot X}$, and U appearing in this RQ factorization, and to discard the remaining rows. Only the selected rows are used to compute an RQ factorization. This later RQ factorization, which has less rows than the original one involving the complete data matrices, will be called the *reduced RQ factorization.* The rows that are selected for the reduced RQ factorization are the ones that contribute most to the matrix Y.

The selection of the rows is basically a subset-selection problem (Lawson and Hanson, 1995, p. 194–196). Roughly speaking, subset selection can be performed by either using a singular value decomposition or an RQ (or QR) factorization (Golub and Van Loan, 1996, p. 590–595). A method based on the RQ factorization is preferred, because it allows to compute the reduced RQ factorization step by step during the selection process. This means that at the end of the selection process, the reduced RQ factorization has already been computed, which obviously results in considerable computational savings.

Reducing the number of rows in this way will of course be at the expense of introducing an error, so that the identified LPV model is only an approximate model. Simulations done by the authors (Verdult, 2002) show that quite often the number of rows can be reduced considerably, without deteriorating the quality of the model too much. However, if the performance of the obtained LPV model is not satisfactory, the model can be improved using a nonlinear optimization-based identification procedure (Lee and Poolla, 1996, 1999; Verdult, 2002).

Let us compare the reduced RQ factorization with the full factorization. For this purpose, let Ω^s contain the rows of Z_X, $Z_{P\odot X}$, and U that are selected for the reduced RQ factorization, and let the matrix Ω^d contain the remaining rows that are discarded. Now, with appropriate definitions of Φ^s and Φ^d, the generalized data equation (32) can be rewritten as

$$Y = \begin{bmatrix}\Phi^s & \Phi^d\end{bmatrix} \begin{bmatrix}\Omega^s \\ \Omega^d\end{bmatrix} + \Phi_V Z_V + \Xi_X + \Xi_{P\odot X}.$$

For simplicity, it is assumed that the matrix

$$\Omega = \begin{bmatrix}\Omega^s \\ \Omega^d\end{bmatrix}$$

has full rank for every N that is larger than the total numbers of rows in Ω.

Let the RQ factorization

$$\begin{bmatrix} \Omega^s \\ \Omega^d \\ Y \end{bmatrix} = \begin{bmatrix} L_{11} & 0 & 0 \\ L_{21} & L_{22} & 0 \\ L_{31} & L_{32} & L_{33} \end{bmatrix} \begin{bmatrix} S_1 \\ S_2 \\ S_3 \end{bmatrix} \tag{51}$$

be equivalent to the RQ factorization Eq. (48). It follows that

$$\Upsilon := \begin{bmatrix} L_{31} & L_{32} \end{bmatrix} \begin{bmatrix} L_{11} & 0 \\ L_{21} & L_{22} \end{bmatrix}^{-1} \begin{bmatrix} \Omega^s \\ \Omega^d \end{bmatrix} \tag{52}$$

is the quantity that needs to be computed. Discarding the rows contained in Ω^d is equivalent to computing a reduced RQ factorization given by

$$\begin{bmatrix} \Omega^s \\ Y \end{bmatrix} = \begin{bmatrix} L_{11} & 0 \\ L_{31} & \overline{L}_{32} \end{bmatrix} \begin{bmatrix} S_1 \\ \overline{S}_3 \end{bmatrix}, \tag{53}$$

which can be used to approximate Υ by

$$\widehat{\Upsilon} := L_{31} L_{11}^{-1} \Omega^s. \tag{54}$$

The error due to this approximation can easily be computed and is given in the following lemma:

Lemma 5. *Given the full RQ factorization (51) and the reduced RQ factorization (53). The matrices Υ and $\widehat{\Upsilon}$ given by (52) and (54), respectively, satisfy*

$$\|\Upsilon - \widehat{\Upsilon}\|_F = \|L_{32}\|_F,$$

where $\|\cdot\|_F$ denotes the Frobenius norm.

Proof. Combining the RQ factorization (51) with Eq. (52) and Eq. (54) it is easy to see that $\Upsilon = L_{31}S_1 + L_{32}S_2$ and $\widehat{\Upsilon} = L_{31}S_1$. Hence, $\|\Upsilon - \widehat{\Upsilon}\|_F = \|L_{32}S_2\|_F = \|L_{32}\|_F$. □

The selection of the rows from the matrices Z_X, $Z_{P \odot X}$, and U is carried out by computing an RQ factorization row by row. The selection of a row is based on its contribution to the matrix Y in a least-squares sense. Let the matrix Z_X^r contain a subset of the rows of the matrix Z_X, let $Z_{P\odot X}^r$ contain a subset of $Z_{P\odot X}$, and U^r a subset of U, such that

$$\Omega^s := \begin{bmatrix} Z_X^r \\ Z_{P\odot X}^r \\ U^r \end{bmatrix}.$$

Let n_z^r denote the number of rows of the matrix Z_X^r. Provided that at least n rows are selected from Z_X ($n_z^r \geq n$), an approximation of the state can be computed, from the singular value decomposition of the matrix

$$L_{31}L_{11}^{-1}Z_X^r,$$

similar to the description in Section 3.2. This state estimate can be used to compute the system matrices in a similar way as described in Section 3.3.

Since the matrix $L_{31}L_{11}^{-1}Z_X^r$ needs to be computed, the matrix Z_X^r needs to be available. Therefore, the selection of the rows from Z_X is separated from the selection of the rows from $Z_{P\odot X}$ and U. Below, it is explained how to select the rows from Z_X. A similar procedure can be used to select rows from $Z_{P\odot X}$ and U. Note that there is no need to separate the selection of rows from $Z_{P\odot X}$ and the selection of rows from U.

The first row that is selected from Z_X is the one that has the largest contribution to the matrix Y. To determine this row, the set of least-squares problems

$$\min_{\psi_i} ||Y - \psi_i Z_X(i,:)||_F^2, \quad i = 1, 2, \ldots, n_z \tag{55}$$

is solved and the corresponding residuals are computed. The row with the smallest residual has the largest contribution to the matrix Y and should hence be selected. To select the next row of Z_X, the effect of the previously selected row needs to be taken into account. This means that the residuals have to be recomputed after which the row that yields the smallest residual can be selected. More specifically, the procedure given below (Verhaegen, 1987) can be used to select the n_z^r most dominant rows from Z_X.

Take $Z^1 := Z_X$ and $Y^1 := Y$. For i ranging from 1 to n_z^r, perform the following steps:

(1) Select the row of Z^i that has the smallest residual when solving

$$\min_{\Psi_q^i} \left|\left|Y^i - \Psi_q^i Z^i(q,:)\right|\right|_F^2, \quad q = i, i+1, \ldots, n_z.$$

(2) Interchange the ith row of Z^i with the row selected in the previous step, by the permutation matrix Π^i.

(3) Compute an orthogonal projection H^i such that

$$\Pi^i \begin{bmatrix} T^i & 0 \\ \overline{T}^i & Z^i \end{bmatrix} H^i = \begin{bmatrix} T^{i+1} & 0 \\ \overline{T}^{i+1} & Z^{i+1} \end{bmatrix},$$

where $T^i \in \mathbb{R}^{(i-1)\times(i-1)}$, and $\overline{T}^i \in \mathbb{R}^{(n_z-i+1)\times(i-1)}$.

(4) Apply H^i to Y^i:

$$\begin{bmatrix} \overline{Y}^i & Y^i \end{bmatrix} H^i = \begin{bmatrix} \overline{Y}^{i+1} & Y^{i+1} \end{bmatrix},$$

where $\overline{Y}^i \in \mathbb{R}^{(i-1)\times(i-1)}$.

Figure 1 gives a graphical representation of this selection algorithm. It shows that in every step the matrix Z^i is reduced by one column and one row, and that the matrix T^i becomes a lower triangular matrix.

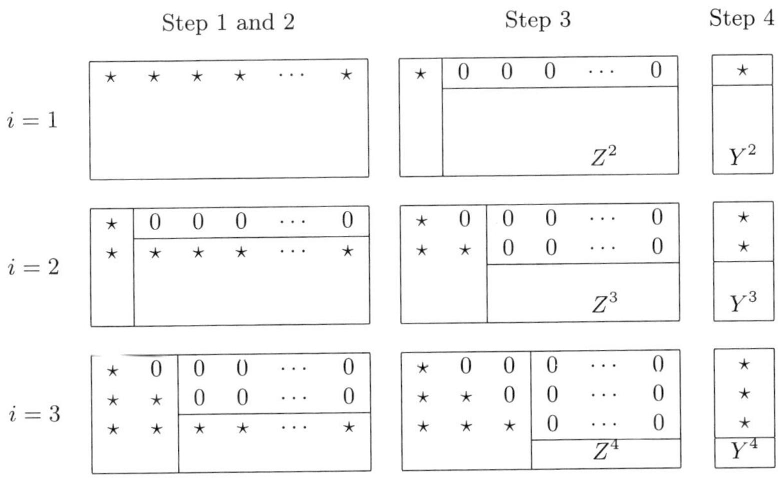

Fig. 1. Graphical representation of the row selection procedure. The row selected in Step 1 is brought to the top of the matrix Z^i by Step 2. Step 3 then computes a rotation such that all the entries of this row become zero, except for the first i entries. The "new" matrix Z^{i+1} is one column and one row smaller than Z^i. In Step 4 the rotation is also applied to Y^i, and this yields Y^{i+1}.

The number of rows that are selected to form Z_X^r determines the approximation error. If the amount of computer memory available is limited, it makes sense to fix this number of rows beforehand. Alternatively, the norm of the residual can be used to decide when to stop the selection process.

Let the matrix $\overline{Z}_X$ be the matrix that is obtained by applying all the computed permutation matrices Π^i, $i = 1, 2, \ldots, n_z^r$ to the matrix Z_X, that is, $\overline{Z}_X := \Pi^{n_z^r} \cdot \Pi^{n_z^r - 1} \cdots \Pi^1 \cdot Z_X$. Since the top n_z^r rows of $\overline{Z}_X$ are now the ones that have the largest contribution to Y_j, these are the rows to construct the matrix Z_X^r from: $Z_X^r = \overline{Z}_X(1\colon n_z^r, :)$. An important motivation for using the RQ factorization in selecting the rows, is that such a selection procedure yields already the first part of the reduced RQ factorization (53):

$$Z_X^r = \begin{bmatrix} L_{11}(1\colon n_z^r, 1\colon n_z^r) & 0 \end{bmatrix} S_1(1\colon n_z^r, :).$$

The next step is to reduce the number of rows in the matrix $Z_{P\odot X}$. For this a similar procedure as for the selection of the rows of Z_X is used. By continuing the application of orthogonal rotations, the remaining part of the matrix L_{11} is computed. Applying all the rotations used in the selection procedures for Z_X and $Z_{P\odot X}$ to the matrix Y yields the matrix L_{31} of (53). Since the matrix $\overline{L}_{32}$ is not needed in the subsequent computations, it is not necessary to complete the reduced RQ factorization.

The selection procedure described above minimizes the Frobenius norm of the approximation error

$$\|\Upsilon - \widehat{\Upsilon}\|_F,$$

for a fixed number of rows n_z^r. To see this, note that

$$\min_{\Phi^s} \|Y - \Phi^s \Omega^s\|_F^2 = \min_{\Phi^s} \|[L_{31}, \overline{L}_{32}] - \Phi^s [L_{11}, 0]\|_F^2 = \|\overline{L}_{32}\|_F^2.$$

Comparing Eq. (51) with Eq. (53) yields

$$\overline{L}_{32}\overline{S}_3 = L_{32}S_2 + L_{33}S_3.$$

Since $S_2 S_3^T = 0$, it follows that

$$\|\overline{L}_{32}\|_F^2 = \|L_{32}\|_F^2 + \|L_{33}\|_F^2.$$

Equation (51) shows that the selection of the rows, that is, which rows go into Ω^s, and which go into Ω^d, does not influence the matrix L_{33}. Therefore, minimizing $\|\overline{L}_{32}\|_F^2$ is equivalent to minimizing $\|L_{32}\|_F^2$. With the result of Lemma 5, the conclusion is that the described selection procedure minimizes the Frobenius norm of the approximation error.

Because of the huge dimensions of the data matrices, an efficient implementation is of paramount importance. Storing the complete matrices Z_X and $Z_{P\odot X}$ into the memory is often not feasible. Therefore, a procedure has been developed that constructs only one row at a time. After evaluating the associated value for the residual, the row is replaced by the next one. In this way, only the residuals need to be stored. If a row is encountered that has the lowest residual thus far, this row is stored into the memory. The Householder rotations associated with the selected rows are also stored. For an efficient implementation, the Householder matrices H^j should not be explicitly formed. It is much more efficient to use and store the associated Householder vectors (Golub and Van Loan, 1996, p. 211).

An algorithm for selecting n_z^r rows from $Z_X \in \mathbb{R}^{n_z \times N}$ is given below. It constructs the matrix $R \in \mathbb{R}^{n_z^r \times n_z^r}$, which is a part of the reduced RQ factorization (53): $R = L_{11}(1\colon n_z^r, 1\colon n_z^r)$.

Define the index set $\mathscr{I}^1 := \{1, 2, \ldots, n_z\}$.
for each $j \in \{1, 2, \ldots, n_z^r\}$

(1) Select the row from $\mathscr{I}^j$ that has the smallest residual:
for each $k \in \mathscr{I}^j$

(a) Construct the kth row of Z_X: $Z_k^1 = Z_X(k, 1\colon N)$.

(b) Apply all previously computed Householder rotations $Z_k^{i+1} = Z_k^i H^i$ for $i \in \{1, 2, \ldots, j-1\}$; this yields Z_k^j.

(c) Compute the residuals

$$r_k^j = \min_{\phi_k^j} \|Y^j(:, j\colon N) - \Psi_k^j Z_k^j(:, j\colon N)\|_F^2.$$

Note that this can be done efficiently by first computing a Householder rotation Q_k^j such that $Z_k^j(:, j\colon N) Q_k^j = [P_k^j,\ 0_{1\times(N-j)}]$ with $P_k^j \in \mathbb{R}$ and then evaluating

$$r_k^j = \|Y^j(:, j\colon N) Q_k^j(:, 2\colon N-j+1)\|_F^2.$$

(d) If $r_k^j \leq r_i^j$ for all $i \in \mathscr{I}^j$ then row k has the smallest residual thus far and we select this row, that is, we make the jth row of Z_X^r equal to Z_k^j and we take $\mathscr{I}^{j+1} = \mathscr{I}^j \setminus \{k\}$ as the new index set.

end

(2) Compute a Householder rotation H^j such that
$Z_X^r(j, :)H^j = [R^j,\ 0_{1\times(N-j)}]$ with $R^j \in \mathbb{R}^{1\times j}$.

(3) Make the jth row of the R matrix equal to $[R^j,\ 0_{1\times(n_z^r - j)}]$.

(4) Apply the rotation H^j to Y^j: $Y^{j+1} = Y^j H^j$.

end

Observe that a further improvement can be obtained by updating the residuals at each step using information from the previous step. After computing the residuals in the first step, the residuals for the following steps can be obtained with the update rule

$$r_k^j = r^{j-1} - \|Y^j(:\,, j\colon N)Q_k^j(:\,, 1)\|_2^2,$$

where r^{j-1} is the residual of the row selected in the previous step.

With this implementation, it is possible to handle systems of moderate dimensions, because the data matrices do not need to be stored. However, the computation time increases drastically. This is mainly due to the need to construct one row at a time. Depending on its place in the data matrix, the construction of a row involves several times N multiplications. It is possible to reduce this computational burden a bit by stepping through the rows in a special order. For example, if we are processing the rows of $U_{j-1|0}$, we could after processing a row in the first part of the matrix (that is, $U_{j-2|0}$), store the row, and use it to first compute the rows in the third part of the matrix (that is, $P_{j-1} \odot U_{j-2|0}$), which can be obtained by element-wise multiplication with a row from P_{j-1}. This requires only N multiplications, in contrast with several times N when the row needs to be constructed from scratch.

6. Implementation by a kernel method

An alternative to deal with the exponentially growing number of rows in the data matrices is the application of a kernel method that only relies on computations with the kernel matrix, which is a square matrix with dimensions equal to the number of samples. The major advantage of such a method is that the computational complexity does no longer depend on the exponentially growing number of rows in the data matrices, but on the (usually much smaller) number of samples available. Besides the reduction of computational complexity, the method also allows to identify LPV systems using only a small number of samples of the input and output data. It is well known that estimation of a large number of parameters using a small number of samples can lead to a large variance error in the estimated parameters (Sjöberg and Ljung, 1995). In such a case it is worthwhile to allow for a small bias in order to reduce the variance error. Such a bias-variance trade-off can be made with a kernel method based on Ridge regression. The kernel method can also be used for the identification of bilinear systems (Verdult and Verhaegen, 2003)

If the matrix Ω has full row rank, the solution to the least squares problem Eq. (35) is given by Eq. (36). If only a limited number of samples N is available, the matrix Ω has much more rows than columns and can thus not be of full row rank. There are not enough data points to determine a unique solution. For such an underdetermined least squares problem, the minimum norm solution is commonly used. This solution is given by

$$\widehat{\Phi} = Y V_a S_a^{-1} U_a^T, \tag{56}$$

based on the SVD of the matrix Ω given by

$$\Omega = \begin{bmatrix} U_a & U_b \end{bmatrix} \begin{bmatrix} S_a & 0 \\ 0 & S_b \end{bmatrix} \begin{bmatrix} V_a^T \\ V_b^T \end{bmatrix}, \tag{57}$$

where S_a contains the dominant singular values.

The minimum norm solution can be computed using a kernel based algorithm in which we only have to perform computations with the kernel matrix $\Omega^T \Omega \in \mathbb{R}^{N \times N}$. Thus, computations in the large dimensional space spanned by the columns of Ω are avoided.

The kernel method can be derived by looking for a solution of the form $\Phi = \Psi \Omega^T$. The least squares problem Eq. (35) becomes

$$\min_{\Psi} \|Y - \Psi \Omega^T \Omega\|_F^2.$$

Using the SVD given by Eq. (57), the minimum norm solution is obtained as:

$$\widehat{\Phi} = \widehat{\Psi} \Omega^T = Y (V_a S_a^2 V_a^T)^{-1} V_a S_a U_a^T = Y V_a S_a^{-1} U_a^T,$$

Now, the state sequence can be estimated from the row space of the matrix $\widehat{\Phi} Z_X = \widehat{\Psi} Z_X^T Z_X$ as discussed in Section 3.2.

An alternative to the minimum norm solution is computing the solution to a Ridge regression problem. In Ridge regression a term that equals the norm of the regression coefficients is added to the cost function as follows:

$$\min_{\Phi} \|\Phi\|_F^2 + \gamma \|Y - \Phi\Omega\|_F^2. \tag{58}$$

The regularization parameter γ determines the relative weight between the two terms. In the limit for $\gamma \to \infty$ we end up with the original least squares problem Eq. (35). The regularization parameter γ can be used to reduce the variance of the solution to the least squares problem, at the expense of introducing some bias. In problems where the number of parameters to estimate in Φ is much larger than the number of data points available, the variance error can be reduced considerably by introducing only a small bias. This improves the predictive capabilities of the models for fresh data sets. Thus, γ can be used to perform a bias variance trade-off. A common way to determine the parameter γ is by cross-validation (Golub and Van Loan, 1996, p. 583). In cross-validation a part of the data available is selected and not used in estimating the parameters Φ. Next, the quality of the estimation is evaluated by computing the residuals for the selected data. By performing this exercise for several different choices of γ, one can get an idea of the bias-variance trade-off.

The regularized least squares problem (58) can be solved using a kernel algorithm, that is, an algorithm that works with the kernel matrix $\Omega^T\Omega$. Clearly, the advantage is that once the kernel matrix has been computed, no further computations are required that involve the huge number of rows present in the matrix Ω.

For the special case were Φ and Y in Eq. (58) are vectors, such a kernel algorithm has been derived for the training of least squares support vector machines (Suykens and Vandewalle, 1999). Support vector machines can be used to solve nonlinear classification and regression problems. They have a firm theoretical foundation in the statistical learning theory developed by Vapnik (1998). The kernel algorithm described below can be seen as a multivariable extension of the least squares support vector machine.

To derive a kernel algorithm, the problem (58) is rewritten as

$$\begin{aligned} &\min_{\Phi, E(i)} \|\Phi\|_F^2 + \gamma \sum_{i=1}^{N} E(i)^T E(i) \\ &\text{subject to} \quad E(i) = Y(i) - \Phi\Omega(i), \quad i = 1, 2, \ldots, N, \end{aligned}$$

where $E(i)$ denotes the ith column of the matrix E, $Y(i)$ the ith column of Y, and $\Omega(i)$ the ith column of Ω. We can define the Lagrangian

$$\mathscr{L} = \|\Phi\|_F^2 + \gamma \sum_{i=1}^{N} E(i)^T E(i) + \sum_{i=1}^{N} \psi(i)^T \Big(Y(i) - \Phi\Omega(i) - E(i)\Big),$$

where the constraints $E(i) = Y(i) - \Phi\Omega(i)$, $i = 1, 2, \ldots, N$ have been incorporated using the vectors $\psi(i) \in \mathbb{R}^{(j+1)n_y}$ containing the Lagrange multipliers. The optimality conditions for the problem (58) are now given by

$$\frac{\partial \mathscr{L}}{\partial \psi(i)} = 0, \tag{59}$$

$$\frac{\partial \mathscr{L}}{\partial E(i)} = 0, \tag{60}$$

$$\frac{\partial \mathscr{L}}{\partial \mathrm{vec}(\Phi)} = 0, \tag{61}$$

where $\mathrm{vec}(\Phi)$ denotes a column vector constructed by stacking the columns of the matrix Φ underneath each other; starting with the left most column.

From condition (59) it follows that

$$Y(i) - \Phi\Omega(i) - E(i) = 0, \quad i = 1, 2, \ldots, N,$$

and from condition (60) it follows that

$$\gamma E(i) - \psi(i) = 0, \quad i = 1, 2, \ldots, N.$$

Combining these two expressions yields

$$Y(i) = \frac{1}{\gamma}\psi(i) + \Phi\Omega(i), \quad i = 1, 2, \ldots, N. \tag{62}$$

Now we turn our attention to condition (61). Observe that

$$\frac{\partial}{\partial \mathrm{vec}(\Phi)} \|\Phi\|_F^2 = \frac{\partial}{\partial \mathrm{vec}(\Phi)} \mathrm{vec}(\Phi)^T \mathrm{vec}(\Phi) = \mathrm{vec}(\Phi),$$

and that if we denote the columns of Φ by $\Phi(i)$, $i = 1, 2, \ldots, p$ we can write

$$\begin{aligned}
&\frac{\partial}{\partial \mathrm{vec}(\Phi)} \psi(i)^T \Phi \Omega(i) \\
&= \frac{\partial}{\partial \mathrm{vec}(\Phi)} \psi(i)^T \left[\Phi(1)\Phi(2) \cdots \; \Phi(p)\right] \Omega(i) \\
&= \Big(I_p \otimes \psi(i)\Big)\Omega(i).
\end{aligned}$$

Therefore, condition (61) yields

$$\mathrm{vec}(\Phi) = \sum_{i=1}^{N} \Big(I_p \otimes \psi(i)\Big)\Omega(i).$$

Using the fact that $\mathrm{vec}(ABC) = (C^T \otimes A)\mathrm{vec}(B)$, and realizing that $\mathrm{vec}(\Omega(i)^T) = \Omega(i)$, we can rewrite this equation as

$$\Phi = \sum_{i=1}^{N} \psi(i)\Omega(i)^T = \Psi\Omega^T, \tag{63}$$

where

$$\Psi := \left[\psi(1) \; \psi(2) \; \cdots \; \psi(N)\right].$$

Substitution of Eq. (63) into Eq. (62) yields

$$Y(i) = \frac{1}{\gamma}\psi(i) + \Psi\Omega^T\Omega(i),$$

or equivalently,

$$Y = \Psi\left(\frac{1}{\gamma} I_N + \Omega^T\Omega\right). \tag{64}$$

Thus, with a proper choice of γ we can determine the matrix Ψ by solving this set of linear equations. If we denote the solution to this set of equations as $\widehat{\Psi}$, we can use Eq. (63) to approximate Y as follows

$$\widehat{Y} = \widehat{\Psi}\Omega^T\Omega \tag{65}$$

Note that in the Eq. (64) and Eq. (65) only the kernel matrix $\Omega^T\Omega$ is needed, and not the matrix Ω. Again the state sequence can be estimated from the row space of the matrix $\widehat{\Psi} Z_X^T Z_X$ as discussed in Section 3.2.

7. Conclusion

Subspace techniques can be used to identify multivariable LPV systems with a time-invariant output equation and an affine time-varying parameter dependence in the state equation. The key point in LPV subspace identification is the reconstruction of the state sequence. Once the state sequence has been reconstructed, the system matrices follow from a linear least-squares problem. The reconstruction of the state sequence is always an approximation, because the effect of the unknown initial state has to be neglected. The state sequence is approximated by a linear combination of the rows of certain data matrices that contain the available measurements of the input, output, time-varying parameter, and products of time-lagged versions of these signals. This linear combination is unknown, but follows by solving a linear least-squares problem. Different identification algorithms can be derived, depending on which combination of data matrices is used to approximate the state.

The dimension of the data matrices in LPV subspace identification can be a major problem. The number of rows in the data matrices grows exponentially with the block size. Since the block size has to be at least equal to the system order, higher-order systems require a huge amount of computer memory.

One way to deal with the large number of rows in the data matrices is to use only the most dominant rows and discard the other ones. Selecting the most dominant rows boils down to solving a subset-selection problem with an RQ factorization. It is possible to perform the selection row by row such that there is no need to store the entire data matrix into the memory. Since certain rows are discarded, an approximation error is introduced. The row selection procedure is an effective way of reducing the dimensions of the data matrices and it makes the LPV subspace identification practically feasible. However, the large memory requirements are partly converted into large computation times.

Another way to deal with the large number of rows in the data matrices is to use a kernel method that uses kernel matrices which are square matrices with dimensions equal to the number of data samples. For multivariable LPV systems the kernel matrices often have much smaller dimensions than the data matrices. The kernel method results in a considerable reduction of computational complexity and memory storage requirements. A regularization parameter is used to control the trade-off between bias and variance in the estimate. The kernel method allows LPV systems to be estimated from small data batches. This is not possible with the original LPV subspace identification methods.

References

Chen, H. and Maciejowski, J. (2000a). An improved subspace identification method for bilinear systems, in *Proceedings of the IEEE Conference on Decision and Control* (Sydney, Australia).

Chen, H. and Maciejowski, J. (2000b). New subspace identification methods for bilinear systems, Tech. Rep. CUED/F-INFENG/TR.357, Department of Engineering, University of Cambridge, Cambridge, United Kingdom, URL `http://www-control.eng.cam.ac.uk/jmm/cuedsid/`.

Favoreel, W. (1999). *Subspace methods for identification and control of linear and bilinear systems*, Ph.D. thesis, Faculty of Engineering, K. U. Leuven.

Golub, G. H. and Van Loan, C. F. (1996). *Matrix computations*, 3rd edn. (The Johns Hopkins University Press, Baltimore, MD, USA).

Jansson, M. and Wahlberg, B. (1998). On consistency of subspace methods for system identification, *Automatica* **34**, 12, pp. 1507–1519.

Katayama, T. (2005). *Subspace methods for system identification* (Springer-Verlag, London, UK).

Khatri, C. G. and Rao, C. R. (1968). Solutions to some functional equations and their applications to characterization of probability distributions, *Sankhya: The Indian Journal of Statistics Series A* **30**, pp. 167–180.

Lawson, C. L. and Hanson, R. J. (1995). *Solving least squares problems*, 2nd edn. (SIAM, Philadelphia, PA, USA).

Lee, L. H. and Poolla, K. (1996). Identification of linear parameter-varying systems via LFTs, in *Proceedings of the IEEE Conference on Decision and Control* (Kobe, Japan).

Lee, L. H. and Poolla, K. (1999). Identification of linear parameter-varying systems using nonlinear programming, *Journal of Dynamic Systems, Measurement and Control* **121**, 1, pp. 71–78.

Ljung, L. (1999). *System identification: Theory for the user*, 2nd edn. (Prentice-Hall, Upper Saddle River, NJ, USA).

Picci, G. and Katayama, T. (1996). Stochastic realization with exogenous inputs and 'subspace-methods' identification, *Signal Processing* **52**, 2, pp. 145–160.

Rugh, W. J. (1996). *Linear system theory*, 2nd edn. (Prentice-Hall, Upper Saddle River, NJ, USA).

Sjöberg, J. and Ljung, L. (1995). Overtraining, regularization and searching for a minimum, with applications to neural networks, *International Journal of Control* **62**, 6, pp. 1391–1407.

Söderström, T. and Stoica, P. (1983). *Instrumental variable methods for system identification* (Springer-Verlag, New York, NY, USA).

Suykens, J. A. K. and Vandewalle, J. (1999). Least squares support vector machine classifiers, *Neural Processing Letters* **9**, 3, pp. 293–300.

Van Overschee, P. and De Moor, B. (1994). N4SID: Subspace algorithms for the identification of combined deterministic and stochastic systems, *Automatica* **30**, 1, pp. 75–93.

Van Overschee, P. and De Moor, B. (1996). *Subspace identification for linear systems; Theory, implementation, applications* (Kluwer Academic Publishers,

Dordrecht, The Netherlands).

van Wingerden, J. W. and Verhaegen, M. (2009). Subspace identification of bilinear and LPV systems for open- and closed loop data, *Automatica* **45**, 2, pp. 372–381.

Vapnik, V. (1998). *Statistical learning theory* (John Wiley & Sons, New York, NY, USA).

Verdult, V. (2002). *Nonlinear system identification: A state-space approach*, Ph.D. thesis, Twente University.

Verdult, V. and Verhaegen, M. (2002). Subspace identification of multivariable linear parameter-varying systems, *Automatica* **38**, 5, pp. 805–814.

Verdult, V. and Verhaegen, M. (2003). A kernel method for subspace identification of multivariable bilinear systems, in *Proceedings of the IEEE Conference on Decision and Control* (Maui, Hawaii, USA).

Verhaegen, M. (1987). The minimal residual QR-factorization algorithm for reliably solving subset regression problems, Tech. Rep. NASA TM 100021, NASA, Ames Research Center, Moffett Field, CA, USA.

Verhaegen, M. and Dewilde, P. (1992). Subspace model identification part 1: The output-error state-space model identification class of algorithms, *International Journal of Control* **56**, 5, pp. 1187–1210.

Verhaegen, M. and Verdult, V. (2007). *Filtering and system identification: A least-squares approach* (Cambridge University Press, Cambridge, UK).

Chapter 7

Subspace identification of MIMO LPV systems

J. W. van Wingerden & M. Verhaegen

Delft Center for Systems and Control
Delft University of Technology
Mekelweg 2, Delft, 2628 CD, The Netherlands
`j.w.vanwingerden@tudelft.nl`
`m.verhaegen@tudelft.nl`

In this chapter we present a novel algorithm to identify LPV systems with affine parameter dependence. This algorithm is applicable for data generated in either open or closed-loop. A factorization is introduced that makes it possible to form a predictor, which is based on past inputs, outputs, and scheduling data. The predictor contains the LPV equivalent of the Markov parameters. Using this predictor, ideas from closed-loop LTI identification are developed to estimate the state sequence from which the LPV system matrices can be constructed. A numerically efficient implementation is presented using the kernel method.

1. Introduction

From a system theoretic point of view the identification and control of Linear Parameter-Varying (LPV) systems has attracted considerable attention in recent years as already highlighted in the other book chapters in this book. In [Felici *et al.* (2007b,a); van Wingerden *et al.* (2007, 2009); van Wingerden (2008)] we developed a set of algorithms that exploit the structure in the scheduling sequence to enable the use of well-known LTI techniques to identify LPV systems. This can be seen as the latest development in the field of LPV system identification. When it is possible to control the scheduling sequence during the data acquisition, this "periodic" approach is appropriate. However, for many applications you can not control the scheduling sequence. For instance for a wind turbine the operational position is strongly correlated with the wind and we all know that the wind is not controllable. For systems where there is no structure is the scheduling sequence we talk about LPV systems with arbitrary scheduling sequences, which is the topic of this chapter.

For the case of arbitrary scheduling there are two approaches with the potential to tackle the identification problem: subspace identification and nonlinear optimization based approaches. Previous work showed that these two methods should be developed simultaneously because for the nonlinear optimization technique good initial guesses are required while for the subspace approaches it is typical to have a biased estimate. In the field of subspace LPV identification a milestone contribution is presented in [Verdult (2002)]. [Verdult (2002)] extends the Bilinear identification scheme of [Favoreel *et al.* (1997, 1999)] and [Favoreel (1999)], and later improved by [Verdult *et al.* (1998)] and [Chen and Maciejowski (2000)], to LPV systems. Compared to the subspace LTI counterpart MOESP (Verhaegen and Dewilde, 1992) this algorithm has the inherent drawback that it estimates the state sequence using a certain past window, possibly leading to biased results. Similar approximations are made in the subspace LTI algorithm: N4SID (Van Overschee and De Moor, 1996), however, by making the past window larger and larger this bias will tend to zero. It turns out that identification of LPV systems with arbitrarily varying scheduling sequences is challenging from a numerical point of view (Verdult and Verhaegen, 2001, 2002); the data matrices involved in this algorithm grow exponentially with the size of the past window. With the introduction of the kernel method, the "curse of dimensionality" was partially solved, however, a different bias was introduced (Verdult and Verhaegen, 2005). Still, they showed that the obtained estimates are good initial guesses for nonlinear optimization approaches (Lee and Poolla, 1999; Verdult *et al.*, 2003). Another disadvantage of the approach presented in [Verdult (2002)] is that the algorithm is only suited for data generated in open loop.

In this chapter we discuss a novel LPV identification algorithm that can cope with data generated in open and closed-loop, which is the first main contribution of this chapter. We present a factorization that makes it possible to formulate a predictor that contains the LPV equivalent of the Markov parameters. The second main contribution is that we present an approach that stays close to the formulations given in [Chiuso (2007)] and that can be seen as an extension of the algorithms presented therein. The computational complexity of the algorithm is significantly smaller than the algorithms in [Verdult (2002)], but still the dimensions grow exponentially. Similar to what is done in [Verdult and Verhaegen (2005)] we present the kernel method to reduce the computational complexity. However, we derive computationally efficient formulations of the kernels, which is the third main contribution of this chapter.

The outline of this chapter is as follows; we start in Section 2 with the problem formulation and assumptions. In Section 3 we present a factorization that separates the unknown system matrices from the known input, output, and scheduling data. In Section 4 the basic idea behind the identification scheme is presented and the "curse of dimensionality" will appear. In Section 5 the kernel method is presented, where compact formulations of the kernels are presented. In Section 6 a simulation example is presented that shows the potential of the proposed algorithm. At the end of this chapter we briefly go to the "smart" rotor concept and present a case study to identify the dynamics of a "smart" airfoil, Section 7. We end this chapter with our conclusions concerning this novel algorithm.

Some results published in this chapter are published elsewhere, see [van Wingerden and Verhaegen (2008a,b, 2009); van Wingerden (2008); van Wingerden *et al.* (2010a)].

2. Problem formulation and assumptions

In this section we present the model structure we consider in this chapter. Furthermore, some assumptions are listed and some notation is introduced.

2.1. *Problem formulation*

For the derivation of the algorithm we consider the LPV model structure defined in (1a)-(1b). However, similar results can be derived for LPV systems with parameter-varying output equation, which can be found in [van Wingerden (2008)]. For now we consider:

$$x_{k+1} = \sum_{i=1}^{m} \mu_k^{(i)} \left(A^{(i)} x_k + B^{(i)} u_k + K^{(i)} e_k \right), \tag{1a}$$

$$y_k = C x_k + D u_k + e_k, \tag{1b}$$

where $x_k \in \mathbb{R}^n$, $u_k \in \mathbb{R}^r$, $y_k \in \mathbb{R}^\ell$, are the state, input, and output vectors. $e_k \in \mathbb{R}^\ell$ denotes the zero-mean white innovation process. The matrices

$A^{(i)} \in \mathbb{R}^{n\times n}$, $B^{(i)} \in \mathbb{R}^{n\times r}$, $C \in \mathbb{R}^{\ell\times n}$, $D \in \mathbb{R}^{\ell\times r}$, $K^{(i)} \in \mathbb{R}^{n\times \ell}$ are the local system, input, output, direct feedthrough, and the observer matrices; and $\mu_k^{(i)} \in \mathbb{R}$ the local weights. The index m is referred to as the number of local models or scheduling parameters. Note that the system, input, and the observer matrices depend linearly on the time-varying scheduling vector. The time-varying system matrix is now given by:

$$A_k = \sum_{i=1}^{m} \mu_k^{(i)} A^{(i)},$$

and a similar thing can be done for the other system matrices. We assume that we have an affine dependence and the scheduling is given by:

$$\mu_k = \left[1,\, \mu_k^{(2)}, \cdots, \mu_k^{(m)}\right]^T.$$

We can rewrite (1a)-(1b) in the predictor form as:

$$x_{k+1} = \sum_{i=1}^{m} \mu_k^{(i)} \left(\tilde{A}^{(i)} x_k + \tilde{B}^{(i)} u_k + K^{(i)} y_k\right), \tag{2a}$$

$$y_k = Cx_k + Du_k + e_k, \tag{2b}$$

with

$$\tilde{A}^{(i)} = A^{(i)} - K^{(i)}C, \qquad \tilde{B}^{(i)} = B^{(i)} - K^{(i)}D.$$

It is well-known that an invertible linear transformation of the state does not change the input-output behavior of a state-space system. Therefore, we can only determine the system matrices up to a similarity transformation $T \in \mathbb{R}^{n\times n}$: $T^{-1}A^{(i)}T$, $T^{-1}B^{(i)}$, $T^{-1}K^{(i)}$, CT, and D.

The identification problem can now be formulated as: Given the input sequence u_k, the output sequence y_k, and the scheduling sequence μ_k over a time interval $k = \{0, \ldots, N-1\}$; find, if they exist, the LPV system matrices $A^{(i)}$, $B^{(i)}$, $K^{(i)}$, C, and D up to a global similarity transformation.

2.2. *Assumptions and notation*

First we define the transition matrix for discrete time time-varying systems (Rugh, 1996) and this is given by:

$$\phi_{j,k} = \tilde{A}_{k+j-1} \cdots \tilde{A}_{k+1}\tilde{A}_k. \tag{3}$$

To make the notation more transparent we define:

$$z_k = \begin{bmatrix} u_k, \\ y_k \end{bmatrix},$$
$$\breve{B}_k = \left[\tilde{B}_k, K_k \right],$$
$$\overline{B}^{(i)} = \left[\tilde{B}^{(i)}, K^{(i)} \right].$$

Similar as in [Jansson (2005)] and [Chiuso (2007)] we define a past window denoted by p. This window is used to define the following stacked vector:

$$\overline{z}_k^p = \begin{bmatrix} z_k \\ z_{k+1} \\ \vdots \\ z_{k+p-1} \end{bmatrix},$$

which is a vector containing past and future data. We assume that the state sequence:

$$X = \left[x_{p+1}, \cdots, x_N \right],$$

has full row rank and the matrix:

$$\Gamma^f = \begin{bmatrix} C \\ C\tilde{A}^{(1)} \\ \vdots \\ C\left(\tilde{A}^{(1)}\right)^{f-1} \end{bmatrix}, \tag{4}$$

has full column rank and where f is referred to as the future window. This last matrix can be interpreted as the extended observability matrix of the first local model. For persistency of excitation it is also required that the scheduling sequence satisfies the following relation:

$$\text{rank}\left(\left[\mu_0, \mu_1, \cdots, \mu_{N-p-f} \right]\right) = m,$$

and $N - p - f + 1 > m$. Furthermore, we assume that the feedback problem is well-posed. So, there is either a delay in the feedback controller or in the system. The problem formulation so far does not require any assumptions on the correlation between the input and noise sequence, which opens the possibility to apply the algorithm in closed-loop.

These definitions and assumptions are used in Section 4 but first we define a factorization to extend the predictor-based subspace identification approaches (see [Chiuso (2007)] for an overview) to LPV systems.

3. Factorization of the LPV controllability matrix

In this section we define a fundamental factorization in which we separate the unknown system matrices from the known weighting sequence. To be more precise we will factorize the time-varying extended controllability matrix, which is defined in the following definition.

Definition 7.1. Given the transition matrix in (3) the time-varying extended controllability matrix is given by:

$$\overline{\mathscr{K}}^p_k = \left[\phi_{p-1,k+1}\breve{B}_k, \cdots, \phi_{1,k+p-1}\breve{B}_{k+p-2}, \breve{B}_{k+p-1}\right].$$

To illustrate this definition see the following example:

Example 1. $\overline{\mathscr{K}}^p_k$ For $m = 2$, $k = 0$, and $p = 3$ one obtains:

$$\overline{\mathscr{K}}^3_0 = \Big[\left(\tilde{A}^{(1)} + \tilde{A}^{(2)} \otimes \mu_2^{(2)}\right)\left(\tilde{A}^{(1)} + \tilde{A}^{(2)} \otimes \mu_1^{(2)}\right)\left(\overline{B}^{(1)} + \overline{B}^{(2)} \otimes \mu_0^{(2)}\right), \cdots$$
$$\left(\tilde{A}^{(1)} + \tilde{A}^{(2)} \otimes \mu_2^{(2)}\right)\left(\overline{B}^{(1)} + \overline{B}^{(2)} \otimes \mu_1^{(2)}\right), \left(\overline{B}^{(1)} + \overline{B}^{(2)} \otimes \mu_2^{(2)}\right)\Big].$$

From this example we clearly see that this time-varying controllability matrix is depending on the known scheduling sequence and the LPV system matrices. The next step is to factorize the time-varying extended controllability matrix in a matrix containing only the scheduling terms and a constant matrix, which depends only on the system matrices $\tilde{A}^{(i)}$, and $\overline{B}^{(i)}$. Before we formulate this factorization in a lemma we have to introduce a number of definitions. We start with the following definition:

Definition 7.2. We define the matrix:

$$\mathscr{L}_j = \left[\tilde{A}^{(1)}\mathscr{L}_{j-1}, \cdots, \tilde{A}^{(m)}\mathscr{L}_{j-1}\right],$$

with

$$\mathscr{L}_1 = \left[\overline{B}^{(1)}, \cdots, \overline{B}^{(m)}\right].$$

To illustrate this definition see the following example:

Example 2. $\mathscr{L}_j$ For $m = 2$ one obtains:

$$\begin{aligned}
\mathscr{L}_1 &= \left[\overline{B}^{(1)}, \overline{B}^{(2)}\right], \\
\mathscr{L}_2 &= \left[\tilde{A}^{(1)}\overline{B}^{(1)}, \tilde{A}^{(1)}\overline{B}^{(2)}, \tilde{A}^{(2)}\overline{B}^{(1)}, \tilde{A}^{(2)}\overline{B}^{(2)}\right], \\
\mathscr{L}_3 &= \Big[\, \tilde{A}^{(1)}\tilde{A}^{(1)}\overline{B}^{(1)}, \tilde{A}^{(1)}\tilde{A}^{(1)}\overline{B}^{(2)}, \tilde{A}^{(1)}\tilde{A}^{(2)}\overline{B}^{(1)}, \tilde{A}^{(1)}\tilde{A}^{(2)}\overline{B}^{(2)}, \cdots \\
&\qquad \tilde{A}^{(2)}\tilde{A}^{(1)}\overline{B}^{(1)}, \tilde{A}^{(2)}\tilde{A}^{(1)}\overline{B}^{(2)}, \tilde{A}^{(2)}\tilde{A}^{(2)}\overline{B}^{(1)}, \tilde{A}^{(2)}\tilde{A}^{(2)}\overline{B}^{(2)}\Big].
\end{aligned}$$

The number of block-columns grows exponentially as m^j. Using this definition we define the matrix $\mathscr{K}^p$, which we refer to as LPV extended controllability matrix.

Definition 7.3. The operator $\mathscr{L}_j$ is used to define the LPV extended controllability matrix:

$$\mathscr{K}^p = \left[\mathscr{L}_p, \mathscr{L}_{p-1}, \cdots, \mathscr{L}_1\right] \quad \in \mathbb{R}^{n\times\tilde{q}},$$

with $\tilde{q} = (r+\ell)\sum_{j=1}^{p} m^j$.

To present the factorized expression of the time-varying extended controllability matrix in Lemma 1, we still need the following two definitions:

Definition 7.4. We define the matrix:

$$P_{p|k} = \mu_{k+p-1} \otimes \cdots \otimes \mu_k \otimes I_{r+\ell},$$

with $P_{p|k} \in \mathbb{R}^{m^p(r+\ell)\times(r+\ell)}$ and $\otimes$ represents the Kronecker product (Brewer, 1978).

Now we define:

Definition 7.5. With Definition 7.4 we can define:

$$N_k^p = \begin{bmatrix} P_{p|k} & & & 0 \\ & P_{p-1|k+1} & & \\ & & \ddots & \\ 0 & & & P_{1|k+p-1} \end{bmatrix},$$

with $N_k^p \in \mathbb{R}^{\tilde{q}\times p(r+\ell)}$.

To illustrate this, see the following example:

Example 3. N_k^p For $m = 2$, $k = 0$, and $p = 3$ one obtains:

$$N_0^3 = \begin{bmatrix} 1 & 0 & 0 \\ \mu_0^{(2)} & 0 & 0 \\ \mu_1^{(2)} & 0 & 0 \\ \mu_1^{(2)}\mu_0^{(2)} & 0 & 0 \\ \mu_2^{(2)} & 0 & 0 \\ \mu_2^{(2)}\mu_0^{(2)} & 0 & 0 \\ \mu_2^{(2)}\mu_1^{(2)} & 0 & 0 \\ \mu_2^{(2)}\mu_1^{(2)}\mu_0^{(2)} & 0 & 0 \\ 0 & 1 & 0 \\ 0 & \mu_1^{(2)} & 0 \\ 0 & \mu_2^{(2)} & 0 \\ 0 & \mu_2^{(2)}\mu_1^{(2)} & 0 \\ 0 & 0 & 1 \\ 0 & 0 & \mu_2^{(2)} \end{bmatrix} \otimes I_{\ell+r}.$$

Now we can state the following lemma:

Lemma 1. *Given the model structure in (2a)-(2b) we use Definition 7.3, and 7.5 to obtain:*

$$\overline{\mathscr{K}}_k^p = \mathscr{K}^p N_k^p,$$

where $\overline{\mathscr{K}}_k^p$ *is the time-varying extended controllability matrix, which equals Definition 7.1,* N_k^p *depends on the known scheduling sequence, and* $\mathscr{K}^p$ *is an unknown matrix defined in Definition 7.3. Note that the number of columns of* $\mathscr{K}^p$ *(rows of* N_k^p*), denoted by* $\tilde{q}$*, increases exponentially with* p *according to the relation* $\tilde{q} = (r + \ell) \sum_{j=1}^{p} m^j$.

Proof. Proof follows through straightforward computations. □

Using the examples in this section this definition can be illustrated.

4. LPV predictor-based subspace identification

With the factorization defined in Section 3 we now can present the core of this chapter. We will use similar ideas as discussed in [Chiuso (2007)] but now we introduce the LPV counterpart.

4.1. *Predictors*

The first objective of the algorithm is to reconstruct the state sequence up to a similarity transformation. The state x_{k+p} is given by:

$$x_{k+p} = \phi_{p,k} x_k + \mathscr{K}^p N_k^p \bar{z}_k^p,$$

where $\phi_{p,k}$ is the transition matrix given in (3), $\mathscr{K}^p$ is the time-invariant extended LPV controllability matrix, and the matrix N_k^p is a matrix solely depending on the scheduling sequence. The key approximation in this algorithm is that we assume that $\phi_{j,k} \approx 0$ for all $j \geq p$. Similar as in the LTI case it can be shown that if the system in (2a)-(2b) is uniformly exponentially stable the approximation error can be made arbitrarily small by making p large (Verdult and Verhaegen, 2002). With this assumption the state x_{k+p} is approximately given by:

$$x_{k+p} \approx \mathscr{K}^p N_k^p \bar{z}_k^p. \tag{5}$$

In a number of LTI subspace methods it is well known to make this step (Van Overschee and De Moor, 1996; Jansson, 2005; Chiuso, 2007). The input-output behavior is now approximately given by:

$$\begin{cases} y_{k+p} & \approx C\mathscr{K}^p N_k^p \bar{z}_k^p + Du_{k+p} + e_{k+p} := y_{k+p}^{(p)} \\ y_{k+p+1} & \approx C\mathscr{K}^{p+1} N_k^{p+1} \bar{z}_k^{p+1} + Du_{k+p+1} + e_{k+p+1} \\ \quad\vdots & \\ y_{k+p+f-1} & \approx C\mathscr{K}^{p+f-1} N_k^{p+f-1} \bar{z}_k^{p+f-1} + Du_{k+p+f-1} + e_{k+p+f-1} \end{cases}. \tag{6}$$

Now we define the stacked matrices U_i, Y_i, and Z_i:

$$U_i = \begin{bmatrix} u_{p+i}, & \cdots, & u_{N-f+i+1} \end{bmatrix}, \tag{7a}$$

$$Y_i = \begin{bmatrix} y_{p+i}, & \cdots, & y_{N-f+i+1} \end{bmatrix}, \tag{7b}$$

$$Z_i = \begin{bmatrix} N_0^{p+i} \bar{z}_0^{p+i}, & \cdots, & N_{N-p-f}^{p+i} \bar{z}_{N-p-f}^{p+i} \end{bmatrix}, \tag{7c}$$

for all $i \in \{0, \cdots, f-1\}$. If the matrix $\left[Z_i^T, U_i^T \right]^T$ has full row rank for all $i \in \{0, \cdots, f-1\}$, the matrices $C\mathscr{K}^{p+i}$ and D can be estimated by solving the following linear regression problem:

$$\min_{C\mathscr{K}^{p+i}, D} \|Y_i - C\mathscr{K}^{p+i} Z_i - DU_i\|_F^2, \tag{8}$$

for all $i \in \{0, \cdots, f-1\}$ and where $\|\cdots\|_F$ represents the Frobenius norm (Golub and van Loan, 1996). For finite p the solution of this linear problem will be biased due the approximation made in (5). In the LTI literature a number of papers appeared that studied the effect of the window size and although they proved the asymptotic properties of the algorithms (if $p \to \infty$ the bias disappears) it is hard to quantify the effect for finite p (Knudsen, 2001; Chiuso and Picci, 2005; Chiuso, 2007). In Section 6 we demonstrate that a rather large approximation does not directly imply a large bias in the estimate of the system matrices.

4.2. *Extended observability times controllability matrix*

The product $\mathscr{K}^p Z_0$, which represents by definition the state sequence, X, cannot directly be estimated. In the LTI literature it is common practice to use the estimate of the linear-time invariant equivalent of $C\mathscr{K}^{p+i}$ to construct the extended observability matrix times the extended controllability matrix. For the LPV situation a similar approach can be followed. However, in this case we look at the product $\Gamma^f \mathscr{K}^p$, where Γ^f is defined in (4). This matrix can be constructed with Definition 7.2 to equal the following matrix (we assume $f = p$):

$$\Gamma^f \mathscr{K}^p = \begin{bmatrix} C\mathscr{L}_p & C\mathscr{L}_{p-1} & \cdots & C\mathscr{L}_1 \\ C\tilde{A}^{(1)}\mathscr{L}_p & C\tilde{A}^{(1)}\mathscr{L}_{p-1} & \cdots & C\tilde{A}^{(1)}\mathscr{L}_1 \\ \vdots & \vdots & \ddots & \vdots \\ C\left(\tilde{A}^{(1)}\right)^{f-1}\mathscr{L}_p & C\left(\tilde{A}^{(1)}\right)^{f-1}\mathscr{L}_{p-1} & \cdots & C\left(\tilde{A}^{(1)}\right)^{f-1}\mathscr{L}_1 \end{bmatrix}. \tag{9}$$

This particular matrix is constructed in the LPV-PBSID algorithm. The following matrix is used in the LPV-PBSID$_{\text{opt}}$ algorithm (we assume $f = p$):

$$\Gamma^f \mathscr{K}^p \approx \begin{bmatrix} C\mathscr{L}_p & C\mathscr{L}_{p-1} & \cdots & C\mathscr{L}_1 \\ 0 & C\tilde{A}^{(1)}\mathscr{L}_{p-1} & \cdots & C\tilde{A}^{(1)}\mathscr{L}_1 \\ & & \ddots & \vdots \\ 0 & & & C\left(\tilde{A}^{(1)}\right)^{f-1}\mathscr{L}_1 \end{bmatrix}. \tag{10}$$

The zeros appear in this equation based on the approximation that $\phi_{j,k} \approx 0$ for all $j \geq p$. This matrix represents the extended observability matrix of the first local model times the LPV controllability matrix. Observe that from the linear regression problems formulated in (8) we only need the solution for $i = 0$ to construct the matrix given in (10). With the factorizations presented in Definition 7.2 we can construct this product between the extended observability matrix of the first local model and the extended LPV controllability matrix solely from the elements of $C\mathscr{K}^p$. To illustrate this we refer to Table 1 where an example is given.

For the LPV-PBSID algorithm we have to construct the matrix given in (9) or alternatively given by:

$$\Gamma^f \mathscr{K}^p = \begin{bmatrix} C\mathscr{K}^p \\ C\left(\tilde{A}^{(1)}\right)\mathscr{K}^p \\ \vdots \\ C\left(\tilde{A}^{(1)}\right)^{f-1}\mathscr{K}^p \end{bmatrix}.$$

Observe that from the linear problems formulated in (8) and the factorizations presented in Definition 7.2 we can construct the $(i+1)^{th}$ block row of $\Gamma^f \mathscr{K}^p$ from the estimate of $C\mathscr{K}^{p+i}$. From which we can construct $\Gamma^f \mathscr{K}^p Z_0$, which equals by definition the extended observability matrix times the state sequence, $\Gamma^f X$. The construction of $\Gamma^f \mathscr{K}^p$ is a rather cumbersome task. However, when we introduce the kernel method the construction of this matrix simplifies significantly.

From the constructed matrix $\Gamma^f \mathscr{K}^p$, either the LPV-PBSID or LPV-PBSID$_{\text{opt}}$ way, we can compute $\Gamma^f \mathscr{K}^p Z_0$, which equals by definition the extended observability matrix times the state sequence, $\Gamma^f X$. By computing a Singular Value Decomposition (SVD) of this estimate we can estimate the state sequence and the order of the system. We will use the following SVD:

$$\widehat{\Gamma^f \mathscr{K}^p} Z_0 = \begin{bmatrix} \mathscr{U} & \mathscr{U}_\perp \end{bmatrix} \begin{bmatrix} \Sigma_n & 0 \\ 0 & \Sigma \end{bmatrix} \begin{bmatrix} \mathscr{V} \\ \mathscr{V}_\perp \end{bmatrix}, \tag{11}$$

where Σ_n is the diagonal matrix containing the n largest singular values and $\mathscr{V}$ is the corresponding row space. Note that we can find the largest singular values by detecting a gap between the singular values (Verhaegen and Verdult, 2007). The state is now estimated by:

$$\widehat{X} = \Sigma_n \mathscr{V}. \tag{12}$$

Table 1. Example of the matrix $\Gamma^p \mathscr{K}^p$ for $p = 3$, $m = 2$, and constant input matrix; $B^{(i)} = 0$ for $i > 1$.

$$\Gamma^3 \mathscr{K}^3 \approx C \left[\begin{array}{cccc|cc|c} \tilde{A}^{(1)}\tilde{A}^{(1)}\overline{B}^{(1)} & \tilde{A}^{(1)}\tilde{A}^{(2)}\overline{B}^{(1)} & \tilde{A}^{(2)}\tilde{A}^{(1)}\overline{B}^{(1)} & \tilde{A}^{(2)}\tilde{A}^{(2)}\overline{B}^{(1)} & \tilde{A}^{(1)}\overline{B}^{(1)} & \tilde{A}^{(2)}\overline{B}^{(1)} & \overline{B}^{(1)} \\ 0 & 0 & 0 & 0 & \tilde{A}^{(1)}\tilde{A}^{(1)}\overline{B}^{(1)} & \tilde{A}^{(1)}\tilde{A}^{(2)}\overline{B}^{(1)} & \tilde{A}^{(1)}\overline{B}^{(1)} \\ 0 & 0 & 0 & 0 & 0 & 0 & \tilde{A}^{(1)}\tilde{A}^{(1)}\overline{B}^{(1)} \end{array}\right].$$

Table 2. Total number of rows in the matrix Z_i for $m = 4$ and $r = \ell = 1$.

	$i = 0$	$i = 1$	$i = 2$	$i = 3$	$i = 4$
$p = 2$	40	168	680	2728	10920
$p = 3$	168	680	2728	10920	43688
$p = 4$	680	2728	10920	43688	174760
$p = 5$	2728	10920	43688	174760	699048

It is well known that when the state, input, output, and scheduling sequence are known the system matrices can be estimated (Nemani *et al.*, 1995; Lovera, 1997; Verdult and Verhaegen, 2002). First we use (1b), which is now a linear relation in C and D and where e_k represents white noise. From this equation an estimate can be found of the C and D matrix while also the noise sequence can be estimated. The estimated noise sequence is used to transform (1a) into a linear expression depending on $A^{(i)}$, $B^{(i)}$, and $K^{(i)}$ and consequently all the system matrices can be estimated.

4.3. *"Curse of dimensionality"*

Similar as in [Verdult and Verhaegen (2002)] the presented method suffers from the "curse of dimensionality". The number of rows of Z_i grow exponentially with the past and future window. The number of rows of Z_i is given by:

$$\rho_{Z_i} = (r + \ell) \sum_{j=1}^{p+i} m^j.$$

In Table 2 the "curse of dimensionality" is demonstrated. Observe that the growth of the dimensions is almost similar to the work of [Verdult and Verhaegen (2002)] if we consider the LPV-PBSID algorithm (see Table 1 in [Verdult and Verhaegen (2002)]). However, we would like to stress that for the LPV-PBSID$_{\text{opt}}$ algorithm we only need Z_0 and consequently resulting is a smaller set of regression parameters that we have to estimate. So, we can conclude that the computational complexity of the LPV-PBSID is similar to the algorithm presented in [Verdult and Verhaegen (2002)] but is much higher compared to the LPV-PBSID$_{\text{opt}}$ case. Still, the "curse of dimensionality" plays an important role. In the next section the kernel method is derived for the LPV-PBSID$_{\text{opt}}$ and the LPV-PBSID algorithm.

4.4. *Summary of the LPV algorithms*

The predictor-based subspace algorithms are presented in a conceptual way. No explicit algorithm recipe is given. This is due to the fact that the construction of the matrices (9) and (10) is quite cumbersome. In the next section we will derive explicit formulas for the algorithms. A sketch of the algorithms is now given below:

LPV-PBSID$_{\text{opt}}$ The LPV-PBSID$_{\text{opt}}$ algorithm can be summarized as follows:

(1) Create the matrices U_i, Y_i, and Z_i using (7a), (7b) and (7c) for $i = 0$,
(2) Solve the linear problem given in (8) for $i = 0$,
(3) Construct $\Gamma^f \mathscr{K}^p Z_0$ using (7c) and (10),
(4) Compute the state sequence using (11) and (12),
(5) With the estimated state, use the linear relations (1a)-(1b) to obtain the system matrices.

LPV-PBSID The LPV-PBSID algorithm can be summarized as follows:

(1) Create the matrices U_i, Y_i, and Z_i using (7a), (7b) and (7c) for all $i \in \{0, \cdots, f-1\}$,
(2) Solve the linear problem given in (8) for all $i \in \{0, \cdots, f-1\}$,
(3) Construct $\Gamma^f \mathscr{K}^p Z_0$ using (7c) and (9),
(4) Compute the state sequence using (11) and (12),
(5) With the estimated state, use the linear relations (1a)-(1b) to obtain the system matrices.

---

5. Kernel method

The LPV identification methods presented in the previous paragraph suffer from the "curse of dimensionality". However, similar as in [Verdult and Verhaegen (2005)] we can use the kernel method to overcome this drawback. In Section 5.1 we present the kernel method for the proposed LPV identification schemes. In Section 5.2, a computationally efficient formula is presented for the proposed model structure. The kernel method is normally ill-conditioned, but in Section 5.3 regularization is proposed to overcome this drawback. In the last subsection a summary of the algorithms with kernels is given.

5.1. *Kernel method*

The LPV identification approach presented in the previous section resulted in a linear problem formulated in (8), from now on also referred to as primal problem. This equation can be solved by using traditional Least Squares (LS). However, the data matrices grow exponentially with the past window for LPV-PBSID$_{\text{opt}}$ and with the past and future window for LPV-PBSID. In [Verdult and Verhaegen (2005)] it was shown that the solution of this linear problem is equal to the solution of the dual problem if the solution with the minimum-norm is considered (Golub and van Loan, 1996). In this subsection we show how the kernel method can be exploited for the presented LPV identification schemes.

The linear problem in (8) has a unique solution if the matrix:

$$\begin{bmatrix} Z_i \\ U_i \end{bmatrix},$$

has full row rank and this solution is given by:

$$\begin{bmatrix} \widehat{C\mathcal{K}^{p+i}} \; \widehat{D} \end{bmatrix} = Y_i \begin{bmatrix} Z_i^T, U_i^T \end{bmatrix} \left(\begin{bmatrix} Z_i \\ U_i \end{bmatrix} \begin{bmatrix} Z_i^T, U_i^T \end{bmatrix} \right)^{-1}.$$

When the matrix $\begin{bmatrix} Z_i^T, U_i^T \end{bmatrix}^T$ has missing row rank the solution is not unique. This will occur when the past window is large. However, the solution with the smallest norm, $\min \|[C\mathcal{K}^{p+i}, D]\|_F^2$, can still be computed by using the SVD of the matrix:

$$\begin{bmatrix} Z_i \\ U_i \end{bmatrix} = \begin{bmatrix} \mathcal{U}_i, \mathcal{U}_{\perp,i} \end{bmatrix} \begin{bmatrix} \Sigma_{n,i} & 0 \\ 0 & 0 \end{bmatrix} \begin{bmatrix} \mathcal{V}_i^T \\ \mathcal{V}_{\perp,i}^T \end{bmatrix},$$

where $\Sigma_{n,i}$ is the diagonal matrix containing the non-zero singular values and $\mathcal{V}_i^T$ and $\mathcal{U}_i$ are the corresponding row and column space, respectively. The solution with the minimum-norm is now given by:

$$\begin{bmatrix} \widehat{C\mathcal{K}^{p+i}}, \widehat{D} \end{bmatrix} = Y\mathcal{V}_i\Sigma_{n,i}^{-1}\mathcal{U}_i^T.$$

The computations take place in a large dimensional space spanned by the columns of Z_i. If we consider the minimum-norm solution of (8) the dual problem avoids computations in this large dimensional space (Suykens *et al.*, 2002). The dual problem results in:

$$\min_{\alpha_i} \|\alpha_i\|_F^2 \quad \text{with} \quad Y_i - \alpha_i \left[Z_i^T Z_i + U_i^T U_i \right] = 0, \tag{13}$$

where α_i are the Lagrange Multipliers and $\left[Z_i^T Z_i + U_i^T U_i \right]$ is referred to as the kernel matrix. If the matrix $\left[Z_i^T, U_i^T \right]^T$ has full column rank the solution to this dual problem is given by:

$$\begin{aligned} \widehat{\alpha}_i &= Y_i \left(\left[Z_i^T Z_i + U_i^T U_i \right] \right)^{-1}, \\ &= Y_i \mathscr{V}_i \Sigma_{n,i}^{-2} \mathscr{V}_i^T. \end{aligned}$$

The estimate of $\left[C\mathscr{K}^{p+i}, D \right]$ is now given by:

$$\begin{aligned} \left[\widehat{C\mathscr{K}^{p+i}}, \widehat{D} \right] &= \widehat{\alpha}_i \left[Z_i^T, U_i^T \right], \\ &= Y_i \mathscr{V}_i \Sigma_{n,i}^{-1} \mathscr{U}_i^T. \end{aligned}$$

The construction of the matrix $\widehat{C\mathscr{K}^{p+i}}$ from the dual problem requires the matrix Z_i, explicitly. However, due to the "curse of dimensionality" this can lead to dimension problems. For the construction of $C\mathscr{K}^{p+i} Z_i$ we do not need the matrix Z_i explicitly, we only have to construct $Z_i^T Z_i$ and $U_i^T U_i$ for the computation of α_i and an estimate of $C\mathscr{K}^{p+i} Z_i$ is given by:

$$\widehat{C\mathscr{K}^{p+i}} Z_i = \hat{\alpha} Z_i^T Z_i. \tag{14}$$

With this above we can not reconstruct the extended controllability matrix times the state sequence directly. However, in the previous section we used these estimates to build the matrix given in (9) and (10). We mentioned that this was a rather cumbersome task. However, in the next two lemma's things clear up:

Lemma 2 (Extended observability times X (LPV-PBSID$_{\text{opt}}$)). *Given the model structure in (1a)-(1b) and Definition 7.4, we can define the following matrices:*

$$Z^{i,j} = \left[P_{p-j|j-i} z_{j-i}, \cdots, P_{p-j|\overline{N}+j-i-1} z_{\overline{N}+j-i-1} \right],$$

with $Z^{i,j} \in \mathbb{R}^{m^{p-j+1}(r+\ell)\times \overline{N}}$ and $\overline{N} = N - p - f + 1$. Now we have:

$$C \left(\tilde{A}^{(1)} \right)^i \mathscr{L}_{p-j} = \alpha_0 \left(Z^{i,j} \right)^T,$$

and then we can construct the matrix $\Gamma^f \mathscr{K}^p Z_0$ as follows:

$$\Gamma^f \mathscr{K}^p Z_0 = \begin{bmatrix} \alpha_0 \sum_{j=0}^{p-1} \left(Z^{0,j}\right)^T Z^{0,j} \\ \alpha_0 \sum_{j=1}^{p-1} \left(Z^{1,j}\right)^T Z^{0,j} \\ \vdots \\ \alpha_0 \sum_{j=p-1}^{p-1} \left(Z^{f-1,j}\right)^T Z^{0,j} \end{bmatrix}. \tag{15}$$

Proof. The proof follows from the derivation of the dual problem. □

Lemma 3 (Extended observability times X (LPV-PBSID)). *Given the model structure in (2a)-(2b) and the Definition 7.4. Now we have:*

$$C\left(\tilde{A}^{(1)}\right)^i \mathscr{K}^p = \alpha_i Z_0^T.$$

So with $\alpha = \left[\alpha_0^T, \alpha_1^T, \cdots, \alpha_{f-1}^T\right]^T$. *We can construct the matrix* $\Gamma^f \mathscr{K}^p Z_0$ *as follows*

$$\Gamma^f \mathscr{K}^p Z_0 = \alpha Z_0^T Z_0. \tag{16}$$

Proof. The proof follows from the derivation of the dual problem. □

With one of the previous lemma's we can go back to the original problem and compute an SVD of this estimate to find the state sequence. Again it is important to stress that we do not require Z_i but we only need $Z_i^T Z_i$. This observation makes it possible to derive a computationally more efficient implementation.

5.2. *Computation of the kernel matrices*

In the previous subsection it was already stressed that we do not have to compute Z_i but we only need $Z_i^T Z_i$. In this section an analytical expression is given that does not require the calculation of Z_i. First we define the matrix

$$\tilde{N} = \left[0, 1, \cdots, N-f-p\right],$$

and the following lemma:

Lemma 4. *Given the vectors* $\lambda_1, \lambda_2, \cdots, \lambda_\nu \in \mathbb{R}^{\kappa \times 1}$ *and* $\theta_1, \theta_2, \cdots, \theta_\nu \in \mathbb{R}^{\kappa\times 1}$ *the product*

$$\left(\lambda_1 \otimes \lambda_2 \otimes \cdots \otimes \lambda_\nu\right)^T \left(\theta_1 \otimes \theta_2 \otimes \cdots \otimes \theta_\nu\right),$$

is given by:

$$\prod_{j=1}^{\nu} \lambda_j^T \theta_j. \tag{17}$$

Proof. With the properties of the Kronecker product ($\otimes$) defined in [Brewer (1978)], $(A \otimes B)(C \otimes D) = AC \otimes BD$, we can rewrite (4) as: $\lambda_1^T \theta_1 \otimes \cdots \otimes \lambda_\nu^T \theta_\nu$ and observing that all the elements between the Kronecker products are scalers results in (17). □

With Lemma 4 we can define the kernels for the model structure given in (1a)-(1b) for both LPV-PBSID$_{\text{opt}}$ and LPV-PBSID.[a]

Theorem 1 (Kernels for LPV-PBSID$_{\text{opt}}$). *Given Lemma 4 and the model structure given in (1a)-(1b) we have for $j \geq i$*

$$\left(Z^{i,j}\right)^T Z^{0,j} = \left(\prod_{v=0}^{p-j-1} \mu_{\tilde{N}+v+j-i}^T \mu_{\tilde{N}+v+j}\right) \left(z_{\tilde{N}+j-i}^T z_{\tilde{N}+j}\right)$$

and

$$Z_0^T Z_0 = \sum_{j=0}^{p-1} \left(Z^{0,j}\right)^T Z^{0,j}.$$

We can solve (13) for $i = 0$ and construct (15).

Proof. Using Lemma 4 the proof follows by straightforward computations. □

Theorem 2 (Kernels for LPV-PBSID). *Given Lemma 4 and the model structure given in (1a)-(1b) we have:*

$$Z_i^T Z_i = \sum_{j=0}^{p+i-1} \left(\left(\prod_{v=j}^{p+i-1} \mu_{\tilde{N}+v}^T \mu_{\tilde{N}+v}\right) \left(z_{\tilde{N}+j}^T z_{\tilde{N}+j}\right)\right)$$

We can solve (13) for all $i \in \{0, \cdots, f-1\}$ and construct (16).

[a] We define $\mu_{\tilde{N}+v} = \left[\mu_v, \cdots, \mu_{N-f-p+v}\right]$.

Proof. Using Lemma 4 the proof follows by straightforward computations. □

For $N \gg p$ and $N \gg m$ the computational complexity of the indirect computation, so first constructing Z_i and then computing the product $\left(Z_i^T Z_i\right)$, is of order $\mathcal{O}(N^2 \hat{q})$ with $\hat{q} = \sum_{j=1}^{p} m^j$ and direct construction of $\left(Z_i^T Z_i\right)$ is of the order $\mathcal{O}(N^2)$, which illustrates the computational efficiency of the algorithm.

The kernels are valid for the model structure given in (1a)-(1b). The whole derivation of the kernel matrices can be repeated for different model structures as well and these kernels can be found in [van Wingerden (2008)].

5.3. *Regularization*

The kernel $Z_i^T Z_i$ described in the previous paragraph is square and has the size of the number of data points available. This normally leads to an ill-conditioned set of equations. This conditioning problem can be circumvented through regularization. There are a number of regularization techniques (for a detailed overview see [Sima (2006)]). In [Verdult and Verhaegen (2005)] a simulation study is performed to select the optimal regularization technique and corresponding regularization parameter selection method. In this study they conclude that Tikhonov regularization (Tikh) with Generalized Cross Validation (GCV) regularization parameter selection gives the best result. In this work we will adopt these settings.

5.4. *Summary of the algorithm*

We end this section with the summary of the closed-loop kernel LPV identification algorithms.

LPV-PBSID$_{\text{opt}}$ (kernel) The algorithm can be summarized as follows:

(1) Create the matrices $\left(Z^{i,j}\right)^T Z^{i,j}$ and $Z_0^T Z_0$ using Theorem 1 and compute $U_0^T U_0$ using (7a) for $i = 0$,
(2) Solve the linear problem given in (13) for $i = 0$. If desired regularized,
(3) Construct $\Gamma^f \mathcal{K}^p Z_0$ using (15),
(4) Compute the state sequence using (11) and (12),
(5) Obtain the system matrices using the estimated state and (1a)-(1b).

LPV-PBSID (kernel) The algorithm can be summarized as follows:
(1) Create the matrices $Z_i^T Z_i$ using Theorem 2 and compute $U_0^T U_0$ using (7a) for all $i \in \{0, \cdots, f-1\}$,
(2) Solve the linear problem given in (13) for all $i \in \{0, \cdots, f-1\}$. If desired regularized,
(3) Construct $\Gamma^p \mathscr{K}^p Z_0$ using (16),
(4) Compute the state sequence using (11) and (12),
(5) Obtain the system matrices using the estimated state and (1a)-(1b).

6. Simulation examples

In this section we show some features of the novel algorithm on an academic simulation example.

6.1. *Closed-loop LPV identification*

In this section we demonstrate the operation of the algorithm with data collected in closed-loop, we compare the LPV-PBSID$_{\text{opt}}$ and LPV-PBSID, and we exploit the structure in the scheduling sequence. We use the model described in [Felici *et al.* (2007b)], which might represent the flapping dynamics of a wind turbine blade (Eggleston and Stoddard, 1987). To apply state feedback we now assume that this model has an identity output matrix. We take a past and future window of 8 and limit ourselves to 2000 data points. To show the potential of the closed-loop setting a time-varying state feedback is used. The controller is synthesized using a discrete time periodic Riccati equation (Hench and Laub, 1994; Varga, 2005). Process noise is added with a variance of $\sqrt{0.1}$.

An LPV model was identified using the kernel version of the LPV-PBSID$_{\text{opt}}$ and the LPV-PBSID algorithm and we exploit the fact that the scheduling sequence contains structure, it is periodic [van Wingerden (2008)]. We take the first M rows of $Z_i^T Z_i$ into account. As mentioned earlier there are smarter ways to select these rows, however, this will illustrate the effectiveness of the approach. In Figure 1 and 2 the mean VAF value as a function of M is presented for the regularized and the un-regularized LPV-PBSID$_{\text{opt}}$ and LPV-PBSID algorithm, respectively. We observe that we can select an optimum of 100 rows and we see that regularization is not required anymore. Furthermore, we observe that for this particular simulation example the LPV-PBSID$_{\text{opt}}$ and LPV-PBSID algorithm have a similar performance.

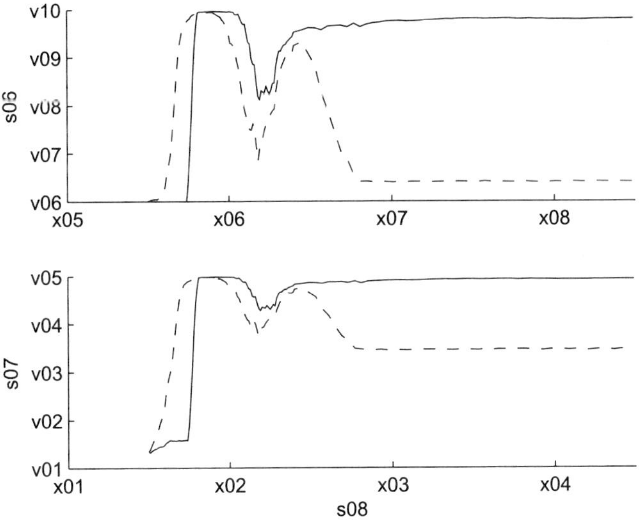

Fig. 1. The VAF for 100 Monte Carlo simulations as a function of the numbers of rows of $Z_i^T Z_i$ that taken into account. The dashed line and the solid line represent the regularized and the un-regularized version of the LPV-PBSID$_{\mathrm{opt}}$ algorithm.

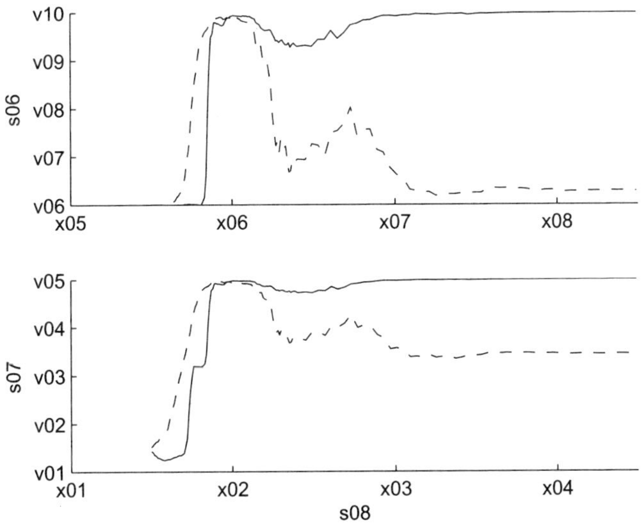

Fig. 2. The VAF for 100 Monte Carlo simulations as a function of the numbers of rows of $Z_i^T Z_i$ that taken into account. The dashed line and the solid line represent the regularized and the un-regularized version of the LPV-PBSID algorithm.

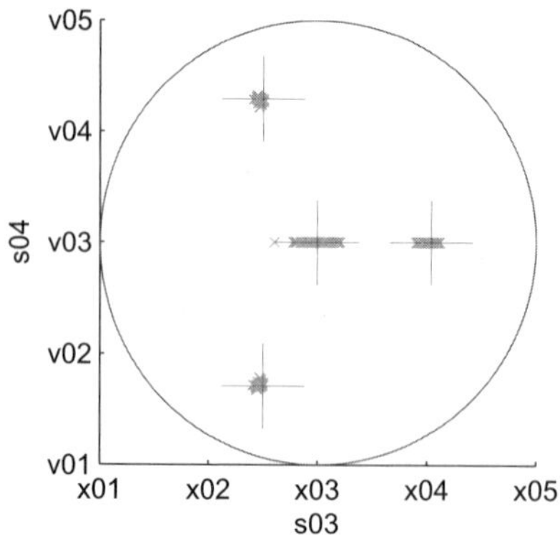

Fig. 3. Eigenvalues of the estimated $A^{(1)}$ and $A^{(2)}$ in one plot for the LPV-PBSID$_{\text{opt}}$ algorithm, for 100 experiments and $M = 100$. The big crosses correspond to the real values of the eigenvalues of the matrices.

In Figure 3 the eigenvalues of the estimated models are compared with their true values. As expected the closed-loop algorithm gives consistent results but a small bias arises due to the approximation made in the algorithm.

7. Case study: A "smart" airfoil

In this section we apply the proposed LPV identification approach on a "smart" 2-D airfoil. This is an airfoil with at the trailing edge a control surface, trailing edge flap. The lay-out of such an airfoil is illustrated in Figure 4. A similar model is used for flutter control and detection and a detailed description of this model can be found in a large number of scientific papers (see for instance [Ko *et al.* (1997); Zeng and Singh (1998); Block and Strganac (1998); Lind and Baldelli (2005); Lee and Singh (2007)]). The control of a "smart" airfoil and flutter control is already a research topic for several years (see also [van Wingerden *et al.* (2008, 2010b)]). The main issue is that the dynamics is strongly dependent on the free stream wind speed. The system can become unstable when the wind speed reaches a certain wind speed; flutter limit. In [Lau and Krener (1999)] and [Barker and Balas (2000)] LPV controller synthesis is proposed to solve this problem. In this section we show that the algorithms presented in this chapter can be used to obtain a model for LPV control.

In this example we first derive an analytical LPV model of the system under consideration. In Section 7.2 we show that due to discretization we lose the

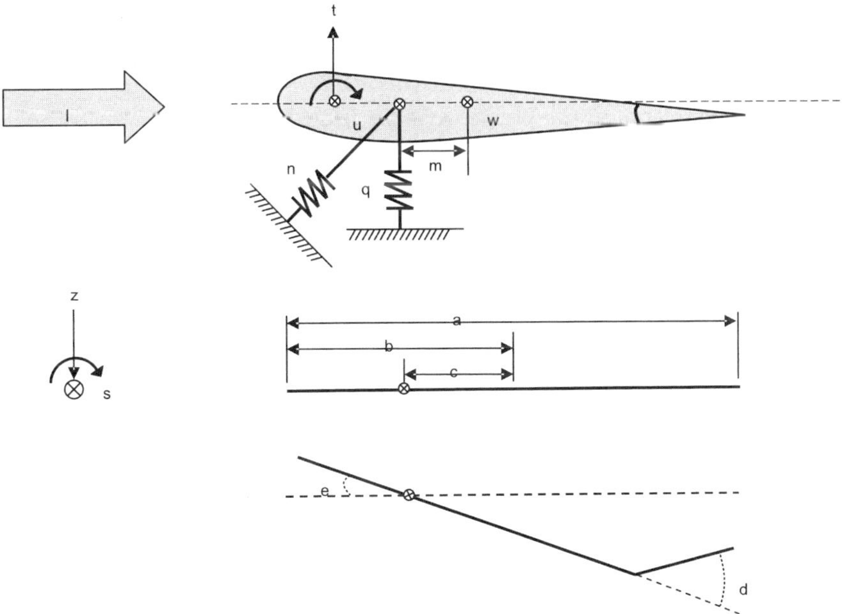

Fig. 4. Schematical representation of the "smart" airfoil model used in this simulation example.

"nice" affine LPV structure. Then we derive a time-varying state-feedback control law that stabilizes the system over the whole trajectory. On the controlled flutter model we apply our closed-loop identification scheme.

7.1. *Analytical LPV modeling*

There are a large number of 2-D "smart" airfoil models available in the literature. The main difference between these models is the number of mechanical degrees of freedom that are taken into account and the unsteadiness of the aerodynamic model. The general mechanical degrees of freedom are the plunge, h, the pitch, α, and the trailing edge flap, β. The trailing edge flap is normally seen as the control variable and if this control surface is considered to be infinitely stiff the corresponding dynamics can be ignored. This is also the approach we will follow in this example. The equations of motion are now given by:

$$\begin{bmatrix} m_t & m_w x_\alpha b \\ m_w x_\alpha b & I_\alpha \end{bmatrix} \begin{bmatrix} \ddot{h} \\ \ddot{\alpha} \end{bmatrix} + \begin{bmatrix} c_h & 0 \\ 0 & c_\alpha \end{bmatrix} \begin{bmatrix} \dot{h} \\ \dot{\alpha} \end{bmatrix} + \begin{bmatrix} k_h & 0 \\ 0 & k_\alpha \end{bmatrix} \begin{bmatrix} h \\ \alpha \end{bmatrix} = \begin{bmatrix} -L \\ M \end{bmatrix},$$

where m_w is the mass of the wing, m_t is the total mass, b is the semi-chord of the wind, I_α is the moment of inertia, x_α is the non-dimensionalized distance of the center of mass from the elastic axis, the thicknesses and damping coefficients of the plunge and pitch degree of freedom are k_h, k_α, c_h, and c_α, respectively. The inputs L and M are the aerodynamic force and moment, respectively. This set of second-order differential equations can also be written as a set of first-order differential equations:

$$\begin{bmatrix} \dot{h} \\ \dot{\alpha} \\ \ddot{h} \\ \ddot{\alpha} \end{bmatrix} = \begin{bmatrix} 0 & 0 & 1 & 0 \\ 0 & 0 & 0 & 1 \\ \frac{I_\alpha}{q_1} k_h & \frac{-m_w x_\alpha b}{q_1} k_\alpha & \frac{I_\alpha}{q_1} c_h & \frac{-m_w x_\alpha b}{q_1} c_\alpha \\ \frac{-m_w x_\alpha b}{q_1} k_h & \frac{m_t}{q_1} k_\alpha & \frac{-m_w x_\alpha b}{q_1} c_h & \frac{m_t}{q_1} c_\alpha \end{bmatrix} \begin{bmatrix} h \\ \alpha \\ \dot{h} \\ \dot{\alpha} \end{bmatrix} + \begin{bmatrix} 0 & 0 \\ 0 & 0 \\ \frac{-I_\alpha}{q_1} & \frac{m_w x_\alpha b}{q_1} \\ \frac{m_w x_\alpha b}{q_1} & \frac{-m_t}{q_1} \end{bmatrix} \begin{bmatrix} -L \\ M \end{bmatrix}, \tag{18}$$

with $q_1 = -m_t I_\alpha + m_w^2 x_\alpha^2 b^2$.

The aerodynamic forces can be modeled in several ways. With unsteady aerodynamics the dynamics of the airflow around the blade is modeled. A number of models are available to describe this dynamic behavior (see [Theodorsen (1935); Leishman (2002); Gaunaa (2006)]) leading to an additional set of differential equations. In this section we assume quasi-steady aerodynamics and that means that the unsteadiness is ignored and we only have static relations between the motion degrees of freedom and the aerodynamic forces. This is a valid assumption if the reduced frequency is considered to be small. For the quasi-steady aerodynamic model we have the following aerodynamic forces:

$$L = q_2 \left[0,\ V^2,\ V,\ q_6 V\right] \begin{bmatrix} h \\ \alpha \\ \dot{h} \\ \dot{\alpha} \end{bmatrix} + q_3 V^2 \beta, \tag{19a}$$

$$M = q_4 \left[0,\ V^2,\ V,\ q_6 V\right] \begin{bmatrix} h \\ \alpha \\ \dot{h} \\ \dot{\alpha} \end{bmatrix} + q_5 V^2 \beta, \tag{19b}$$

with:

$$q_2 = \rho b s_p c_{l\alpha},$$
$$q_3 = \rho b c_{l\beta} s_p,$$
$$q_4 = \rho b^2 s_p c_{m\alpha},$$
$$q_5 = \rho b^2 c_{m\beta} s_p,$$
$$q_6 = \left[\frac{1}{2} - a\right] b,$$

with V the free stream velocity, ρ the air density, a is the nondimensionalized distance from the midchord to the elastic axis, s_p is the span, $c_{l\alpha}$ and $c_{m\alpha}$ are the lift and moment coefficients per angle of attack, and $c_{l\beta}$ and $c_{m\beta}$ are lift and moment coefficients per control surface deflection β.

If we substitute (19a) and (19b) in (18) we obtain a model that describes the aeroelastics of a "smart" rotor blade. The dynamics is depending on the free stream velocity. If we assume that we can measure the plunge and pitch motion of the blade we obtain the following continuous time LPV model:

$$\begin{bmatrix} \dot{h} \\ \dot{\alpha} \\ \ddot{h} \\ \ddot{\alpha} \end{bmatrix} = \left(A_1 + A_2 V + A_3 V^2\right) \begin{bmatrix} h \\ \alpha \\ \dot{h} \\ \dot{\alpha} \end{bmatrix} + B_3 V^2 \beta \tag{20a}$$

$$\begin{bmatrix} h \\ \alpha \end{bmatrix} = C_1 \begin{bmatrix} h \\ \alpha \\ \dot{h} \\ \dot{\alpha} \end{bmatrix} \tag{20b}$$

with:

$$A_1 = \begin{bmatrix} 0 & 0 & 1 & 0 \\ 0 & 0 & 0 & 1 \\ \frac{I_\alpha}{q_1} k_h & \frac{-m_w x_\alpha b}{q_1} k_\alpha & \frac{I_\alpha}{q_1} c_h & \frac{-m_w x_\alpha b}{q_1} c_\alpha \\ \frac{-m_w x_\alpha b}{q_1} k_h & \frac{m_t}{q_1} k_\alpha & \frac{-m_w x_\alpha b}{q_1} c_h & \frac{m_t}{q_1} c_\alpha \end{bmatrix},$$

$$A_2 = \begin{bmatrix} 0\ 0 & 0 & 0 \\ 0\ 0 & 0 & 0 \\ 0\ 0 & \frac{I_\alpha}{q_1} q_2 + \frac{m_w x_\alpha b}{q_1} q_4 & \frac{I_\alpha}{q_1} q_2 q_6 + \frac{m_w x_\alpha b}{q_1} q_4 q_6 \\ 0\ 0 & \frac{-m_w x_\alpha b}{q_1} q_2 + \frac{-m_t}{q_1} q_4 & \frac{-m_w x_\alpha b}{q_1} q_2 q_6 + \frac{-m_t}{q_1} q_4 q_6 \end{bmatrix},$$

$$A_3 = \begin{bmatrix} 0 & 0 & 0 \; 0 \\ 0 & 0 & 0 \; 0 \\ 0 & \frac{I_\alpha}{q_1} q_2 + \frac{m_w x_\alpha b}{q_1} q_4 & 0 \; 0 \\ 0 & \frac{-m_w x_\alpha b}{q_1} q_2 + \frac{-m_t}{q_1} q_4 & 0 \; 0 \end{bmatrix},$$

$$B_3 = \begin{bmatrix} 0 \\ 0 \\ \frac{I_\alpha}{q_1} q_3 + \frac{m_w x_\alpha b}{q_1} q_5 \\ \frac{-m_w x_\alpha b}{q_1} q_3 + \frac{-m_t}{q_1} q_5 \end{bmatrix},$$

$$C_1 = \begin{bmatrix} 1 \; 0 \; 0 \; 0 \\ 0 \; 1 \; 0 \; 0 \end{bmatrix}.$$

The parameters used are given in Table 3.

Table 3. Numerical values of the model parameters of the simple flutter model (Lee and Singh, 2007).

parameter	**value**	**parameter**	**value**
b	0.135 m	a	−0.6847 -
k_h	2844.4 N/m	k_α	2.84 N/m
c_h	27.43 Ns/m	c_α	0.036 × Ns/m
$c_{m\alpha}$	−1.1599 -	$c_{l\alpha}$	6.28 -
$c_{m\beta}$	−0.635 -	$c_{l\beta}$	3.358 -
m_w	2.049 kg	m_t	12.387 kg
I_α	0.0558 kgm^2	x_α	0.3314 -
ρ	1.225 kg/m^3	s_p	1 m

7.2. *Discretization*

The LPV system given in (20a) and (20b) is used to obtain the input, output, and scheduling sequence for the identification algorithm. We assume that the scheduling variable V can be chosen arbitrarily. The LPV identification algorithms introduced in this chapter are valid for discrete time systems and the modeling so far is done in continuous time. The "nice" affine LPV structure that we have in continuous time will be lost if we convert it to discrete time as highlighted in [Tóth (2008)]. In this example we use Tustin discretization, which for this model is given by:

$$A_d\left(V, V^2, \frac{1}{V}, \cdots\right) = \left(I + \frac{T_2}{2}A(V, V^2)\right)\left(I - \frac{T_s}{2}A(V, V^2)\right)^{-1},$$

$$B_d\left(V, V^2, \frac{1}{V}, \cdots\right) = \sqrt{T_s}\left(I - \frac{T_s}{2}A(V, V^2)\right)^{-1} B(V^2),$$

$$C_d\left(V, V^2, \frac{1}{V}, \cdots\right) = \sqrt{T_s}C\left(I - \frac{T_s}{2}A(V, V^2)\right)^{-1},$$

$$D_d\left(V, V^2, \frac{1}{V}, \cdots\right) = \frac{T_s}{2}C\left(I - \frac{T_s}{2}A(V, V^2)\right)^{-1} B(V^2),$$

where T_s is the sample time. There are two important things we would like to stress at this point. The first is that in the continuous time representation of the system the output equation is parameter-invariant. After discretization, the output equation is depending on the scheduling because in the discretization formula the continuous time matrices A and C are used for computing the discrete time system matrices: C_d and D_d. The second issue is that we lose the affine LPV structure. This is caused by the fact that in the discretization formula we have an inverse of the parameter dependent A matrix. The main conclusion is that we completely lose the "nice" affine LPV structure with parameter-invariant output equation. However, still we have the freedom to force an affine model on the data we obtain. In the sequel we try to identify the LPV model with an affine LPV model with parameter-varying output equation on data generated by the Tustin discretized model.

In this example we use a sampling time of 0.08 s. The pole trajectory in the continuous time and discrete time, for a wind speed going from

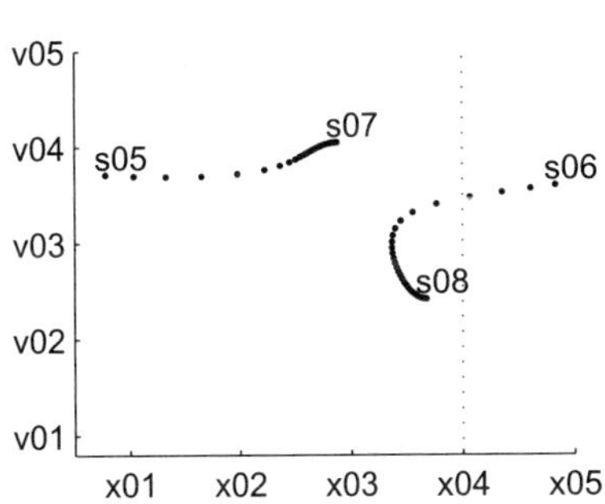

Fig. 5. Poles of the continuous time LPV model for different wind speeds.

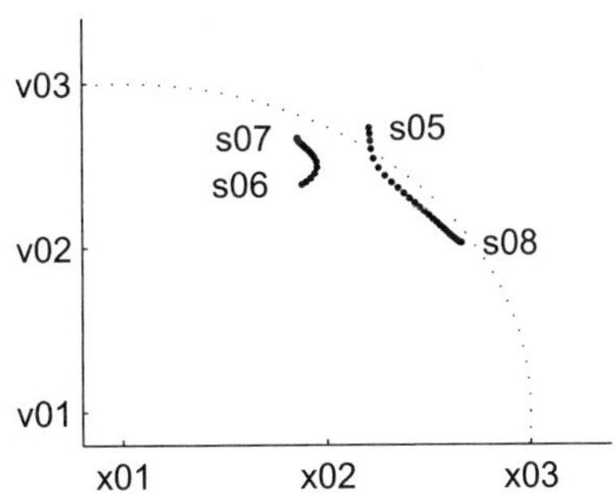

Fig. 6. Poles of the discrete time LPV model for different wind speeds.

0 m/s to 14 m/s, is given in Figure 5 and Figure 6. In these figures we clearly see that the system becomes unstable. Consequently, to do system identification we have to apply feedback control. The focus of this chapter is on identification and not on control that is why we use a finite-horizon state-feedback controller (Ogata, 1995) that stabilizes the system.

7.3. *Simulation results*

For the identification algorithm we use the following wind signal:

$$V_k = 5 + 3\sin\left(\frac{2\pi}{10}kT_s\right) + 2\cos\left(\frac{12\pi}{10}kT_s\right) + w_k,$$

with w_k a white noise with *var(1)*. On the actuator, the flap β, we apply a white noise sequence with unit variance. With these two signals we can simulate the dynamic system given by (20a)-(20b) and discretized using the described Tustin discretization method. To investigate the sensitivity of the identification algorithm with respect to output and process noise, a Monte-Carlo simulation with 100 runs was carried out. For each of the 100 simulations a different realization of the input u_k and scheduling μ_k is used. We added process and measurement noise with a Signal-to-Noise Ratio (SNR) of 40dB. We collect 2500 data points for each trial.

For the identification procedure we assume to have the following affine scheduling sequence:

$$\mu_k = \left[1, V_k, V_k^2\right]^T.$$

All the signals are scaled such that they have a variance of 1. Concerning the identification, we observed that we obtain the best results if we assume parameter-varying A, B, D, and K matrices. We used the kernel version of the LPV-PBSID$_{\text{opt}}$ algorithm with $p = 3$ and we select the first 140 rows of $Z_0^T Z_0$. The collected scaled data u_k, y_k, and μ_k is then used in the proposed identification algorithm.

For this example we do not use the VAF but we look at the Bode plots for a frozen parameter because for meaningful VAF values the system under consideration must be asymptotically stable, otherwise a small error will give low VAF values due to the increasing or decreasing characteristic of the outputs. In Figure 7 and 8 the Bode plots for the pitch position and plunge position, respectively, are presented for 4 frozen values of the free stream velocity, $V = 2$, $V = 6$, $V = 10$, and $V = 14$. In these figures the dashed

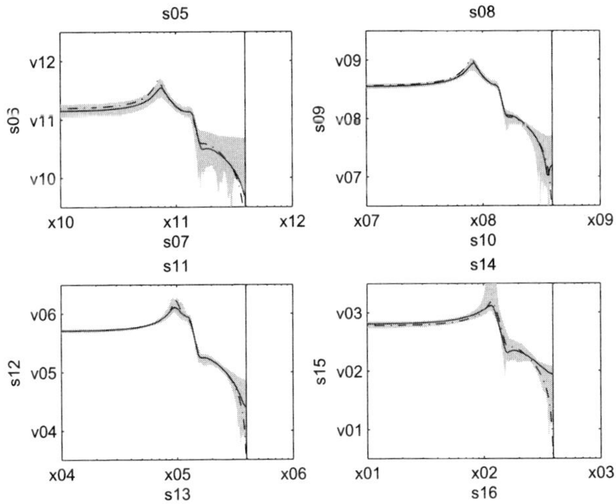

Fig. 7. Bode diagrams of the original transfer functions (dashed) and the identified transfer functions of the experiment with the highest mean correlation coefficients (solid). The transfer functions of the other 99 experiments are within the gray confidence region. The determine the Bode diagram, the free stream velocity is fixed at the values $V = \{2, 6, 10, 14\}$.

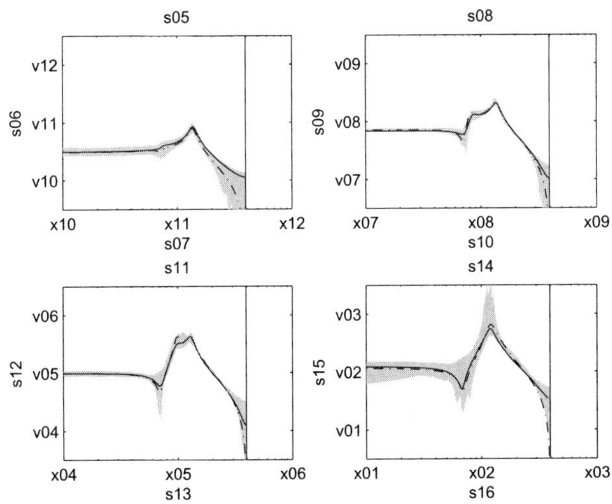

Fig. 8. Bode diagrams of the original transfer functions (dashed) and the identified transfer functions of the experiment with the highest mean correlation coefficients (solid). The transfer functions of the other 99 experiments are within the gray confidence region. The determine the bode diagram, the free stream velocity is fixed at the values $V = \{2, 6, 10, 14\}$.

line represents the "real" system and the solid line the fit with the highest correlation coefficients (Seber, 1984; Krzanowski, 1988) between the "real" states and the identified states. The gray area represents the interval in which the other 99 estimates are located. We see that we have "good" fits with a small variance (the gray area is small).

8. Conclusion

In this chapter we presented an open-loop and closed-loop LPV subspace identification method, which is an extension of Predictor-Based Subspace IDentification (PBSID) algorithms. This methodology from closed-loop LTI subspace identification is used to formulate the input-output behavior of an LPV system. From this input-output behavior the LPV equivalent of the Markov parameters can be estimated. We showed that with this estimate the product between the extended observability matrix and state sequence can be reconstructed and an SVD can be used to estimate the state sequence and consequently the system matrices. The "curse of dimensionality" in subspace LPV identification appeared and the kernel method was proposed. A computationally efficient representation of the kernel is presented, which makes the approach numerical attractive. The algorithm was illustrated with an academic simulation example. Furthermore, we applied the algorithm on a simple "smart" rotor model.

References

Barker, J. M. and Balas, G. J. (2000). Comparing linear parameter-varying gain-scheduled control techniques for active flutter suppression, *Journal of Guidance, Control, and Dynamics* **23**, 5, pp. 948–955.

Block, J. J. and Strganac, T. W. (1998). Applied active control for a nonlinear aeroelastic structure, *Journal of Guidance, Control and Dynamics* **21**, 6, pp. 838–845.

Brewer, J. W. (1978). Kronecker products and matrix calculus in system theory, *IEEE Transactions on Circuits and Systems* **25**, 9, pp. 772–781.

Chen, H. and Maciejowski, J. (2000). An improved subspace identification method for bilinear systems, in *Proceedings of the IEEE Conference on Decision and Control* (Sydney, Australia).

Chiuso, A. (2007). The role of vector auto regressive modeling in predictor based subspace identification, *Automatica* **43**, 6, pp. 1034–1048.

Chiuso, A. and Picci, G. (2005). Consistency analysis of certain closed-loop subspace identification methods, *Automatica* **41**, 3, pp. 377–391.

Eggleston, D. M. and Stoddard, F. S. (1987). *Wind turbine engineering design* (Van Nostrand Reinhold, New York City, NY, USA).

Favoreel, W. (1999). *Subspace methods for identification and control of linear and bilinear systems*, Ph.D. thesis, Katholieke Universiteit Leuven, Belgium.

Favoreel, W., De Moor, B. and Van Overschee, P. (1997). A bilinear extension of subspace identification for systems subject to white noise, in *Proceedings of the American Control Conference* (Albuquerque, NM, USA).

Favoreel, W., De Moor, B. and Van Overschee, P. (1999). Subspace identification of bilinear systems subject to white inputs, *IEEE Transactions on Automatic Control* **44**, 6, pp. 1157–1165.

Felici, F., van Wingerden, J. W. and Verhaegen, M. (2007a). Dedicated periodic scheduling sequences for LPV system identification, in *Proceedings of the European Control Conference* (Kos, Greece).

Felici, F., van Wingerden, J. W. and Verhaegen, M. (2007b). Subspace identification of MIMO LPV systems using a periodic scheduling sequence, *Automatica* **43**, 10, pp. 1684–1697.

Gaunaa, M. (2006). Unsteady 2D potential-flow forces on a thin variable geometry airfoil undergoing arbitrary motion, Tech. Rep. Risø-R-1478(EN), Risø.

Golub, G. H. and van Loan, C. F. (1996). *Matrix computations* (The John Hopkins University Press, Baltimore, MD, USA).

Hench, J. J. and Laub, A. (1994). Numerical solution of the discrete-time periodic riccati equation, *IEEE Transactions on Automatic Control* **39**, 6, pp. 1197–1210.

Jansson, M. (2005). A new subspace identification method for open and closed loop data, in *Proceedings of the IFAC World Congress* (Prague, Czech Republic).

Knudsen, T. (2001). Consistency analysis of subspace identification methods based on a linear regression approach, *Automatica* **37**, 1, pp. 81–89.

Ko, J., Kurdila, A. J. and Strganac, T. W. (1997). Nonlinear control of a prototypical wing section with torsion nonlinearity, *Journal of Guidance, Control and Dynamics* **20**, 6, pp. 1181–189.

Krzanowski, J., W (1988). *Principles of Multivariate analysis: A user's perspective* (Oxford University Press, Oxford, UK).

Lau, E. and Krener, A. J. (1999). LPV control of two dimensional wing flutter, in *Proceedings of the IEEE Conference on Decision and Control* (Phoenix, USA, AZ).

Lee, K. W. and Singh, S. N. (2007). Global robust control of an aeroelastic system using output feedback, *Journal of Guidance, Control, and Dynamics* **30**, 1, pp. 271–275.

Lee, L. H. and Poolla, K. (1999). Identification of linear parameter-varying systems using non-linear programming, *Journal of Dynamic Systems, Measurements and control* **121**, 1, pp. 71–78.

Leishman, J. G. (2002). Challenges in modelling the unsteady aerodynamics of wind turbines, *Wind Energy* **5**, 2–3, pp. 85–132.

Lind, R. and Baldelli, D. H. (2005). Identifying parameter-dependent volterra kernels to predict aeroelastic instabilities, *AIAA journal* **43**, 12, pp. 2496–

2502.

Lovera, M. (1997). *Subspace identification methods: Theory and applications*, Ph.D. thesis, Politecnico Di Milano, Milano, Italy.

Nemani, M., Ravikanth, R. and Bamieh, B. A. (1995). Identification of linear parametrically varying systems, in *Proceedings of the IEEE Conference on Decision and Control* (New Orleans, LA, USA).

Ogata, K. (1995). *Discrete-Time Control Systems: Second edition* (Prentice Hall, Upper Saddle River, NJ, USA).

Rugh, W. (1996). *Linear System Theory* (Prentice Hall, Upper Sadle River, NJ, USA).

Seber, G. A. F. (1984). *Multivariate observations* (John Wiley & Sons, New York City, NY, USA).

Sima, D. M. (2006). *Regularization techniques in model fitting and parameter estimation*, Ph.D. thesis, Katholieke Universiteit Leuven, Belgium.

Suykens, J. A., Van Gestel, T., De Brabantar, J., De Moor, B. and Vandewalle, J. (2002). *Least squares support vector machines* (World Scientific, Singapore).

Theodorsen, T. (1935). General theory of aerodynamic instability and the mechanism of flutter, Report 496, N.A.C.A.

Tóth, R. *et al.*. (2008). Crucial aspects of zero-order-hold LPV state-space system discretization, in *Proceedings of the IFAC world congress* (Seoul, South-Korea).

Van Overschee, P. and De Moor, B. (1996). *Subspace Identification for linear systems* (Kluwer Academic Publishers, Boston, MA, USA).

van Wingerden, J. W. (2008). *Control of Wind Turbines with "Smart" Rotors: Proof of Concept & LPV Subspace Identification*, Ph.D. thesis, Delft University of Technology, The Netherlands.

van Wingerden, J. W., Felici, F. and Verhaegen, M. (2007). Subspace identification of MIMO LPV systems using piecewise constant scheduling sequence with hard/soft switching, in *Proceedings of the European Control Conference* (Kos, Greece).

van Wingerden, J. W., Gebraad, P. M. O. and Verhaegen, M. (2010a). LPV identification of an aeroelastic flutter model, in *Proceedings of the IEEE Conference on Decision and Control* (Atlanta, GA, USA).

van Wingerden, J. W., Houtzager, I., Felici, F. and Verhaegen, M. (2009). Closed-loop identification of the time-varying dynamics of variable-speed wind turbines, *International Journal of Robust and Nonlinear Control,* Special issue on Wind turbines: New challenges and advanced control solutions **19**, 1, pp. 4–21.

van Wingerden, J. W., Hulskamp, A., Barlas, T., Houtzager, I., Bersee, H., Van Kuik, G. A. M. and Verhaegen, M. (2010b). Two-degree-of-freedom active vibration control of a prototyped "smart" rotor, *IEEE Transactions on Control System Technologies* **19**, 2, pp. 284–296.

van Wingerden, J. W., Hulskamp, A., Barlas, T., Marrant, B., Van Kuik, G. A. M., Molenaar, D.-P. and Verhaegen, M. (2008). On the proof of concept of a "smart" wind turbine rotor blade for load alleviation, *Wind Energy* **11**, 3, pp. 265–280.

van Wingerden, J. W. and Verhaegen, M. (2008a). Subspace identification of MIMO LPV systems: the PBSID approach, in *Proceedings of the IEEE Conference on Decision and Control* (Cancun, Mexico).

van Wingerden, J. W. and Verhaegen, M. (2008b). Subspace identification of multivariable LPV systems: a novel approach, in *Proceedings of the IEEE Multi-conference on Systems and Control* (San Antonio, TX, USA).

van Wingerden, J. W. and Verhaegen, M. (2009). Subspace identification of Bilinear and LPV systems for open and closed loop data, *Automatica* **45**, 2, pp. 372–381.

Varga, A. (2005). On solving discrete-time periodic Riccati equations, in *Proceedings of the IFAC World Congress* (Prague, Czech Republic).

Verdult, V. (2002). *Nonlinear system identification; A state-space approach*, Ph.D. thesis, University of Twente.

Verdult, V., Bergboer, N. and Verhaegen, M. (2003). Identification of fully parameterized linear and nonlinear state-space systems by projected gradient search. in *Proceedings of the IFAC Symposium on System Identification* (Rotterdam, The Netherlands).

Verdult, V., Verhaegan, M. H., Chou, C. T. and Lovera, M. (1998). Efficient and systematic identification of MIMO bilinear state-space models, in *Proceedings of the IEEE Conference on Decision and Control* (Tampa, FL, USA).

Verdult, V. and Verhaegen, M. (2001). Identification of multivariable bilinear state-space systems based on subspace techniques and separable least squares optimization, *International Journal of Control* **74**, 18, pp. 1824–1836.

Verdult, V. and Verhaegen, M. (2002). Subspace identification of multivariable linear parameter-varying systems, *Automatica* **38**, 5, pp. 805–814.

Verdult, V. and Verhaegen, M. (2005). Kernel methods for subspace identification of multivariable LPV and bilinear systems, *Automatica* **41**, 9, pp. 1557–1565.

Verhaegen, M. and Dewilde, P. (1992). Subspace model identification part 1: The output-error state-space model identification class of algorithms, *International Journal of Control* **56**, 5, pp. 1187–1210.

Verhaegen, M. and Verdult, V. (2007). *Filtering and System Identification: A least squares approach* (Cambridge University Press, Cambridge, UK).

Zeng, Y. and Singh, S. N. (1998). Output feedback variable structure adaptive control of an aeroelastic system, *Journal of Guidance, Control and Dynamics* **21**, 6, pp. 830–837.

Chapter 8

Subspace identification of continuous-time state-space LPV models

M. Bergamasco & M. Lovera

Dipartimento di Elettronica e Informazione, Politecnico di Milano
Piazza Leonardo da Vinci 32, 20133 Milano, Italy
bergamasco@elet.polimi.it
lovera@elet.polimi.it

LPV model identification has attracted a significant attention in the last few years, in view of the increasing interest for the analysis and the design of gain-scheduled control systems based on parameter-varying models. In particular, a number of reliable algorithms are now available for the estimation of state-space and input-output LPV models both in a local and in a global perspective. Most of the methods proposed so far in the literature, however, deal with the identification of discrete-time models. This can be viewed as a limitation, for the following reasons: first, most LPV control design methods are formulated in continuous-time and therefore assume that a continuous-time model of the plant to be controlled is available; in addition, the conversion of LPV models from discrete-time to continuous-time is not as straightforward and well understood as its LTI counterpart; finally, when it comes to the choice of model structure (i.e., the definition of the parameter vector scheduling the dynamics of the LPV model), the intuition associated with continuous-time variables generally plays a major role and might be lost when performing the modelling exercise in discrete-time. In view of the above issues, in this work the problem of deriving continuous-time LPV models in state-space form from sampled measurements of input, output and scheduling variables is considered, in the framework of a local approach. The proposed method is formulated in the subspace model identification framework and builds on previous, recent results on the subspace-based identification of continuous-time LTI models derived by the Authors.

1. Introduction

In many control engineering applications a single control system must be designed to guarantee stability and performance requirements for a given plant in many different operating conditions. The gain scheduling approach, which has been part of the engineering practice for decades, can be roughly summarised as follows: find one or more *scheduling variables* which can completely parameterise the operating space of interest for the system to be controlled; define a parametric *family* of linearised models for the plant associated with the set of operating points of interest; finally, design a *parametric* controller which can both ensure the desired control objectives in each operating point and an acceptable behaviour during (slow) transients between one operating condition and the other. As is well known, a wide body of design techniques is now available for this problem (see, e.g., [Apkarian and Adams (1998); Kaminer *et al.* (1995); Shamma and Athans (1990)]), which can be reliably solved, provided that a suitable model in parameter-dependent form has been derived.

This modelling problem can be formulated in a number of different perspectives, which consequently lead to a number of different approaches. More precisely, two broad classes of methods can be identified in the literature (see, e.g., [Lovera *et al.* (2011)]): *analytical* methods based on the availability of (relatively) reliable nonlinear equations for the dynamics of the plant, from which suitable control-oriented representations can be derived (see [Leith and Leithead (2000); Rugh and Shamma (2000)] and [Marcos and Balas (2004)]); *experimental* methods based entirely on identification, i.e., aiming at deriving Linear Parameter-Varying (LPV) models for the plant directly from input/output data. The methods belonging to the first class aim at developing LPV models for the plant to be controlled by resorting, broadly speaking, either to suitable extensions of the notion of linearisation, developed in order to take into account off-equilibrium operation of the system, or to interpolation schemes applied to local numerical linearisations. As far as identification methods are concerned, a number of algorithms has been proposed in the literature in the last ten years or so, aiming at the estimation of the parameters for both input/output and state-space models, the latter either in affine or Linear Fractional Transformation (LFT) form [Lee and Poolla (1999); Nemani *et al.* (1995); Lovera *et al.* (1998); Bamieh and Giarré (2002); Verdult (2002); Previdi and Lovera (2003); Tóth (2010); van Wingerden and Verhaegen (2009)].

Most LPV identification techniques are based on the assumption that the identification procedure can rely on one *global* identification experiment in which both the control input and the scheduling variables are (persistently) excited in a simultaneous way. This assumption may not be a reasonable one in many applications, in which it would be desirable to try and derive a parameter-dependent model on the basis of *local* experiments only, i.e., experiments in which the scheduling variable is held constant and only the control input is excited.

Such a viewpoint has been considered in [Steinbuch *et al.* (2003); van Helvoort *et al.* (2004); Groot Wassink *et al.* (2005); Paijmans *et al.* (2008)], where numerical procedures for the construction of parametric models for gain scheduling on the basis of local experiments and for the interpolation of local controllers have been proposed. In [Steinbuch *et al.* (2003); van Helvoort *et al.* (2004)], i) input-output local models are identified using a frequency-domain approach; ii) the identified models are written in state-space representation using a canonical controllability form; iii) an interpolation step for the elements of the state-space matrices completes the identification procedure. In [Groot Wassink *et al.* (2005)], a similar approach is adopted, followed by a conversion of the identified model to LFT form. Finally, in [Paijmans *et al.* (2008); De Caigny *et al.* (2009)] methods for the identification of LPV SISO models are proposed, based on the decomposition of the system in first and second order subsystems and the subsequent interpolation of the parameter-dependent gain, zeros and poles of each subsystem. An extension to the MIMO case has been presented in [De Caigny *et al.* (2011)]. All the above methods are clearly based on the idea of following a two stage procedure, which consists in deriving a set of local models and performing some kind of interpolation between them. A number of issues deserve attention as far as this general procedure is concerned. First of all, performing the interpolation on input-output models directly might lead to significant problems if the obtained LPV model is to be eventually used in a state-space representation (see, e.g., [Tóth (2010)] for a detailed discussion of the equivalence between input-output and state-space representations for LPV systems). Similarly, numerical issues might turn out to be critical in the interpolation problem if an ill-conditioned parameterisation for the model class is chosen (e.g., numerator and denominator coefficients in a high order transfer function). Finally, when dealing with the problem of interpolating black-box local models in state-space form, the challenging problem of realizing all the models with respect to the same state variables comes into play. Indeed, interpolating such state-space models without taking this issue into account may lead to inaccurate LPV models, even if the local LTI models are consistent. Quantifying the modelling errors associated with this aspect is, to our best knowledge, an open problem.

In view of the above discussion, the aim of this paper is to further elaborate on the existing approaches and on preliminary results presented in [Lovera and Mercére (2007)], in order to provide a number of extensions, namely:

- the identification of local models is performed using recently developed subspace techniques of the predictor-based class for continuous-time models (see, e.g., [Bergamasco and Lovera (2010, 2011)]), so enabling the straightforward treatment of MIMO as well as SISO modelling problems, in state-space form;
- in addition, the ability to deal with data generated in closed-loop is obtained as a benefit of the predictor-based framework.

In the last twenty years or so, Subspace Model Identification (SMI) algorithms have been developed, which have proven extremely successful in dealing with the estimation of discrete-time state space models for MIMO systems. Not surprisingly, the problem of extending SMI methods to the identification of continuous-time systems has been studied in a number of contributions. In [Van Overschee and De Moor (1996)] a frequency-domain approach to subspace identification of continuous-time models was proposed, while a time-domain SMI algorithm able to identify a continuous-time model from sampled input-output data was first proposed in [Haverkamp (2001)], building on the framework introduced in [Johansson *et al.* (1999)]. More precisely, in the cited thesis the MIMO Output-Error State sPace (MOESP, see [Verhaegen (1994)]) class of SMI algorithms was extended to the identification of continuous-time models through the use of Laguerre filters: this allowed the development of a method that deals with noise in a similar way as its discrete-time counterparts. More recently, in [Ohsumi *et al.* (2002)] the version of the MOESP algorithm presented in [Verhaegen and Dewilde (1992a,b)] is adopted and a discrete-time algebraic equation is derived starting from sampled input-output data by describing derivatives of stochastic processes in the distribution sense, while in [Bastogne *et al.* (2001); Mercére *et al.* (2008)] the combination of the MOESP algorithm with filtering methods to avoid the need to compute numerical derivatives of input-output signals was proposed. In [Ohta and Kawai

(2004)] a novel approach to the problem of continuous-time SMI has been presented, based on the adoption of orthonormal basis functions to arrive, again, at a MOESP-like data equation for a continuous-time system.

All the above mentioned contributions, however, assume that the system under study is operating in open-loop. This assumption is frequently restrictive in practice and is typically violated, for example, in aerospace applications, in which partial loop closures must be retained during identification experiments, primarily for safety issues. The problem of closed-loop SMI has been studied extensively in recent years due to its high relevance for practical applications (see, e.g., [Ljung and McKelvey (1996); Chou and Verhaegen (1997); Chiuso and Picci (2005); Chiuso (2007); Huang *et al.* (2005)] and the references therein). The present state-of-the-art is represented by the so-called Predictor-Based Subspace IDentification (PBSID) algorithm (see, again, [Chiuso and Picci (2005); Chiuso (2007)]) which, under suitable assumptions, can provide consistent estimates of the state space matrices for a discrete-time, linear time-invariant system operating under feedback. To the best knowledge of the authors, the problem of closed-loop subpace identification in continuous-time has been only considered in the literature in [Mohd-Moktar and Wang (2008)], where the application of the errors-in-variables approach of [Chou and Verhaegen (1997)] is proposed to deal with correlation in a continuous-time setting.

In the light of the above discussion, in this contribution the identification of the local models will be carried out using novel continuous-time SMI schemes, first proposed in [Bergamasco and Lovera (2010, 2011)], based on the derivation of PBSID-like algorithms within the all-pass domain proposed in [Ohta and Kawai (2004)] and relying on Laguerre projections of the sampled input-output data. Note that even though the proposed approach is capable of dealing with some of the limitations of the existing methods, the state-space basis issue remains open and will be dealt with in the present study only by means of simulations at the model validation stage.

2. Definitions

In the following sections a number of definitions will be used, which are summarised hereafter for the sake of clarity. See, e.g., [Zhou *et al.* (1996); Johansson *et al.* (1999); Ohta (2005)] for further details.

Let $\mathscr{L}_2(0,\infty)$ denote the space of square integrable and Lebesgue measurable functions of time $0 < t < \infty$, with the inner product defined as $< f, g >= \int_0^\infty f(t)g(t)dt$, for $f, g \in \mathscr{L}_2(0,\infty)$; the space $\mathscr{H}_2$ is the closed subspace of $\mathscr{L}_2(i\mathbb{R})$ with functions analytic in the open right half plane, with norm

$$\|U\|_2^2 = \sup_{\sigma>0} \frac{1}{2\pi} \int_{-\infty}^{\infty} |U(\sigma + j\omega)|^2 d\omega = \frac{1}{2\pi} \int_{-\infty}^{\infty} |U(j\omega)|^2 d\omega. \quad (1)$$

In view of Parseval's relation, the spaces $\mathscr{L}_2(0,\infty)$ and $\mathscr{H}_2$ are related by the isometric isomorphism defined by the bilateral Fourier transform, if $U \in \mathscr{H}_2$, the inverse Fourier transform $u = \mathscr{F}^{-1}[U]$ is in $\mathscr{L}_2(0,\infty)$ and $\|U\|_2 = \|u\|_2$.

A scalar transfer function $w(s)$ is called inner if it is a bounded analytic function in the open right half plane (i.e., $w(j\omega) \in \mathscr{H}_\infty$), such that $|w(j\omega)| = 1$ or $w^\sim(j\omega)w(j\omega) = 1$ almost everywhere on the imaginary axis, where $w^\sim(j\omega) = w^T(-j\omega)$ is the para-conjugate (i.e., w is an all-pass transfer function). We further denote by Λ_w the multiplication operator $\mathscr{L}_2(0,\infty) \mapsto \mathscr{L}_2(0,\infty)$ defined as

$$\Lambda_w u(t) = \mathscr{F}^{-1}[w\mathscr{F}[u(t)]]. \quad (2)$$

In the following the focus will be on the first order inner function

$$w(s) = \frac{s-a}{s+a}, \quad (3)$$

$a > 0$, together with the associated realisation

$$w(s) = \frac{c_w b_w}{s - a_w} + d_w, \quad (4)$$

where $a_w = -a$, $b_w = -\sqrt{2a}$, $c_w = \sqrt{2a}$, $d_w = 1$. Then, it can be shown that $w(s)\mathscr{H}_2$ is a proper closed subspace of $\mathscr{H}_2$, the orthogonal complement of which is denoted as $S = \mathscr{H}_2 \ominus w(s)\mathscr{H}_2$, that

$$\mathscr{L}_0(s) = \frac{c_w}{s+a} = \frac{\sqrt{2a}}{s+a} \quad (5)$$

is a basis of the (one-dimensional) subspace S and that the set

$$\{\mathscr{L}_0, w\mathscr{L}_0, \ldots, w^i\mathscr{L}_0, \ldots\} \quad (6)$$

is an orthonormal basis of $\mathscr{H}_2$, i.e., $\mathscr{H}_2 = \bigoplus_{i=0}^{\infty} w^i S$. Equivalently, letting $\ell_0 = \mathscr{F}^{-1}[\mathscr{L}_0]$, the set

$$\{\ell_0, \Lambda_w \ell_0, \ldots, \Lambda_w^i \ell_0, \ldots\} \quad (7)$$

is an orthonormal basis of $\mathscr{L}_2(0,\infty)$, i.e., $\mathscr{L}_2(0,\infty) = \bigoplus_{i=0}^{\infty} \Lambda_w^i S$.

Finally, the transfer function of the i-th (order $i+1$) Laguerre filter is then defined as

$$\mathscr{L}_i(s) = w^i(s)\mathscr{L}_0(s) = \sqrt{2a}\frac{(s-a)^i}{(s+a)^{i+1}}. \tag{8}$$

3. Problem statement

The system identification problem considered in this paper can be summarised as follows. Consider the MIMO linear parametrically-varying system given by

$$\dot{x} = A(\theta)x + B(\theta)u \tag{9}$$

$$y = C(\theta)x + D(\theta)u \tag{10}$$

where $u \in \mathbb{R}^{n_u}$, $y \in \mathbb{R}^{n_y}$, $x \in \mathbb{R}^{n_x}$ and $\theta \in \mathbb{R}^{n_p}$ and assume that the results of a number of N_θ identification experiments are available, associated with the operation of the system near N_θ different values of the parameter vector θ. The aim of the identification procedure proposed in this paper is to determine a set of parameter dependent matrices $\hat{A}(\theta)$, $\hat{B}(\theta)$, $\hat{C}(\theta)$ and $\hat{D}(\theta)$ either in affine or LFT form which can provide a good approximation of the system (9)-(10) over the considered range of operating points.

Note that in (9)-(10) no noise terms have been taken into account, as the above equations are just intended to illustrate the parameter dependent structure of the considered system. Since the actual identification experiments will be performed near operating points characterised by a constant value of the parameter vector, i.e., assuming that the system can be considered as LTI, the relevant noise assumptions will be introduced in Section 4 when presenting the considered algorithms for the identification of the local models.

4. A balanced subspace approach to identification for gain scheduling

The approach to the problem of identification for gain scheduling proposed in this paper can be summarised in the following steps:

- linear continuous-time state-space models are estimated for each operating point, using a continuous-time approach (see [Bergamasco and Lovera (2010, 2011)]),
- The identified models are balanced using the numerical algorithm of [Laub *et al.* (1987)] (as implemented in the Matlab Control Toolbox).
- Finally, the parameter-dependent model is obtained by direct interpolation of the state-space matrices of the local models, made possible by the unique properties of balanced realizations (see [Laub *et al.* (1987)]), and can eventually be converted to LFT form using results from the LPV identification literature [Lee and Poolla (1999)].

Each of the above mentioned steps will be described in detail in the following sections.

5. Continuous-time predictor-based subspace identification

Consider the linear, time-invariant continuous-time system

$$\begin{aligned} dx(t) &= Ax(t)dt + Bu(t)dt + dw(t),\ x(0) = x_0 \\ dz(t) &= Cx(t)dt + Du(t)dt + dv(t) \\ y(t)dt &= dz(t) \end{aligned} \tag{11}$$

where $x \in \mathbb{R}^{n_x}$, $u \in \mathbb{R}^{n_u}$ and $y \in \mathbb{R}^{n_y}$ are, respectively, the state, input and output vectors and $w \in \mathbb{R}^{n_x}$ and $v \in \mathbb{R}^{n_y}$ are the process and the measurement noise, respectively, modelled as Wiener processes with incremental covariance given by

$$E\left\{\begin{bmatrix} dw(t) \\ dv(t) \end{bmatrix}\begin{bmatrix} dw(t) \\ dv(t) \end{bmatrix}^T\right\} = \begin{bmatrix} Q & S \\ S^T & R \end{bmatrix} dt.$$

The system matrices A, B, C and D, of appropriate dimensions, are such that (A, C) is observable and $(A, [B, Q^{1/2}])$ is controllable. Assume that a dataset $\{u(t_i), y(t_i)\}$, $i \in [1, N]$ of sampled input/output data (possibly associated with a non equidistant sequence of sampling instants) obtained from system (11) is available. Then, the problem is to provide a recursive estimator of the state space matrices A, B, C and D (up to a similarity transformation) on the basis of the available data.

The continuous-time algorithms discussed in this paper are based on the results first presented in [Ohta and Kawai (2004); Ohta (2005)], and further expanded in [Kinoshita and Ohta (2010); Ohta (2011)], which allow to obtain a discrete-time equivalent model starting from the continuous-time system (11), along the following lines.

First note that under the previously stated assumptions (11) can be written in innovation form as

$$\begin{aligned} dx(t) &= Ax(t)dt + Bu(t)dt + Kde(t) \\ dz(t) &= Cx(t)dt + Du(t)dt + de(t) \\ y(t)dt &= dz(t), \end{aligned} \tag{12}$$

and it is possible to apply the results of [Ohta and Kawai (2004)] to derive a discrete-time equivalent model, as follows. Consider the first order inner function $w(s)$ and apply to the input u, the output y and the innovation e of (12) the transformations

$$\begin{aligned} \tilde{u}(k) &= \int_0^\infty \Lambda_w^k \ell_0(t)u(t)dt \\ \tilde{y}(k) &= \int_0^\infty \Lambda_w^k \ell_0(t)y(t)dt \\ \tilde{e}(k) &= \int_0^\infty \Lambda_w^k \ell_0(t)de(t), \end{aligned} \tag{13}$$

where $\tilde{u}(k) \in \mathbb{R}^{n_u}$, $\tilde{e}(k) \in \mathbb{R}^{n_y}$ and $\tilde{y}(k) \in \mathbb{R}^{n_y}$. Then (see [Ohta and Kawai (2004)] for details) the transformed system has the state space representation

$$\begin{aligned} \xi(k+1) &= A_o\xi(k) + B_o\tilde{u}(k) + K_o\tilde{e}(k),\ \xi(0) = 0 \\ \tilde{y}(k) &= C_o\xi(k) + D_o\tilde{u}(k) + \tilde{e}(k) \end{aligned} \tag{14}$$

where the state space matrices are given by

$$\begin{aligned} A_o &= (A - aI)^{-1}(A + aI) \\ B_o &= \sqrt{2a}(A - aI)^{-1}B \\ C_o &= -\sqrt{2a}C(A - aI)^{-1} \\ D_o &= D - C(A - aI)^{-1}B. \end{aligned} \tag{15}$$

Starting from system (12), the continuous-time PBSID algorithm proposed in [Bergamasco and Lovera (2010, 2011)] for the estimation of the state space matrices A_o, B_o, C_o, D_o, K_o is outlined in the following. Considering the sequence of sampling instants t_i, $i = 1, \ldots, N$, the input u, the output

y and the innovation e of (12) are subjected to the transformations

$$\begin{aligned}\tilde{u}_i(k) &= \int_0^\infty (\Lambda_w^k \ell_0(\tau))u(t_i+\tau)d\tau \\ \tilde{e}_i(k) &= \int_0^\infty (\Lambda_w^k \ell_0(\tau))de(t_i+\tau) \\ \tilde{y}_i(k) &= \int_0^\infty (\Lambda_w^k \ell_0(\tau))y(t_i+\tau)d\tau\end{aligned} \tag{16}$$

(or to the equivalent ones derived from (13)), where $\tilde{u}_i(k) \in \mathbb{R}^{n_u}$, $\tilde{e}_i(k) \in \mathbb{R}^{n_y}$ and $\tilde{y}_i(k) \in \mathbb{R}^{n_y}$. Then (see [Ohta and Kawai (2004)] for details) the transformed system has the state space representation

$$\begin{aligned}\xi_i(k+1) &= A_o\xi_i(k) + B_o\tilde{u}_i(k) + K_o\tilde{e}_i(k),\ \xi_i(0) = x(t_i) \\ \tilde{y}_i(k) &= C_o\xi_i(k) + D_o\tilde{u}_i(k) + \tilde{e}_i(k)\end{aligned} \tag{17}$$

where the state space matrices are given by (15).

Letting now

$$\tilde{z}_i(k) = \left[\tilde{u}_i^T(k)\ \tilde{y}_i^T(k)\right]^T$$

and

$$\begin{aligned}\bar{A}_o &= A_o - K_oC_o \\ \bar{B}_o &= B_o - K_oD_o \\ \widetilde{B}_o &= \left[\bar{B}_o\ K_o\right],\end{aligned}$$

system (17) can be written as

$$\begin{aligned}\xi_i(k+1) &= \bar{A}_o\xi_i(k) + \widetilde{B}_o\tilde{z}_i(k),\ \xi_i(0) = x(t_i) \\ \tilde{y}_i(k) &= C_o\xi_i(k) + D_o\tilde{u}_i(k) + \tilde{e}_i(k),\end{aligned} \tag{18}$$

to which the PBSID$_{opt}$ algorithm can be applied to compute estimates of the state space matrices A_o, B_o, C_o, D_o, K_o. To this purpose note that iterating $p-1$ times the projection operation (i.e., propagating $p-1$ forward in the index k the first of equations (18), where p is the so-called past window length) one gets

$$\begin{aligned}\xi_i(k+2) &= \bar{A}_o^2\xi_i(k) + \left[\bar{A}_o\widetilde{B}_o\ \widetilde{B}_o\right]\begin{bmatrix}\tilde{z}_i(k) \\ \tilde{z}_i(k+1)\end{bmatrix} \\ &\ \ \vdots \\ \xi_i(k+p) &= \bar{A}_o^p\xi_i(k) + \mathscr{K}^p Z_i^{0,p-1}\end{aligned} \tag{19}$$

where

$$\mathscr{K}^p = \left[\bar{A}_o^{p-1}\widetilde{B}_0 \ldots \widetilde{B}_o\right] \tag{20}$$

is the extended controllability matrix of the system in the transformed domain and

$$Z_i^{0,p-1} = \begin{bmatrix} \tilde{z}_i(k) \\ \vdots \\ \tilde{z}_i(k+p-1) \end{bmatrix}.$$

Under the considered assumptions, $\bar{A}_o$ has all the eigenvalues inside the open unit circle, so the term $\bar{A}_o^p \xi_i(k)$ is negligible for sufficiently large values of p and we have that

$$\xi_i(k+p) \simeq \mathscr{K}^p Z_i^{0,p-1}.$$

As a consequence, the input-output behaviour of the system is approximately given by

$$\begin{aligned} \tilde{y}_i(k+p) &\simeq C_o \mathscr{K}^p Z_i^{0,p-1} + D_o \tilde{u}_i(k+p) + \tilde{e}_i(k+p) \\ &\vdots \\ \tilde{y}_i(k+p+f) &\simeq C_o \mathscr{K}^p Z_i^{f,p+f-1} + D_o \tilde{u}_i(k+p+f) + \\ &\quad + \tilde{e}_i(k+p+f), \end{aligned} \tag{21}$$

so that introducing the vector notation

$$\begin{aligned} Y_i^{p,f} &= \left[\tilde{y}_i(k+p)\ \tilde{y}_i(k+p+1) \ldots \tilde{y}_i(k+p+f)\right] \\ U_i^{p,f} &= \left[\tilde{u}_i(k+p)\ \tilde{u}_i(k+p+1) \ldots \tilde{u}_i(k+p+f)\right] \\ E_i^{p,f} &= \left[\tilde{e}_i(k+p)\ \tilde{e}_i(k+p+1) \ldots \tilde{e}_i(k+p+f)\right] \\ \Xi_i^{p,f} &= \left[\xi_i(k+p)\ \xi_i(k+p+1) \ldots \xi_i(k+p+f)\right] \\ \bar{Z}_i^{p,f} &= \left[Z_i^{0,p-1}\ Z_i^{1,p} \ldots Z_i^{f,p+f-1}\right] \end{aligned} \tag{22}$$

equations (19) and (21) can be rewritten as

$$\begin{aligned} \Xi_i^{p,f} &\simeq \mathscr{K}^p \bar{Z}_i^{p,f} \\ Y_i^{p,f} &\simeq C_o \mathscr{K}^p \bar{Z}_i^{p,f} + D_o U_i^{p,f} + E_i^{p,f}. \end{aligned} \tag{23}$$

Considering now the entire dataset for $i = 1, \ldots, N$, the data matrices become

$$Y^{p,f} = \left[\tilde{y}_1(k+p) \ldots \tilde{y}_N(k+p) \ldots \tilde{y}_1(k+p+f) \ldots \tilde{y}_N(k+p+f)\right], \tag{24}$$

and similarly for $U_i^{p,f}$, $E_i^{p,f}$, $\Xi_i^{p,f}$ and $\bar{Z}_i^{p,f}$. The data equations (23), in turn, are given by

$$\begin{aligned} \Xi^{p,f} &\simeq \mathscr{K}^p \bar{Z}^{p,f} \\ Y^{p,f} &\simeq C_o \mathscr{K}^p \bar{Z}^{p,f} + D_o U^{p,f} + E^{p,f}. \end{aligned} \tag{25}$$

From this point on, the algorithm can be developed along the lines of the discrete-time PBSID$_{opt}$ method, i.e., by carrying out the following steps. Considering $p = f$, estimates for the matrices $C_o\mathscr{K}^p$ and D_o are first computed by solving the least-squares problem

$$\min_{C_o\mathscr{K}^p, D_o} \|Y^{p,p} - C_o\mathscr{K}^p \bar{Z}^{p,p} - D_o U^{p,p}\|_F. \tag{26}$$

Defining now the extended observability matrix Γ^p as

$$\Gamma^p = \begin{bmatrix} C_o \\ C_o\bar{A}_o \\ \vdots \\ C_o\bar{A}_o^{p-1} \end{bmatrix} \tag{27}$$

and noting that the product of Γ^p and $\mathscr{K}^p$ can be written as

$$\Gamma^p \mathscr{K}^p \simeq \begin{bmatrix} C_o\bar{A}^{p-1}\widetilde{B}_o & \dots & C_o\widetilde{B}_o \\ 0 & \dots & C_o\bar{A}\widetilde{B}_o \\ \vdots & & \\ 0 & \dots & C_o\bar{A}^{p-1}\widetilde{B}_o \end{bmatrix}, \tag{28}$$

such product can be computed using the estimate $\widehat{C_o\mathscr{K}^p}$ of $C_o\mathscr{K}^p$ obtained by solving the least squares problem (26).

Recalling now that

$$\Xi^{p,p} \simeq \mathscr{K}^p \bar{Z}^{p,p} \tag{29}$$

it also holds that

$$\Gamma^p \Xi^{p,p} \simeq \Gamma^p \mathscr{K}^p \bar{Z}^{p,p}. \tag{30}$$

Therefore, computing the singular value decomposition

$$\Gamma^p \mathscr{K}^p \bar{Z}^{p,p} = U\Sigma V^T \tag{31}$$

an estimate of the state sequence can be obtained as

$$\widehat{\Xi}^{p,p} = \Sigma_{n_x}^{1/2} V_{n_x}^T = \Sigma_{n_x}^{-1/2} U_{n_x}^T \Gamma^p \mathscr{K}^p \bar{Z}^{p,p}, \tag{32}$$

from which, in turn, an estimate of C_o can be computed by solving the least squares problem

$$\min_{C_o} \|Y^{p,p} - \widehat{D}_o U^{p,p} - C_o \widehat{\Xi}^{p,p}\|_F. \tag{33}$$

The final steps consist of the estimation of the innovation data matrix $E^{p,p}$

$$E^{p,p} = Y^{p,p} - \widehat{C}_o \widehat{\Xi}^{p,p} - \widehat{D}_o U^{p,p} \tag{34}$$

and of the entire set of the state space matrices for the system in the transformed domain, which can be obtained by solving the least squares problem

$$\min_{A_o, B_o, K_o} \|\widehat{\Xi}^{p+1,p} - A_o \widehat{\Xi}^{p,p-1} - B_o U^{p,p-1} - K_o E^{p,p-1}\|_F. \tag{35}$$

6. Balancing of the identified models

Once N_θ local models have been obtained, the problem of recovering the whole parameter-dependent system has to be faced. It turns out that this task is greatly simplified if the local models are converted to a suitable balanced form. In order to clarify this point, the main properties of internally balanced realizations have to be briefly summarised (see [Moore (1981); Moonen and Ramos (1993)]).

Definition 8.1. Consider the continuous-time system (11) and the matrices

$$\mathscr{W}_o = \int_0^\infty e^{A^T t} C^T C e^{At} dt, \quad \mathscr{W}_c = \int_0^\infty e^{At} B B^T e^{A^T t} dt$$

which are respectively known as the observability and controllability Gramians of the (deterministic part of the) state-space system (11). The state-space realization is internally balanced if

$$\mathscr{W}_o = \mathscr{W}_c = \Sigma = \operatorname{diag}(\sigma_1 \ \cdots \ \sigma_n), \tag{36}$$

where $\{\sigma_i\}_{i=1}^n$ are the singular values of $\mathscr{W}_c \mathscr{W}_o$.

In order to improve the numerical performance of the subsequent interpolation algorithm, it is proposed to fix the state-space basis of the estimated matrices by finding a similarity transformation T such that $\left\{T^{-1}\hat{A}T, T^{-1}\hat{B}, \hat{C}T, \hat{D}\right\}$ is in internally balanced form. To this purpose, we use the algorithm first derived in [Laub *et al.* (1987)], which consists of the following steps:

(1) Compute the Gramians $\mathscr{W}_o$ and $\mathscr{W}_c$ of the identified model;

(2) apply a Cholesky factorisation to obtain $\mathscr{W}_o = L_o L_o^T$, $\mathscr{W}_c = L_c L_c^T$, where L_o and L_c are the lower triangular factors of the Gramians;

(3) compute the SVD $L_o^T L_c = U \Sigma V^T$.

The balancing transformation is then defined by $T = L_c V \Sigma^{-1/2}$, $T^{-1} = \Sigma^{-1/2} U^T L_o^T$.

The most interesting property of balanced realizations, as far as this contribution is concerned, is associated with the *uniqueness* of the balancing transformation (see [Laub *et al.* (1987)] and [Moore (1981)] for details). Essentially, the question turns on the eigenvalues (real and nonnegative) of the product of the reachability and observability Gramians, and their corresponding eigenvectors, i.e., the columns of the balancing transformation T. If these eigenvalues are distinct, then the corresponding eigenvectors are uniquely determined to within sign, i.e., T is essentially unique. If, on the other hand, two or more eigenvalues are repeated, then their corresponding eigenvectors can be rotated arbitrarily in the corresponding eigenspace. This, in turn, implies that as long as the eigenvalues are distinct, if the true system exhibits a smooth dependence from the scheduling parameter p, then the overall parameter dependent model can be directly reconstructed from the identified local models.

Since, in general, the assumption of distinct eigenvalues cannot be guaranteed a priori, as far as our identification problem is concerned, the following situations may arise:

- If the elements of the state-space matrices of the identified local models exhibit a smooth variation as a function of the parameter θ, then the overall parameter-dependent model can be directly recovered using the interpolation techniques outlined in the following.

- If, instead, the behaviour of the elements of the state-space matrices exhibit abrupt sign changes (associated with sign changes in the columns of the balancing transformations), then one should check for situations in which the eigenvalues of the balanced Gramians are not distinct. If signs changes are present but the eigenvalues are distinct, then a "mild" nonuniqueness is occurring, which can be corrected by adjusting the signs of the "stray" matrix elements, and the proposed procedure is applicable.

7. Model interpolation

In this section the interpolation of the local models, in order to obtain an LPV model, is discussed. The problem of interpolating local models derived from jacobian linearisation of a nonlinear simulator is a subject of active research, see for example [Groot Wassink *et al.* (2005); Pfifer and Hecker (2011)]. Once the elements of the state-space matrices of the system have been estimated following the above steps, and set of N_θ local linear time-invariant state space models has been obtained, a number of options are available as far as the derivation of the actual parameter dependent model is concerned. The first, and simplest, would be to directly fit to the system matrices of the local models using suitable regressors formed from the scheduling parameter $\theta \in \mathbb{R}^{n_p}$. This would directly yield a parameter dependent model in so-called *affine* form, i.e.,

$$\begin{aligned} \dot{x} &= A(\gamma)x + B(\gamma)u \\ y &= C(\gamma)x + D(\gamma)u \end{aligned} \tag{37}$$

where $\gamma \in \mathbb{R}^{n_\gamma}$ is the vector of regressors (formed from linear or nonlinear combinations of the elements of θ) such that the parameter dependent matrices can be written as

$$A(\gamma) = A_0 + A_1\gamma_1 + A_2\gamma_2 + ... + A_{n_\gamma}\gamma_{n_\gamma}$$

and similarly for $B(\gamma)$, $C(\gamma)$ and $D(\gamma)$. The state space matrices are represented as transfer functions

$$G_k(s) = C(\gamma(k))(sI - A(\gamma(k)))^{-1}B(\gamma(k)) + D(\gamma(k)), \quad k = 1, \ldots, N_\theta$$

where k is the scheduling index of the kth identified local model. The matrices are also aggregated in the following form

$$F(\gamma(k)) = \begin{bmatrix} A(\gamma(k)) & B(\gamma(k)) \\ C(\gamma(k)) & D(\gamma(k)) \end{bmatrix}, \; k = 1, \ldots, N_\theta. \tag{38}$$

The polynomial interpolation of the elements of the matrix F is performed by solving a least squares problem as $Y = \Phi\Theta$, where Y contains the elements f_{ij} of F, Φ contains the regressors γ, and Θ contains the polynomial coefficients. The root means square error (RMS_e) is defined as

$$RMS_e = \frac{\|Y - \Phi\hat{\Theta}\|_2}{\|Y\|_2}, \tag{39}$$

and it expresses the fitting error.

In order to reduce the model complexity, as suggested in [Pfifer and Hecker (2011)] and reported here for the sake of completeness, the fitting procedure is divided in two steps: the sensitivity analysis and the polynomial fitting.

The H_∞-norm sensitivity (note that in [Pfifer and Hecker (2011)] the ν-gap metric is instead used as a measure of sensitivity) is computed for each element f_{ij} of the matrix $F(\gamma(k))$ and it is defined as

$$S_{ij} = \max_k \frac{\|G_k - G_{k_{ij}}\|_\infty}{\|G_k\|_\infty}, \tag{40}$$

where $G_{k_{ij}}$ is equal to G_k except for the element f_{ij} that is substituted with the mean of the N_θ values $f_{ij}(\gamma(k))$. If the influence of an element f_{ij} on the transfer functions is lower than a (user-defined) bound $\underline{S} \in [0,1]$, its mean value is used instead of performing a polynomial fitting. It is also possible to define an upper bound $\overline{S}$ beyond which the maximum polynomial degree is used. If S_{ij} is in the interval $[\underline{S}, \overline{S}]$ a linear function of the influence is considered in order to obtain the most reasonable trade-off between model complexity, i.e., polynomial degree, and fitting error. In other words, the coefficient ζ_{ij} is defined for each (i,j) as a function of the sensitivity S_{ij} as follows

$$\zeta_{ij} = \begin{cases} \infty & S_{ij} < \underline{S} \\ \frac{K_e - 1}{\underline{S} - \overline{S}} S_{ij} + \frac{\underline{S} - K_e \overline{S}}{\underline{S} - \overline{S}} & \underline{S} \leq S_{ij} \leq \overline{S} \\ 1 & S_{ij} > \overline{S}, \end{cases} \tag{41}$$

where $K_e > 1$ is the value of ζ_{ij} when S_{ij} is equal to $\underline{S}$. The value for this parameter suggested in [Pfifer and Hecker (2011)] is 2, which corresponds to twice the absolute and the relative tolerances than their original values, and in this way the low sensitivity of the parameter (i,j) is taken into account. The absolute and the relative tolerances of the mean square error are respectively $\zeta_{ij}\varepsilon_a$ and $\zeta_{ij}\varepsilon_r$.

The algorithm increases the polynomial degree of the approximation and computes the new polynomial coefficients by solving the least squares problem until the RMS_e drops below the absolute tolerance $\zeta_{ij}\varepsilon_a$ or the improvement in the RMS_e becomes less than the relative tolerance $\zeta_{ij}\varepsilon_r$. The interpolation algorithm is summarized in Table 2.

Remark 8.1. This algorithm can suffer from the curse of dimensionality, because of the dimension of γ, n_γ, that is

$$n_\gamma = \binom{M + n_p}{M} \tag{42}$$

where M is the maximum degree of the polynomial and n_p is the dimension of θ.

Interpolation Algorithm

- Input parameters: $\underline{S}$, $\overline{S}$, ε_a, ε_r, K_e
- For each $i = 1, \ldots, n_x + n_y$ $j = 1, \ldots, n_x + n_u$

(1) Compute S_{ij} according to (40)
(2) According to (41) compute ζ_{ij}
(3) Set the polynomial order equal to 0
(4) Solve the least squares problem obtaining an estimate of the polynomial coefficients and compute the RMS_e according to (39)
(5) Increase the polynomial order by 1
(6) If $RMS_e > \zeta_{ij}\varepsilon_a$ and the difference between two sequential RMS_e is larger than $\zeta_{ij}\varepsilon_r$ then return to the step 4

Finally, in some applications it is desirable to obtain as output of the system identification procedure a model in so-called LPV-LFT form, i.e., in which the state-space matrices are expressed as a linear fractional transformation over a suitable linear operator representing time-varying parameters:

$$\dot{x} = Ax + B_0 w + B_1 u \tag{43}$$

$$z = C_0 x + D_{00} w + D_{01} u \tag{44}$$

$$y = C_1 x + D_{10} w + D_{11} u \tag{45}$$

$$w = \Gamma z, \quad \Gamma = \mathrm{diag}(\gamma_1 I_{r_1} \ldots \gamma_{n_\gamma} I_{r_{n_\gamma}}) \tag{46}$$

and $w, z \in \mathbb{R}^r$, $r = r_1 + \ldots + r_{n_\gamma}$.

As is well known in the robust control literature and also pointed out in [Lee and Poolla (1999); Lovera *et al.* (1998)], LPV-A and LPV-LFT representations are related to each other. Denoting the composition of the system matrices

$$M(\gamma) = \begin{bmatrix} A(\gamma) & B(\gamma) \\ C(\gamma) & D(\gamma) \end{bmatrix} = M_0 + M_1\gamma_1 + \ldots + M_{n_\gamma}\gamma_{n_\gamma},$$

by expressing each of the $M_i's$, $i = 1, ..., n_\gamma$ by means of a rank r_i decomposition as $M_i = U_i V_i$ one can write $M(\gamma)$ as $M(\gamma) = M_0 + U\Gamma V$, where $U = [U_1 \ldots U_{n_\gamma}]$, $V = [V_1^T \ldots V_{n_\gamma}^T]^T$ and Γ is given by equation (46). The obtained form for the system matrices coincides with the one which is obtained in the special case of a linear fractional transformation characterised by having $D_{00} = 0$, hence the transformation between the two forms.

8. Comments and discussion

As mentioned in the Introduction, the proposed approach seems to offer a useful trade-off between the two classes of solutions to this problem which have been so far proposed in the literature.

On one hand, subspace methods for the identification of LPV state-space models such as the ones in [Verdult (2002); van Wingerden and Verhaegen (2009)] provide a very general way of dealing with the problem, at the cost of critical requirements on the experimental conditions which might not be easily realizable in many applications.

On the other hand, the techniques of [Steinbuch *et al.* (2003); van Helvoort *et al.* (2004); Groot Wassink *et al.* (2005)] have the advantage of being much closer to the actual practice of system identification but suffer from the numerical drawbacks associated with the adoption of a fixed (generally not well conditioned) basis for the manipulation of state-space models and are restricted to SISO models.

Finally, as far as the methods proposed in [Paijmans *et al.* (2008); De Caigny *et al.* (2009)] are concerned: the numerical issues associated with the use of the canonical observability form for the local models are circumvented by splitting the local models into first and second order submodels; however, a significant level of manual intervention is expected from the user, in order to choose the structure of the local submodels and formulate the interpolation problem, even more so when moving from SISO to

MIMO models. The main advantage of this class of methods is the general applicability, as no specific preconditions must be met in order to use them.

In this respect, the proposed method presents two advantages and one limitation. The advantages are that it is directly applicable to both SISO and MIMO systems (thanks to the use of subspace methods for the local identification) with a minimal level of manual intervention to deal with sign changes in the elements of the state space matrices (thanks to the adoption of the balanced form, see the following section for an example). The limitation is due to the need to have distinct eigenvalues of the balanced Gramians of the local models in order to ensure that the interpolation problem is well posed. Note, in passing, that this issue has also been resolved by manual intervention on the local models in a few examples, but no systematic approach to it is available at present.

9. Simulation examples

In this section the results obtained by applying the above described approach to a numerical example are presented . More precisely, both open-loop and closed-loop operating conditions are considered and the performance obtained in both situations by using either the continuous-time PB-SID algorithm presented in Section 5 or the equivalent approach based on the PO-MOESP algorithm (see [Ohta and Kawai (2004)]) is discussed.

The parameter-dependent system used in this example (considered in [De Filippi and Lovera (2011)] in an LPV control design framework) is given by the equations

$$\begin{aligned} \dot{x}_1 &= (-2+\theta)x_1 - 2x_2 + d + 3u \\ \dot{x}_2 &= x_1 - 2x_2 + d + u \\ z_1 &= x_1 \\ z_2 &= x_2 \\ z_3 &= x_1 + x_2 + u \end{aligned} \tag{47}$$

where $\theta \in [\theta_{min}, \theta_{max}]$, with $\theta_{min} = 0$ and $\theta_{max} = 5$. The eigenvalues of the system strongly depend on θ. For the system the $N_\theta = 8$ local operating points corresponding to the values $\{0, 0.4, 0.8, 1.2, 1.6, 2, 2.4, 2.8\}$ have been considered. Simulated data has been collected near each of the considered local operating points by applying to the system a sequence of filtered white Gaussian noise as input, for a duration of 10 s. White Gaussian noise of increasing variance has been added to the output in order to assess the influence of decreasing signal-to-noise ratio on the quality of the computed estimates. For the input and output variables two values of the sampling interval Δt have been considered, i.e., $\Delta t = 0.001$ s and $\Delta t = 0.005$ s. Similar choices for the p and f parameters have been made in the two algorithms, namely $p = f = 10$. As mentioned above, both open-loop and closed-loop operating conditions have been considered. In the latter case, the system is operating under feedback with a controller given by

$$K(\theta) = K_0 + \theta K_1 \tag{48}$$

where

$$K_0 = \begin{bmatrix} 1.1254 & 1.1878 & 0 \\ 1.1878 & 1.2537 & 0 \end{bmatrix}, \quad K_1 = \begin{bmatrix} -0.0345 & -0.0078 & 0 \\ -0.0078 & 0.6752 & 0 \end{bmatrix}. \tag{49}$$

The singular values of the balanced Gramians for the local dynamics of the system are depicted in Figure 1. As can be seen, they are distinct for all the considered values of θ, so it can be anticipated that no major obstacles to the interpolation step of the procedure should be encountered.

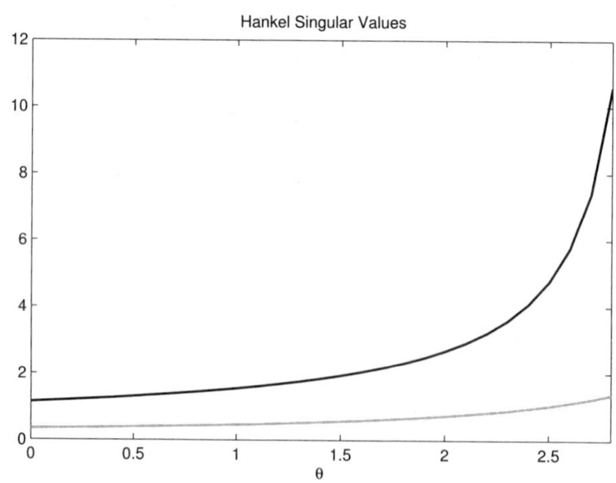

Fig. 1. Eigenvalues of the balanced Gramian matrices, as function of θ.

The results obtained in the validation of the identified model can be summarised as follows, by referring to Figures 2-8 for the achieved values of the Variance-Accounted-For (VAF) in a Monte-Carlo study with 100 simulations for each case.

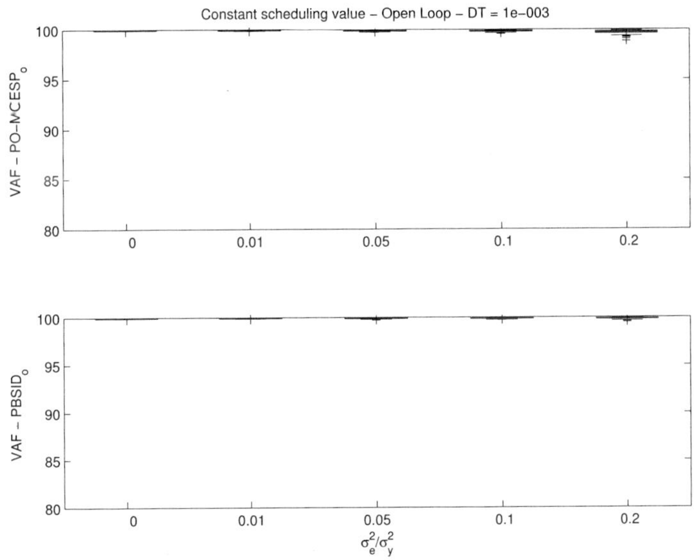

Fig. 2. Open loop VAF - constant scheduling parameter with $DT = 10^{-3}$ s.

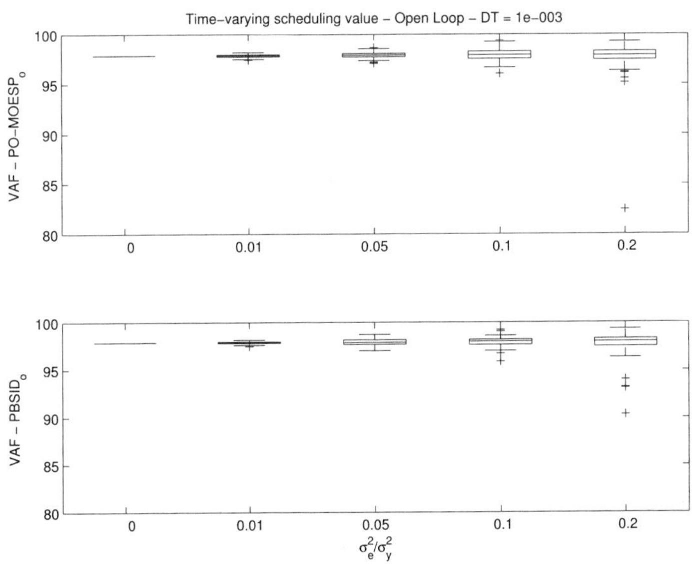

Fig. 3. Open loop VAF - variable scheduling parameter with $DT = 10^{-3}$ s.

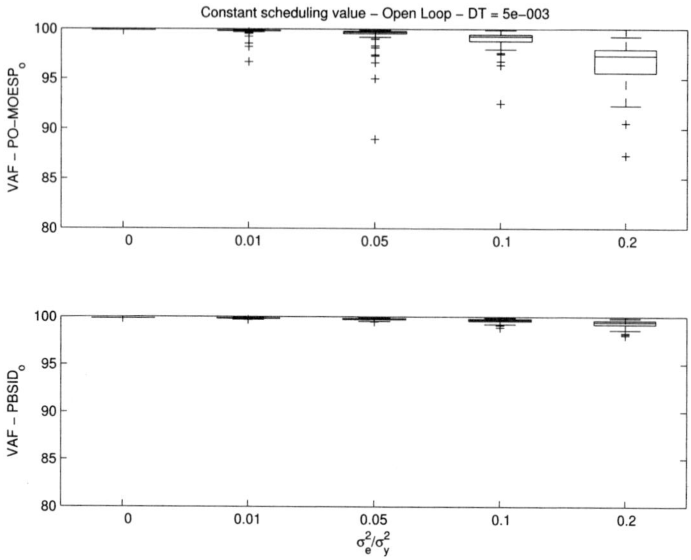

Fig. 4. Open loop VAF - constant scheduling parameter with $DT = 5 \times 10^{-3}$ s.

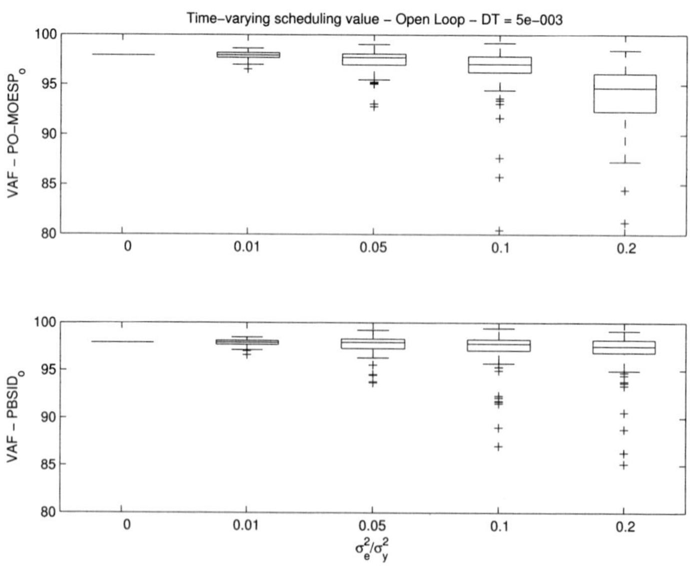

Fig. 5. Open loop VAF - variable scheduling parameter with $DT = 5 \times 10^{-3}$ s.

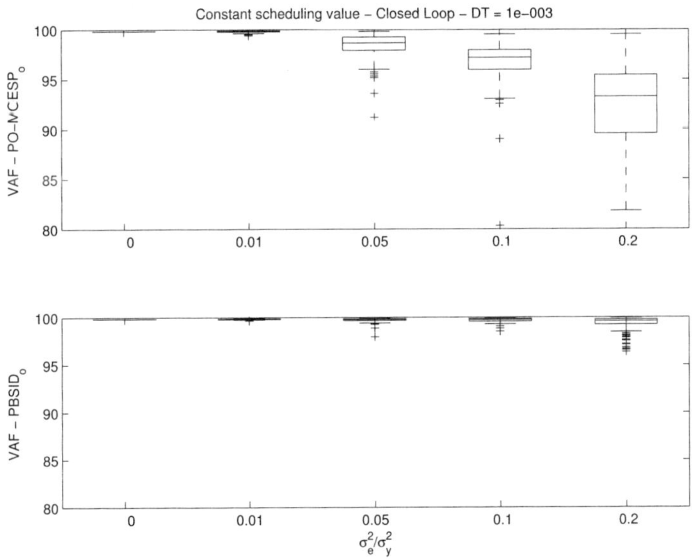

Fig. 6. Closed-loop VAF - constant scheduling parameter with $DT = 10^{-3}$ s.

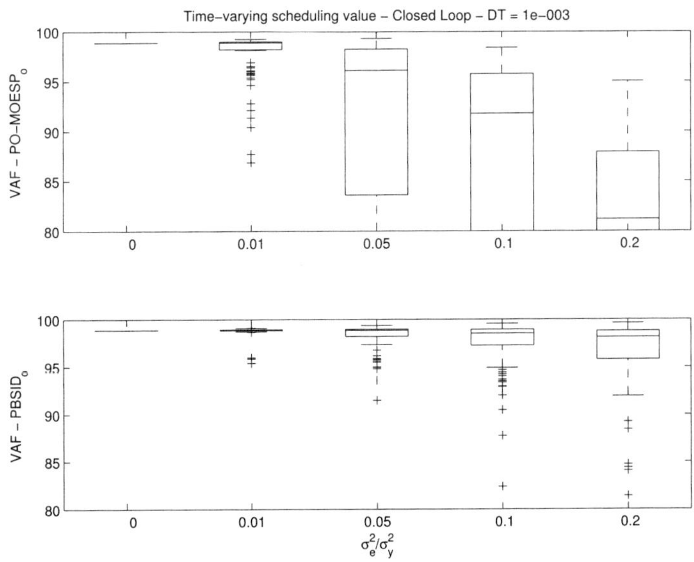

Fig. 7. Closed-loop VAF - variable scheduling parameter with $DT = 10^{-3}$ s.

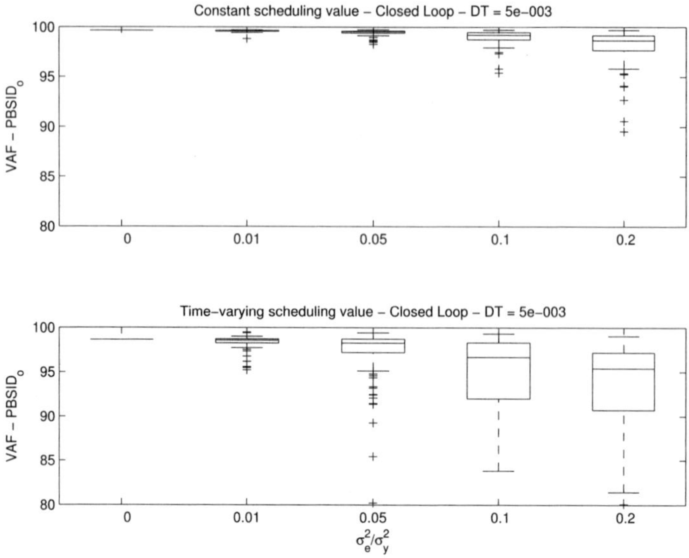

Fig. 8. Closed-loop VAF - constant (top) and varying (bottom) scheduling parameter with $DT = 5 \times 10^{-3}$ s.

Concerning open-loop operation, a direct inspection of the box plots in Figures 2-3 shows that for the case of sampling interval given by $\Delta t = 0.001$ s the two subspace algorithms provide essentially equivalent results both when carrying out the validation for a constant value of the scheduling parameter and by considering a time-varying trajectory for it. A difference in performance becomes apparent when increasing the sampling interval to $\Delta t = 0.005$ s, as depicted in Figures 4-5, from which the greater sensitivity of the continuous-time MOESP becomes visible. Finally, as one can anticipate, when turning to the results obtained in closed-loop operation, comparing the performance of the two approaches for the "fast" sampling case (see Figures 6-7) clearly shows a wide gap in performance; for $\Delta t = 0.005$ s, the local models computed using the continuous-time PO-MOESP are not asymptotically stable, and therefore in Figure 8 only the performance obtained using the continuous-time PBSID is illustrated.

As a conclusion, some representative time-domain validation results are presented in Figures 9-12, to illustrate the ability of the interpolated model to provide a satisfactory performance also in the case of a varying parameter.

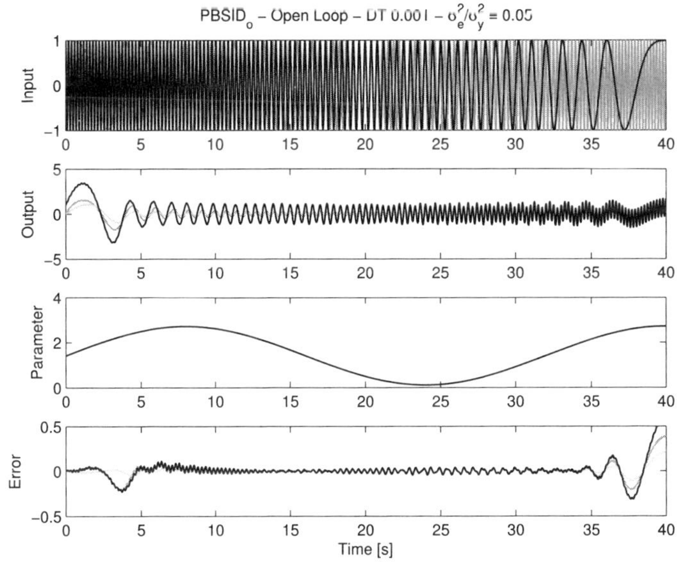

Fig. 9. Validation performance with slowly-varying parameter θ, for PBSID-based models identified in open-loop.

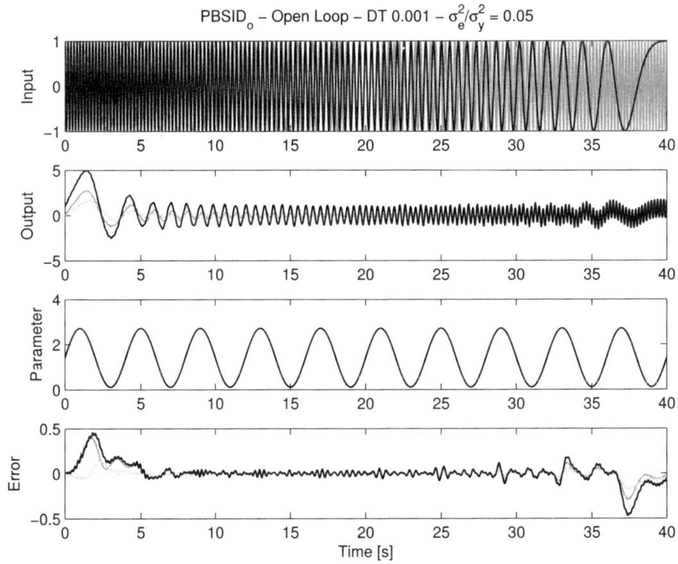

Fig. 10. Validation performance with rapidly-varying parameter θ, for PBSID-based models identified in open-loop.

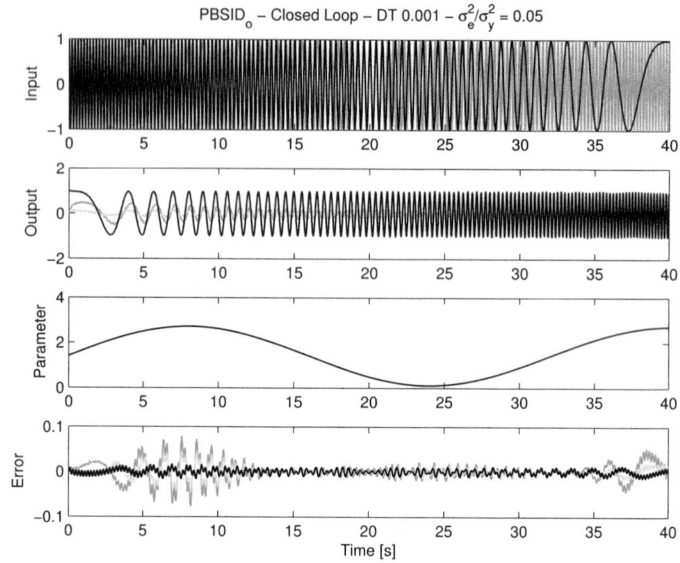

Fig. 11. Validation performance with slowly-varying parameter θ, for PBSID-based models identified in closed-loop.

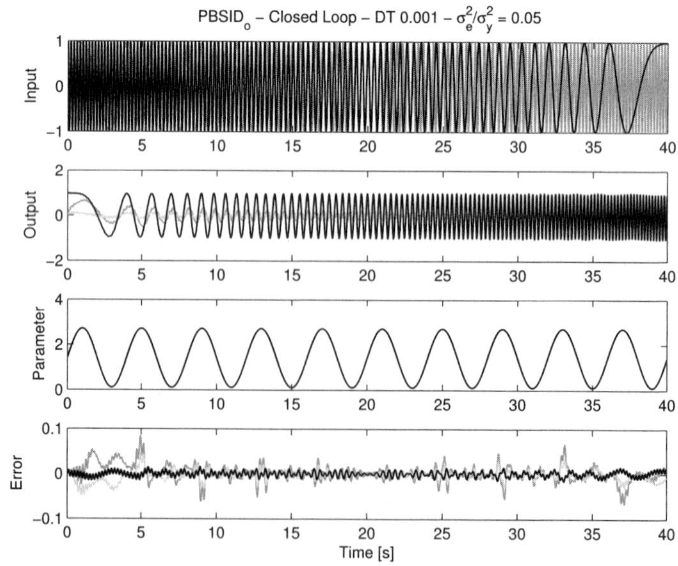

Fig. 12. Validation performance with rapidly-varying parameter θ, for PBSID-based models identified in closed-loop.

10. Conclusions

The problem of deriving MIMO parameter-dependent models for gain-scheduling control design from data generated by *local* identification experiments has been considered and a continuous-time approach based on subspace identification ideas combined with the use of suitable properties of balanced state-space realizations has been presented and discussed.

11. Acknowledgements

This paper has been supported by the Italian MIUR project Identification and Adaptive Control of Industrial Systems.

References

Apkarian, P. and Adams, R. J. (1998). Advanced gain-scheduling techniques for uncertain systems, *IEEE Transactions on Control System Technology* **6**, 1, pp. 21–32.

Bamieh, B. and Giarré, L. (2002). Identification of linear parameter-varying models, *International Journal of Robust and Nonlinear Control* **12**, 9, pp. 841–853.

Bastogne, T., Garnier, H. and Sibille, P. (2001). A PMF-based subspace method for continuous-time model identification. Application to a multivariable winding process, *International Journal of Control* **74**, 2, pp. 118–132.

Bergamasco, M. and Lovera, M. (2010). Continuous-time subspace identification in closed-loop, in *Proceedings of the International Symposium on Mathematical Theory of Networks and Systems* (Budapest, Hungary).

Bergamasco, M. and Lovera, M. (2011). Continuous-time predictor-based subspace identification using Laguerre filters, *IET Control Theory and Applications* Special issue on Continuous-time Model Identification **5**, 2, pp. 856–867.

Chiuso, A. (2007). The role of autoregressive modeling in predictor-based subspace identification, *Automatica* **43**, 3, pp. 1034–1048.

Chiuso, A. and Picci, G. (2005). Consistency analysis of certain closed-loop subspace identification methods, *Automatica* **41**, 3, pp. 377–391.

Chou, C. and Verhaegen, M. (1997). Subspace algorithms for the identification of multivariable dynamic error-in-variables state space models, *Automatica* **33**, 10, pp. 1857–1869.

De Caigny, J., Camino, J. F. and Swevers, J. (2009). Interpolating model identification for SISO Linear Parameter Varying systems, *Mechanical Systems and Signal Processing* **23**, 8, pp. 2395–2417.

De Caigny, J., Camino, J. F. and Swevers, J. (2011). Interpolation-based modelling of MIMO LPV systems, *IEEE Transactions on Control Systems Technology* **19**, 1, pp. 46–63.

De Filippi, P. and Lovera, M. (2011). Linear parameter-varying state feedback synthesis with hierarchical performance requirements, in *Proceedings of the IFAC World Congress* (Milano, Italy).

Groot Wassink, W., van de Wal, M., Scherer, C. and Bosgra, O. (2005). LPV control for a wafer stage: beyond the theoretical solution, *Control Engineering Practice* **13**, pp. 231–245.

Haverkamp, B. R. J. (2001). *State space identification: Theory and practice*, Ph.D. thesis, Delft University of Technology, The Netherlands.

Huang, B., Ding, S. and Qin, S. (2005). Closed-loop subspace identification: an orthogonal projection approach, *Journal of Process Control* **15**, 1, pp. 53–66.

Johansson, R., Verhaegen, M. and Chou, C. (1999). Stochastic theory of continuous-time state-space identification, *IEEE Transactions on Signal Processing* **47**, 1, pp. 41–51.

Kaminer, I., Rascoal, A. M., Karghonekar, P. P. and Coleman, E. E. (1995). A velocity algorithm for the implementation of gain-scheduled controllers, *Automatica* **31**, 8, pp. 1185–1191.

Kinoshita, Y. and Ohta, Y. (2010). Continuous-time system identification using compactly supported filter kernels generated from Laguerre basis functions, in *Proceedings of the IEEE Conference on Decision and Control* (Atlanta, GA, USA).

Laub, A., Heath, M., Paige, C. and Ward, R. (1987). Computation of system balancing transformations and other applications of simultaneous diagonalization algorithms, *IEEE Transactions on Automatic Control* **32**, pp. 115–122.

Lee, L. and Poolla, K. (1999). Identification of linear parameter-varying systems using nonlinear programming, *ASME Journal of Dynamic Systems, Measurement and Control* **121**, 1, pp. 71–78.

Leith, D. J. and Leithead, W. E. (2000). Survey of gain-scheduling analysis and design, *International Journal of Control* **73**, 11, pp. 1001–1025.

Ljung, L. and McKelvey, T. (1996). Subspace identification from closed loop data, *Signal Processing* **52**, 2, pp. 209–215.

Lovera, M. and Mercére, G. (2007). Identification for gain-scheduling: a balanced subspace approach, in *Proceedings of the American Control Conference* (New York City, NY, USA).

Lovera, M., Novara, C., Lopes, P. and Rivera, D. (2011). Guest editorial special issue on Applied LPV modelling and identification, *IEEE Transactions on Control Systems Technology* **19**, 1, pp. 1–4.

Lovera, M., Verhaegen, M. and Chou, C. T. (1998). State space identification of MIMO linear parameter varying models, in *Proceedings of the International Conference on Mathematical Theory of Networks and Systems* (Padova, Italy).

Marcos, A. and Balas, G. (2004). Development of linear-parameter-varying models for aircraft, *Journal of Guidance, Control and Dynamics* **27**, 2, pp.

218–228.

Mercére, G., Ouvrard, R., Gilson, M. and Garnier, H. (2008). Identification de systémes multivariables au temps continu par approche des sous-espaces, *Journal Européen des Systémes Automatisés* **42**, 2-3, pp. 261–285.

Mohd-Moktar, R. and Wang, L. (2008). Continuous-time state space model identification using closed-loop data, in *Proceedings of the Second Asia International Conference on Modelling & Simulation* (Kuala Lumpur, Malaysia).

Moonen, M. and Ramos, J. (1993). A subspace algorithm for balanced state space system identification, *IEEE Transactions on Automatic Control* **38**, 11, pp. 1727–1729.

Moore, B. (1981). Principal component analysis in linear systems: controllability, observability and model reduction, *IEEE Transactions on Automatic Control* **26**, 1, pp. 17–32.

Nemani, M., Ravikanth, R. and Bamieh, B. (1995). Identification of linear parameter-varying systems, in *Proceedings of the IEEE Conference on Decision and Control* (Kobe, Japan).

Ohsumi, A., Kameyama, K. and Yamagushi, K. I. (2002). Subspace identification for continuous-time stochastic systems via distribution-based approach, *Automatica* **38**, 1, pp. 63–79.

Ohta, Y. (2005). Realization of input-output maps using generalized orthonormal basis functions, *Systems & Control Letters* **22**, 6, pp. 437–444.

Ohta, Y. (2011). System transformation of unstable systems induced by a shift-invariant subspace, in *Proceedings of the IEEE Conference on Systems and Control* (Orlando, FL, USA).

Ohta, Y. and Kawai, T. (2004). Continuous-time subspace system identification using generalized orthonormal basis functions, in *Proceedings of the 16th International Symposium on Mathematical Theory of Networks and Systems* (Leuven, Belgium).

Paijmans, B., Symens, W., Brussel, H. V. and Swevers, J. (2008). Identification of interpolating affine LPV models for mechatronic systems with one varying parameter, *European Journal of Control* **14**, 1, pp. 16–29.

Pfifer, H. and Hecker, S. (2011). Generation of optimal linear parametric models for LFT-based robust stability analysis and control design, *IEEE Transactions on Control Systems Technology* **19**, 1, pp. 118–131.

Previdi, F. and Lovera, M. (2003). Identification of a class of nonlinear parametrically varying models, *International Journal on Adaptive Control and Signal Processing* **17**, pp. 33–50.

Rugh, W. and Shamma, J. (2000). Research on gain scheduling, *Automatica* **36**, 10, pp. 1401–1425.

Shamma, J. S. and Athans, M. (1990). Analysis of Gain-Scheduled Control for Nonlinear Plants, *IEEE Transactions on Automatic Control* **35**, 8, pp. 898–907.

Steinbuch, M., van de Molengraft, R. and van der Voort, A. (2003). Experimental modelling and LPV control of a motion system, in *Proceedings of the American Control Conference* (Denver, CO, USA).

Tóth, R. (2010). *Modeling and identification of linear parameter-varying systems,*

Lecture Notes in Control and Information Sciences, Vol. 403 (Springer-Verlag, London, UK).

van Helvoort, J., Steinbuch, M., Lambrechts, P. and van de Molengraft, R. (2004). Analytical and experimental modelling for gain-scheduling of a double scara robot, in *Proceedings of the IFAC Symposium on Mechatronic Systems* (Sydney, Australia).

Van Overschee, P. and De Moor, B. (1996). Continuous time frequency domain subspace system identification, *Signal Processing* **52**, 2, pp. 179–194.

van Wingerden, J. W. and Verhaegen, M. (2009). Subspace identification of bilinear and LPV systems for open- and closed-loop data, *Automatica* **45**, 2, pp. 372–381.

Verdult, V. (2002). *Nonlinear system identification: a state space approach*, Ph.D. thesis, University of Twente, The Netherlands.

Verhaegen, M. (1994). Identification of the deterministic part of MIMO state space models given in innovations form from input-output data, *Automatica* **30**, 1, pp. 61–74.

Verhaegen, M. and Dewilde, P. (1992a). Subspace model identification part 1: output error state space model identification class of algorithms, *International Journal of Control* **56**, pp. 1187–1210.

Verhaegen, M. and Dewilde, P. (1992b). Subspace model identification part 2: analysis of the elementary output error state space model identification algorithm, *International Journal of Control* **56**, pp. 1211–1241.

Zhou, K., Doyle, J. and Glover, K. (1996). *Robust and optimal control* (Prentice Hall).

Chapter 9

Indirect continuous-time LPV system identification through a downsampled subspace approach

P. Lopes dos Santos†, T.-P. Azevedo Perdicoúlis‡, J. A. Ramos* & J. L. Martins de Carvalho†

†*Faculdade de Engenharia da Universidade do Porto*
Rua do Dr Roberto Frias s/n
4200-465 Porto, Portugal
pjsantos@fe.up.pt *and* jmartins@fe.up.pt

‡*ISR—Coimbra & UTAD*
Departamento das Engenharias, UTAD
5000-801 Vila Real, Portugal
tazevedo@utad.pt

**Farquhar College of Arts and Sciences*
Division of Mathematics, Science and Technology
Nova Southeastern University
3301 College Avenue, Fort Lauderdale
FL 33314, USA
jr1284@nova.edu

The successive approximation Linear Parameter Varying systems subspace identification algorithm for discrete-time systems is based on a convergent sequence of linear time invariant deterministic-stochastic state-space approximations. In this chapter, this method is modified to cope with continuous-time LPV state-space models. To do this, the LPV system is discretised, the discrete-time model is identified by the successive approximations algorithm and then converted to a continuous-time model. Since affine dependence is preserved only for fast sampling, a subspace downsampling approach is used to estimate the linear time invariant deterministic-stochastic state-space approximations. A second order simulation example, with complex poles, illustrates the effectiveness of the new algorithm.

1. Introduction

Although the field of linear parameter varying (LPV) systems has evolved mostly in the last 20 years, it is already rather used in the modern control industry. Some of the applications are gain-scheduling based nonlinear controller synthesis [Rugh and Shamma (2000)], optimal control design via $\mu-$synthesis [Zhou and Doyle (1998)], linear matrix inequalities (LMI) based solutions with guaranteed stability margins and performance bounds over the entire operation envelope of the plant [Scheerer (1996)]. More recently, a wide range of applications appeared in a Special Issue of the IEEE Transactions on Control Systems Technology: identification of optimal set membership LPV (SM-LPV) filters for application to vehicle yaw rate estimation [Novara *et al.* (2011)], design of a gain-scheduled controller for the electronic throttle body in ride-by-wire racing motorcycles [Corno *et al.* (2011)], freeway traffic modeling [Luspay *et al.* (2011)], interpolation-based modelling of multiple input multiple output (MIMO) systems for the mechatronic industry [De Caigny *et al.* (2011)], modeling of a cross flow heat exchanger [Mercère *et al.* (2011)], leakage detection in high pressure natural gas transportation networks [Lopes dos Santos *et al.* (2011b)], autonomic web service systems [Tanelli *et al.* (2011)], LPV modeling and validating tools for the aerospace industry [Szabó *et al.* (2011); Pfifer and Hecker (2011); Ferreres (2011)]. Also, worth mentioning is the recent work of Hu and Yurkovitch in the use of LPV models in battery pack management systems for hybrid electric vehicles [Hu and Yurkovitch (2011)].

Besides the applications, many theoretical advances have been registered in the last years. See Tóth for a very good and complete overview in the field [Tóth (2008)]. In spite of this, the estimation of LPV models remains a difficult problem that is still in a state of development and some basic issues of the LPV system theory still need to be resolved. For instance, most dynamic systems encountered in the real world are continuous in time and nonlinear. Also, first principle LPV models of nonlinear systems are also derived in continuous-time form. On the other side, system identification tools are mainly developed in discrete-time, and some structural information about the plant is necessarily lost.

Very few works exist in continuous-time LPV identification. Among those, we can mention [Gaspar *et al.* (2006)], where a gray-box identification problem for LPV continuous-time systems is investigated. The proposed method is based on an adaptive observer scheme designed by applying general nonlinear ideas to the LPV systems. The result is demonstrated through an estimation of the features of the gravity center of heavy vehicles. In [De Caigny *et al.* (2008)], SMILE (state-space model interpolation of local estimates) is presented. This is a new technique to estimate LPV state-space (SS) models for MIMO systems whose dynamics depends on a single varying parameter, called the scheduling. The SMILE technique is based on the interpolation of linear time-invariant (LTI) models that are valid for fixed operating conditions of the system, that is, for constant values of the scheduling parameters. The methodology yields affine LPV models that are numerically well-conditioned and therefore suitable for LPV control synthesis procedures. The underlying interpolation technique is formulated as a nonlinear least-squares optimization problem that can be efficiently solved by standard solvers. Application of the proposed methodology to a vibroacoustic setup, whose dynamics are highly sensitive to the ambient temperature, clearly demonstrates the potential of the SMILE technique.

The availability of several identification methods for discrete-time LPV systems suggests that continuous-time LPV systems could be identified by an indirect approach. But, from our knowledge, there is not any indirect approach reported in the literature. This may be due to the complexity and nonlinearity of the discretization methods for LPV systems. In fact, exact discretization methods for LPV systems introduce a nonlinear dependence on the scheduling signal [Tóth *et al.* (2008); Tóth (2010)]. This nonlinear dependence can be removed by a first order approximation of an exponential term. However, to preserve accuracy, a high sampling rate must be used. It is well known that the identification of high sampling rate discrete-time models of continuous-time LTI systems is an ill conditioned problem. This is due to the clustering of the discrete-time poles and zeros around $1 + j0$. It is expected that similar problems occur for LPV fast sampling systems.

In [Lopes dos Santos *et al.* (2011a)], a new indirect identification method for continuous-time LTI systems able to resolve the problem of fast sampling is presented. To do this, a subspace identification downsampling (SIDDS) approach that takes into consideration the intermediate sampling instants of the input signal is proposed. This is done by partitioning the data set into m subsets, where m is the downsampling factor. Then, the discrete-time model is identified using a subspace discrete-time identification algorithm where the data subsets are fused into a single one. Using the algebraic properties of the system, some of the continuous-time parameters are directly estimated.

The advantages of using subspace methods are that they do not require any type of optimisation technique, neither the need to impose a canonical form on the system. This implies that subspace systems can be equally applied to single input single output (SISO) as well as to MIMO systems. Verdult and Verhagen proposed subspace identification algorithms for LPV systems with very high computational complexity. [Verdult (2002); Verdult and Verhaegen (2002, 2005)]. More recently, new subspace LPV identification algorithms were also developed by van Wingerden and Verhaegen [van Wingerden and Verhaegen (2009)]. The computational complexity of van Wingerden's algorithm is significantly smaller but still uses matrices with dimensions growing exponentially leading to a curse of dimensionality problem. Lopes dos Santos *et al.* used an iterative subspace identification algorithm to circumvented this problem [Lopes dos Santos *et al.* (2007, 2008a,b)]. This algorithm is based on a convergent Picard sequence of LTI approximations estimated by an LTI subspace identification algorithm.

In the work hereby presented, we extended this algorithm to fast sampled LPV continuous-time systems using a downsampling approach similar to the one proposed in [Lopes dos Santos *et al.* (2011a)].

In what follows, the LPV systems are presented in Section 2 and then discretised with a high sampling rate in Section 3. The successive approximations algorithm is briefly described in Section 4. In Section 5 the downsampled identification approach is described. In Section 6 the viability of the method is illustrated through the identification of a fast sampled second-order LPV system.

2. LPV systems

In an LPV system, the parameters are functions of a known time-varying scheduling signal, $p(t) \in \mathbb{R}^{n_p}$. An important class of these systems is described by state-space models where the dependence on the scheduling signal is static, and thence represented as:

$$\dot{x}(t) = \mathscr{A}\,[p(t)]\,x(t) + \mathscr{B}\,[p(t)]\,u(t), \tag{1}$$

$$y(t) = \mathscr{C}\,[p(t)]\,x(t) + \mathscr{D}\,[p(t)]\,u(t) + v(t), \tag{2}$$

where $x(t) \in \mathbb{R}^n, u(t) \in \mathbb{R}^{n_u}$ and $y(t) \in \mathbb{R}^{n_y}$ are the state, input and output vectors at time instant t, respectively. The unknown time-varying parameters are $\mathscr{A}\,[p(t)] \in \mathbb{R}^{n\times n}, \mathscr{B}\,[p(t)] \in \mathbb{R}^{n\times n_u}, \mathscr{C}\,[p(t)] \in \mathbb{R}^{n_y\times n}$, and

$\mathscr{D}[p(t)] \in \mathbb{R}^{n_y \times n_u}$. The measurement noise $v(t) \in \mathbb{R}^{n_y}$ is bandwidth limited zero mean white noise signal.

LPV systems with affine parameter dependence are linear systems in which the system parameters depend on the scheduling signal

$$p(t) = \begin{pmatrix} p_1(t) & \cdots & p_{n_p}(t) \end{pmatrix}^T \in \mathbb{R}^{n_p}$$

as follows:

$$\mathscr{A}[p(t)] = A_0 + \sum_{i=1}^{n_p} p_i(t) A_i, \tag{3}$$

$$\mathscr{B}[p(t)] = B_0 + \sum_{i=1}^{n_p} p_i(t) B_i, \tag{4}$$

$$\mathscr{C}[p(t)] = C_0 + \sum_{i=1}^{n_p} p_i(t) C_i, \tag{5}$$

$$\mathscr{D}[p(t)] = D_0 + \sum_{i=1}^{n_p} p_i(t) D_i, \tag{6}$$

where $A_i \in \mathbb{R}^{n \times n}, B_i \in \mathbb{R}^{n \times n_u}, C_i \in \mathbb{R}^{n_y \times n}, D_i \in \mathbb{R}^{n_y \times n_u}$, for $i = 1, \ldots, n_p$. Using (3)–(6), system (1)–(2) may be written as:

$$\begin{aligned} \dot{x}(t) &= A_0 x(t) + A_p \left(p(t) \otimes x(t)\right) + B_0 u(t) + B_p \left(p(t) \otimes u(t)\right), \\ y(t) &= C_0 x(t) + C_p \left(p(t) \otimes x(t)\right) + D_0 u(t) + D_p \left(p(t) \otimes u(t)\right) + v(t), \end{aligned} \tag{7}$$

where $\otimes$ is the Kronecker product and

$$A_p = \begin{pmatrix} A_1 \; A_2 & \cdots & A_{n_p} \end{pmatrix}, \tag{8}$$

$$B_p = \begin{pmatrix} B_1 \; B_2 & \cdots & B_{n_p} \end{pmatrix}, \tag{9}$$

$$C_p = \begin{pmatrix} C_1 \; C_2 & \cdots & C_{n_p} \end{pmatrix}, \tag{10}$$

$$D_p = \begin{pmatrix} D_1 \; D_2 & \cdots & D_{n_p} \end{pmatrix}. \tag{11}$$

3. Discretisation of LPV systems

Consider the continuous-time LPV system (1)-(2). If both the input and scheduling signals are assumed to be constant between sampling instants, then the system may be described by [Tóth (2010)]

$$x(k+1) = e^{\mathscr{A}[p(k)]T_s} x(k) \\ +\mathscr{A}\left[p(k)\right]^{-1}\left(e^{\mathscr{A}[p(k)]T_s} - I_n\right)\mathscr{B}\left[p(k)\right]u(k), \tag{12}$$

$$y(k) = \mathscr{C}\left[p(k)\right]x + \mathscr{D}\left[p(k)\right]u(k) + v(k), \tag{13}$$

where

$$u(k) = u(kT_s),$$

$$p(k) = p(kT_s),$$

$$x(k) = x(kT_s),$$

$$y(k) = y(kT_s),$$

$$r(k) = r(kT_s).$$

If $v(t)$ is a limited bandwidth zero mean white noise process with bandwidth $\omega_m = \dfrac{\pi}{T_s}$, then $v(k)$ is a zero mean white noise sequence. In order to further analyse (12), we expand $e^{\mathscr{A}[p(k)]T_s}$ in a Taylor series

$$e^{\mathscr{A}[p(k)]T_s} = I_n + \mathscr{A}\left[p(k)\right]T_s + \frac{\left(\mathscr{A}\left[p(k)\right]T_s\right)^2}{2} + \cdots + \frac{\left(\mathscr{A}\left[p(k)\right]T_s\right)^i}{i!} + \dots,$$

and then (12) can be written as:

$$x(k+1) = e^{\mathscr{A}[p(k)]T_s} x(k) + \left(I_n + \cdots + \frac{\left(\mathscr{A}\left[p(k)\right]T_s\right)^i}{(i+1)!} + \dots\right)\mathscr{B}\left[p(k)\right]T_s u(k).$$

If $\mathscr{A}\left[p(k)\right]$ is an affine function of p,

$$\mathscr{A}(p) = A_0 + \sum_{j=1}^{n_p} A_j p_j,$$

then

$$x(k+1) = e^{A_0 T_s + \sum_{j=1}^{n_p} A_j p_j(k) T_s} x(k) + \\ \left(I_n + \cdots + \frac{\left(A_0 T_s + \sum_{j=1}^{n_p} A_j p_j(k) T_s\right)^i}{(i+1)!} + \dots\right)\mathscr{B}\left[p(k)\right]T_s u(k).$$

Next, we add and subtract $e^{A_0 T_s} x(k)$ to the right side of the last equation

$$x(k+1) = e^{A_0 T_s} x(k) + e^{A_0 T_s} \left(e^{\sum_{j=1}^{n_p} A_j p_j(k) T_s} - I_n \right) x(k) + \left(I_n + \cdots + \frac{\left(A_0 T_s + \sum_{j=1}^{n_p} A_j p_j(k) T_s\right)^i}{(i+1)!} + \dots \right) \mathscr{B}\left[p(k)\right] T_s u(k).$$

and we replace $e^{\sum_{j=1}^{n_p} A_j p_j(k) T_s}$ by its Taylor series expansion:

$$x(k+1) = e^{A_0 T_s} x(k) + e^{A_0 T_s} \left(\sum_{j=1}^{n_p} A_j T_s p_j(k) + \cdots + \frac{\left(\sum_{j=1}^{n_p} A_j T_s p_j(k)\right)^i}{i!} + \dots \right) x(k)$$
$$+ \left(I_n + \cdots + \frac{\left(A_0 T_s + \sum_{j=1}^{n_p} A_j p_j(k) T_s\right)^i}{(i+1)!} + \dots \right) \mathscr{B}\left[p(k)\right] T_s u(k).$$

If T_s is small, the solution can be approximated by

$$x(k+1) = e^{A_0 T_s} x(k) + e^{A_0 T_s} \left(\sum_{j=1}^{n_p} A_j T_s p_j(k) \right) x(k) + \mathscr{B}\left[p(k)\right] T_s u(k)$$

or, equivalently, by

$$x(k+1) = A_{d0} x(k) + A_{dp} \left(p(k) \otimes x(k)\right) + \mathscr{B}_d \left[p(k)\right] u(k),$$

where

$$\begin{aligned} A_{d0} &= e^{A_0 T_s}, \\ A_{dp} &= e^{A_0 T_s} \begin{pmatrix} A_1 T_s & \cdots & A_{n_p} T s \end{pmatrix} = A_{d0} A_p T_s, \\ \mathscr{B}_d(p) &= \mathscr{B}(p) T_s. \end{aligned}$$

If $\mathscr{B}(p)$ is also an affine function, i. e.,

$$\mathscr{B}(p) = B_0 + \sum_{j=1}^{n_p} B_j p_j,$$

then

$$x(k+1) = A_{d0} x(k) + A_{dp} \left(p(k) \otimes x(k)\right) + B_{d0} u(k) + B_{dp} \left(p(k) \otimes u(k)\right),$$

where

$$\begin{aligned} B_{d0} &= B_0 T_s, \\ B_{dp} &= \begin{pmatrix} B_1 T_s & \cdots & B_{n_p} T s \end{pmatrix}. \end{aligned}$$

We also consider affine dependencies for $\mathscr{C}(p)$ and $\mathscr{D}(p)$:

$$\mathscr{C}(p) = C_0 + \sum_{j=1}^{n_p} C_j p_j,$$

$$\mathscr{D}(p) = D_0 + \sum_{j=1}^{n_p} D_j p_j.$$

Finally, we end up with the following discrete-time LPV system

$$\begin{aligned} x(k+1) = A_{d0}x(k) + A_{dp}\left(p(k) \otimes x(k)\right) + B_{d0}u(k) \\ +B_{dp}\left(p(k) \otimes u(k)\right), \end{aligned} \tag{14}$$

$$\begin{aligned} y(k) = C_{d0}x(k) + C_{dp}\left(p(k) \otimes x(k)\right) + D_{d0}u(k) \\ +D_{dp}\left(p(k) \otimes u(k)\right) + v(k), \end{aligned} \tag{15}$$

where

$$\begin{aligned} C_{d0} &= C_0, \\ C_{dp} &= \begin{pmatrix} C_1 & \cdots & C_{n_p} \end{pmatrix}, \\ D_{d0} &= D_0, \\ C_{dp} &= \begin{pmatrix} D_1 & \cdots & D_{n_p} \end{pmatrix}. \end{aligned}$$

This system can be identified with the successive approximations algorithm [Lopes dos Santos *et al.* (2008b)] with B_{dp} restricted to $A_{d0}\begin{pmatrix} B_{d1} & \cdots & B_{dn_p} \end{pmatrix}$. The successive approximations algorithm is described in the following section.

4. Successive approximations identification algorithm

The successive approximations identification algorithm is an iterative process where the bilinear signals $A_{dp}\left(p(k) \otimes x(k)\right)$ and $C_{dp}\left(p(k) \otimes x(k)\right)$ are seen as process and measurement noises, respectively, at the first iteration. As a result, system (14)-(15) can be modeled as the LTI system

$$x(k+1) = A_{d0}x(k) + B_{g_0}u_{g_0}(k) + w^{(0)}(k), \tag{16}$$

$$y(k) = C_0 x(k) + D_{g_0}u_{g_0}(k) + v^{(0)}(k), \tag{17}$$

with

$$u_{g_0}(k) = \begin{pmatrix} u(k) \\ p(k) \otimes u(k) \end{pmatrix},$$

$$
\begin{aligned}
w^{(0)}(k) &= A_{dp}\left(p(k)\otimes x(k)\right),\\
v^{(0)}(k) &= C_p\left(p(k)\otimes x(k)\right) + v(k),\\
B_{g_0} &= \begin{pmatrix} B_{d0} & B_{dp} \end{pmatrix},\\
D_{g_0} &= \begin{pmatrix} D_0 & D_p \end{pmatrix}.
\end{aligned}
$$

At the first iteration, this LTI system is identified by a deterministic-stochastic subspace identification algorithm such as the N4SID [Overschee and Moor (1996)], the MOESP [Verhaegen and Dewilde (1992a,b)] or the CVA [Larimore (1990)]. The obtained state sequence is estimated by a Kalman filter which is then used to estimate the bilinear signal $p(k)\otimes x(k)$. In the next iteration the LPV system is described by the LTI model

$$x(k+1) = A_{d0}x(k) + B_e u_e^{(1)}(k) + w^{(1)}(k), \tag{18}$$

$$y(k) = C_{d0}x(k) + C_e u(k) e^{(1)}(k) + v^{(1)}(k), \tag{19}$$

with

$$u_e^{(1)}(k) = \begin{pmatrix} u_{g_0}(k) \\ p(k)\otimes \hat{x}^{(0)}(k) \end{pmatrix}, \tag{20}$$

$$w^{(1)}(k) = A_p\left[p(k)\otimes \tilde{x}^{(0)}(k)\right], \tag{21}$$

$$v^{(1)}(k) = C_p\left[p(k)\otimes \tilde{x}^{(0)}(k)\right] + v(k), \tag{22}$$

and

$$B_e = \begin{pmatrix} B_{g_0} & A_p \end{pmatrix}, \tag{23}$$

$$D_e = \begin{pmatrix} B_{g_0} & C_p \end{pmatrix}. \tag{24}$$

Here $\hat{x}^{(0)}(k)$ and $\tilde{x}^{(0)}(k)$ denote the state and state error estimate. This model is also estimated by a deterministic-stochastic subspace identification algorithm, and the process is repeated until convergence. If $p(k)$ is a zero mean white noise sequence independent of $u(k)$, the signals $p(k)\otimes x(k)$ and $p(k)\otimes\tilde{x}^{(i)}(k)$, $i = 0, 1, \ldots$ are second order zero mean white noise sequences. Under this condition the algorithm converges if, for all iterations, the estimated models fulfill the stationary condition at every iteration [Lopes dos Santos *et al.* (2007, 2008a,b, 2009)] if

$$\rho\left(A_{d0}^{(i)}\otimes A_{d0}^{(i)} + (R_p)_{j\ell}\sum_{j=1}^{s}\sum_{\ell=1}^{s} A_{d\ell}^{(i)}\otimes A_{dj}^{(i)}\right) < 1,$$

where $\rho(M)$ is the spectral radius of M and $(R_p)_{j\ell}$ is the $j\ell^{th}$ entry of matrix

$$R_p = \lim_{N\to\infty} \frac{1}{N} \sum_{k=0}^{N-1} u(k)u^T(k).$$

Unfortunately, no proof exists that the process converges for general scheduling parameter sequences. However, convergence has been observed for several non-white noise scheduling sequences suggesting that this algorithm could converge for general scheduling sequences as well.

As we have seen in the previous section, the continuous-time system must be sampled at a high sampling rate in order to preserve the affine parameter dependence in the LPV discrete-time model. On the other hand, discrete-time system identification algorithms do not cope very well with high sampling rates due to the clustering of the poles and zeros around $1 + j0$, which induces serious numerical problems. In the next section, we present a new subspace LTI discrete-time system identification algorithm that uses a downsampling approach to circumvent this problem.

5. Downsampled LTI discrete-time deterministic-stochastic subspace identification algorithm

Consider

$$x(k+1) = A_d x(k) + B_d u(k) + w(k), \tag{25}$$

$$y(k) = C_d x(k) + D_d u(k) + v(k) \tag{26}$$

as a discretised model of a continuous-time LTI system using zero order hold (ZOH) and a constant sampling period T_s.

We assume that the process and measurement noises $w(k) \in \mathbb{R}^n$ and $v(k) \in \mathbb{R}^{n_y}$ are zero mean white noise sequences, independent of both the input and the scheduling signal, $u(k)$ and $p(k)$, with covariance

$$\mathbb{E}\left\{ \begin{pmatrix} w(k) \\ v(k) \end{pmatrix} \begin{pmatrix} w^T(k) & v^T(k) \end{pmatrix} \right\} = \begin{pmatrix} R_w & R_{wv} \\ R_{wv}^T & R_v \end{pmatrix}, \tag{27}$$

where $R_w \in \mathbb{R}^{n\times n}$, $R_{wv} \in \mathbb{R}^{n\times n_y}$ and $R_v \in \mathbb{R}^{n_y\times n_y}$. It is our objective to estimate the innovation model

$$\begin{aligned} x(k+1) &= A_d x(k) + B_d u(k) + Ke(k), \\ x(k) &= C_d x(k) + D_d u(k) + e(k), \end{aligned}$$

where $e(k)$ is a white noise sequence, using a subspace deterministic-stochastic identification algorithm.

To overcome some of the problems due to fast sampling [Sung *et al.* (2001)], we derive a low rate sampling discrete-time model. Consider the same problem as before but with a larger sampling time, i.e., mT_s, $m > 1$. Thence

$$x\left[(k+1)m\right] = A_d^m x(km) + \sum_{i=0}^{m-1} A_d^{m-i-1} B_d \, u(km+i) + \sum_{i=0}^{m-1} A_d^{m-i-1} w(km+i). \tag{28}$$

Observe that we have m shifted sequences of the same type:

$$\begin{aligned} x\left[((k+1)m+\ell)\right] = A_d^m x(km+\ell) &+ \sum_{i=0}^{m-1} A_d^{m-i-1} B_d \, u(km+\ell+i) \\ &+ \sum_{i=0}^{m-1} A_d^{m-i-1} w(km+\ell+i), \end{aligned} \tag{29}$$

with $\ell = 0, \ldots, m-1$.

For a more compact notation define

$$x_m\left(k,\ell\right) := x\left(km+\ell\right), \tag{30}$$

$$u_m(k,\ell) := \begin{pmatrix} u(km+\ell) \\ u(km+1+\ell) \\ \cdots \\ u\left((k+1)m-1+\ell\right) \end{pmatrix}, \tag{31}$$

$$y_m\left(k,\ell\right) := y\left(km+\ell\right), \tag{32}$$

$$w_m(k,\ell) := \begin{pmatrix} w(km+\ell) \\ w(km+1+\ell) \\ \cdots \\ w\left((k+1)m-1+\ell\right) \end{pmatrix}, \tag{33}$$

$$v_m\left(k,\ell\right) := v\left(km+\ell\right) \tag{34}$$

and

$$A_m := A_d^m, \tag{35}$$

$$B_{m_i} := A_d^{m-i-1} B_d, \tag{36}$$

$$B_m := \left(B_{m_0} \cdots B_{m_{m-1}} \right), \tag{37}$$

$$C_m := C_d, \tag{38}$$

$$D_m := \begin{cases} D_d \ , m = 1 \\ \left(D_d \ 0 \cdots 0 \right) , m > 1. \end{cases} \tag{39}$$

Hence, we end up with the following state-space model:

$$x_m(k+1,\ell) = A_m x_m(k,\ell) + B_m u_m(k,\ell) + w_m(k,\ell), \tag{40}$$

$$y_m(k,\ell) = C_m x_m(k,\ell) + D_m u_m(k,\ell) + v_m(k,\ell), \tag{41}$$

where

$$\mathbb{E}\left\{ \begin{pmatrix} w_m(k,\ell) \\ v_m(k,\ell) \end{pmatrix} \left(w_m^T(k,\ell) \ v_m^T(k,\ell) \right) \right\} = \begin{pmatrix} R_{w_m} & R_{w_m v_m} \\ R_{w_m v_m}^T & R_{v_m} \end{pmatrix} \tag{42}$$

with

$$\begin{aligned} R_{w_m} &= A_d^{m-1} R_w \left(A_d^{m-1} \right)^T + A_d^{m-2} R_w \left(A_d^{m-2} \right)^T \\ &\quad + \cdots + A_d R_w A_d^T + R_w, \end{aligned} \tag{43}$$

$$R_{w_m v_m} = A_d^{m-1} R_{wv}, \tag{44}$$

$$R_{v_m} = R_v. \tag{45}$$

Solving (40)–(41) for $k = 0, 1, \ldots, i-1$, we have

$$\mathbf{y}_i(0,\ell) = \Gamma_{mi} x_m(0,\ell) + \mathscr{H}_i \mathbf{u}_i(0,\ell) + \mathbf{y}_i^s(0,\ell), \tag{46}$$

where

$$\mathbf{y}_i(0,\ell) = \left(y_m^T(0,\ell) \ y_m^T(1,\ell) \ \cdots \ y_m^T(i-1,\ell) \right)^T,$$

$$\mathbf{u}_i(0,\ell) = \left(u_m^T(0,\ell) \ u_m^T(1,\ell) \ \cdots \ u_m^T(i-1,\ell) \right)^T,$$

are input and output sequences, respectively, $\mathbf{y}_i^s(0,\ell)$ is the (stochastic) component of $\mathbf{y}_i(0,\ell)$ due to noises $w_m(0,\ell), \ \ldots, \ w_m(i-2,\ell)$ and

$v_m(0,\ell),\ \ldots, v_m(i-1,\ell)$,

$$\Gamma_{mi} := \begin{pmatrix} C_m \\ C_m A_m \\ C_m A_m^2 \\ \cdots \\ C_m A_m^{i-1} \end{pmatrix} \tag{47}$$

is the extended observability matrix and $\mathscr{H}_i$ is the Töplitz matrix

$$\mathscr{H}_i := \begin{pmatrix} D_m & 0 & \cdots & 0 & 0 \\ C_m B_m & D_m & \cdots & 0 & 0 \\ C_m A_m B_m & C_m B_m & \cdots & 0 & 0 \\ \vdots & & & & \\ C_m A_m^{i-2} B_m & \cdots & \cdots & C_m B_m & D_m \end{pmatrix}.$$

Equation (46) also holds if we replace $\mathbf{y}_i(0,\ell)$, $\mathbf{y}_i^s(0,\ell)$, $\mathbf{u}_i(0,\ell)$ and $x_m(0,\ell)$ by $Y_p = Y_{0|i-1}$, $Y_p^s = Y_{0|i-1}^s$, $U_p = U_{0|i-1}$ and $X_p = X_0$, respectively, where

$$\begin{aligned} Y_{0|i-1} &= \left(\mathbf{y}_{i\times m}(0)\ \mathbf{y}_{i\times m}(1) \cdots \mathbf{y}_{i\times m}(j-1)\right), \\ Y_{0|i-1}^s &= \left(\mathbf{y}_{i\times m}^s(0)\ \mathbf{y}_{i\times m}^s(1) \cdots \mathbf{y}_{i\times m}^s(j-1)\right), \\ U_{0|i-1} &= \left(\mathbf{u}_{i\times m}(0)\ \mathbf{u}_{i\times m}(1) \cdots \mathbf{u}_{i\times m}(j-1)\right), \\ X_0 &= \left(\mathbf{x}_m(0)\ \mathbf{x}_m(1) \ldots \mathbf{x}_m(j-1)\right), \end{aligned} \tag{48}$$

with

$$\begin{aligned} \mathbf{y}_{i\times m}(k) &= \left(\mathbf{y}_i(k,0)\ \mathbf{y}_i(k,1) \cdots \mathbf{y}_i(k,m-1)\right), \\ \mathbf{y}_{i\times m}^s(k) &= \left(\mathbf{y}_i^s(k,0)\ \mathbf{y}_i^s(k,1) \cdots \mathbf{y}_i^s(k,m-1)\right), \\ \mathbf{u}_{i\times m}(k) &= \left(\mathbf{u}_i(k,0)\ \mathbf{u}_i(k,1) \cdots \mathbf{u}_i(k,m-1)\right), \\ \mathbf{x}_m(k) &= \left(x_m(k,0)\ x_m(k,1) \cdots x_m(k,m-1)\right). \end{aligned}$$

In a similar manner, equation (46) also holds for $Y_f = Y_{i|2i-1}$, $Y_f^s = Y_{i|2i-1}^s$, $U_f = U_{i|2i-1}$ and $X_f = X_i$, that is,

$$Y_f \ = \ \Gamma_i X_f \ + \ \mathscr{H}_i U_f + Y_f^s.$$

On the other hand, we can easily see that

$$X_f \ = \ A_m^i X_p \ + \ \mathscr{C}_i^R U_p + \ \mathscr{C}_i^{QR} Q_p,$$

where $\mathscr{C}_i^R$, $\mathscr{C}_i^{QR}$ are the reversed extended accessibility matrices

$$\mathscr{C}_i^R = \left(A_m^{i-1} B_m \cdots A_m B_m\ B_m\right), \tag{49}$$

$$\mathscr{C}_i^{QR} = \left(A_m^{i-1} \; \cdots \; A_m \; I_n \right), \tag{50}$$

and Q_p is defined in a similar manner as U_p with $u(\cdot)$ replaced by $w(\cdot)$.

From these equations, we can estimate the discrete-time state-space model parameters, A_m, B_m, C_m, D_m, using a subspace identification algorithm such as the N4SID [Overschee and Moor (1996)], the MOESP [Verhaegen and Dewilde (1992b,a)], or the CVA [Larimore (1990)]. These algorithms estimate the extended observability matrix Γ_{mi} defined in (47) as explained in [Overschee and Moor (1996)]. That is,

$$\Gamma_{mi} = W_1^{-1} U_1 S_1^{1/2} \tag{51}$$

with U_1 and V_1 being defined by the singular value decomposition

$$W_1 \mathscr{O}_i W_2 = \left(U_1 \; U_2 \right) \begin{pmatrix} S_1 & 0 \\ 0 & S_2 \end{pmatrix} \begin{pmatrix} V_1^T \\ V_2 \end{pmatrix} \tag{52}$$

where the singular values in S_2 are significantly smaller than the ones in S_1 ($S_2 = 0$ if there is no noise) and

$$\mathscr{O}_i = Y_f /_{U_f} W_p. \tag{53}$$

Here $A/_C B$ denotes the oblique projection of the row space of A along the row space of B on the row space of C and

$$W_p = \begin{pmatrix} U_p \\ Y_p \end{pmatrix}.$$

The main difference between these algorithms resides on the used weighting matrices W_1 and W_2:

- CVA: $\begin{cases} W_1 = \left(Y_f \Pi_{U_f^\perp} Y_f^T \right)^{-1/2} \\ W_2 = \Pi_{U_f^\perp} \end{cases}$

- MOESP: $\begin{cases} W_1 = I_{n_y i} \\ W_2 = \Pi_{U_f^\perp} \end{cases}$

- N4SID $\begin{cases} W_1 = I_{n_y i} \\ W_2 = I_j \end{cases}$

where

$$\Pi_{U_f^\perp} = I_j - U f^T \left(U_f U_f^T \right)^\dagger U_f.$$

A_m and C_m may be computed by exploiting the shift invariance of the extended observability matrix, i.e, making C_m equal to the n_y first rows of Γ_{im} and $A_m = \overline{\Gamma}^{\dagger}_{mi}\underline{\Gamma}_{mi}$, where $\overline{\Gamma}_{mi}$ and $\underline{\Gamma}_{mi}$ are built from Γ_{mi} by removing the first and the last n_y rows, respectively. Alternatively, these matrices are part of the least squares solution of

$$\left(\begin{array}{c} \underline{\Gamma}^{\dagger}_{mi}\mathscr{Z}_{i+1} \\ \hline Y_{i|i} \end{array}\right) = \left(\begin{array}{c} A_m \\ \hline C_m \end{array}\right) \Gamma^{\dagger}_{mi}\mathscr{Z}_i + \mathscr{K} U_f + \left(\begin{array}{c} \rho_w \\ \hline \rho_v \end{array}\right) \tag{54}$$

with

$$\mathscr{K} = \left(\left(\begin{array}{c|c} B_m & \underline{\Gamma}^{\dagger}_{mi}\mathscr{H}_{i-1} \\ \hline D_m & 0 \end{array}\right) - \left(\begin{array}{c} A_m \\ \hline C_m \end{array}\right) \Gamma^{\dagger}_{mi}\mathscr{H}_i \right)$$

and

$$\mathscr{Z}_i = Y_f / \begin{pmatrix} W_p \\ U_f \end{pmatrix}, \tag{55}$$

$$\mathscr{Z}_{i+1} = Y_f^- / \begin{pmatrix} W_p^+ \\ U_f^- \end{pmatrix}, \tag{56}$$

$$\tag{57}$$

where A/B denotes the orthogonal projection of the rowspace of A on the rowspace of B, $W_p^+ = \begin{pmatrix} U_{0|i+1} \\ Y_{0|i+1} \end{pmatrix}$, $Y_f^- = Y_{i+1|2i-1}$ and $U_f^- = U_{i+1|2i-1}$. From (38), we have that

$$C_d = C_m$$

and from (35)

$$A_d = e^{\frac{1}{m}\ln A_m}.$$

Notice that $\mathscr{K}$, the other solution of (54), is a linear function of B_m and D_m and, consequently, of A_d and B_d. As a result, these matrices may be estimated by a least squares estimator. To formulate this estimator let us define

$$\mathscr{M} = \left(\begin{array}{c|c} 0 & \underline{\Gamma}^{\dagger}_{mi} \\ \hline I_{n_y} & 0 \end{array}\right) - \left(\begin{array}{c} A_m \\ \hline C_m \end{array}\right) \Gamma^{\dagger}_{mi} \in \mathbb{R}^{(n+n_y)\times n_y i}.$$

Partition $\mathscr{K}$ and $\mathscr{M}$ as

$$\mathscr{K} = \left(\mathscr{K}_{1,1} \cdots \mathscr{K}_{1,m}\ \mathscr{K}_{2,1} \cdots \mathscr{K}_{2,m} \cdots \mathscr{K}_{i,1} \cdots \mathscr{K}_{i,m} \right), \tag{58}$$

$$\mathscr{M} = \left(\mathscr{M}_1\ \mathscr{M}_2 \cdots \mathscr{M}_i \right), \tag{59}$$

with $\mathscr{K}_{\ell,q} \in \mathbb{R}^{(n+n_y)\times n_u}$, $\ell = 1,\ldots,i$, $q = 1,\ldots,m$ and $\mathscr{M}_\ell \in \mathbb{R}^{(n+n_y)\times n_y}$, $\ell = 1,\ldots,i$. Now, if we define

$$\mathscr{N}_1 = \begin{pmatrix} \mathscr{M}_1 \\ \mathscr{M}_2 \\ \cdots \\ \mathscr{M}_i \end{pmatrix} \in \mathbb{R}^{(n+n_y)i\times n_y} \tag{60}$$

and

$$\mathscr{N}_2 = \begin{pmatrix} \mathscr{M}_2 & \mathscr{M}_3 & \cdots & \mathscr{M}_{i-1} & \mathscr{M}_i \\ \mathscr{M}_3 & \mathscr{M}_4 & \cdots & \mathscr{M}_i & 0 \\ \vdots & \vdots & \vdots & \vdots & \vdots \\ \mathscr{M}_i & 0 & \cdots & 0 & 0 \\ 0 & 0 & \cdots & 0 & 0 \end{pmatrix} \underline{\Gamma}_{mi} \in \mathbb{R}^{(n+n_y)i\times n}. \tag{61}$$

We end up with

$$\begin{pmatrix} \mathscr{K}_{1,1} \\ \vdots \\ \mathscr{K}_{i,1} \\ \mathscr{K}_{1,2} \\ \vdots \\ \mathscr{K}_{i,2} \\ \vdots \\ \mathscr{K}_{1,m} \\ \vdots \\ \mathscr{K}_{i,m} \end{pmatrix} = \left(\begin{array}{c|c} \mathscr{N}_1 & \mathscr{N}_2 A_d^{m-1} \\ 0 & \mathscr{N}_2 A_d^{m-2} \\ \vdots & \vdots \\ 0 & \mathscr{N}_2 \end{array}\right) \left(\frac{D_d}{B_d}\right), \tag{62}$$

and $\left(\frac{D_d}{B_d}\right)$ is the least-squares solution of this equation.

In [Overschee and Moor (1996)], it is shown that $\mathscr{Z}_i$, as defined in (55), is given by

$$\mathscr{Z}_i = \Gamma_{mi}\hat{X}_i + \mathscr{H}_i U_f$$

where $\hat{X}_i$ is Kalman filter state sequence. If $\Gamma_{mi}^{\perp} \in \mathbb{R}^{n_y i\times(n_y i-n)}$ spans the orthogonal complement of the column-space of Γ_{mi}, then

$$\left(\Gamma_{mi}^{\perp}\right)^T \mathscr{Z}_i = \left(\Gamma_{mi}^{\perp}\right)^T \mathscr{H}_i U_f.$$

Now we can define

$$\mathscr{K} = \left(\Gamma_{mi}^{\perp}\right)^T \mathscr{H}_i = \left(\Gamma_{mi}^{\perp}\right)^T \mathscr{Z}_i U_f^{\dagger},$$
$$\mathscr{M} = \left(\Gamma_{mi}^{\perp}\right)^T$$

and equation (62) holds for $\mathscr{K}_{\ell,q}$, $\mathscr{M}_\ell$, N_1 and N_2 defined in (58)-(61) with $\begin{pmatrix} D_d \\ \hline B_d \end{pmatrix}$ being its solution in the least squares sense.

Moreover, R_{w_m}, $R_{w_m v_m}$ and R_{v_m} can be estimated from

$$\begin{pmatrix} R_{w_m} & R_{w_m v_m} \\ R_{w_m v_m}^T & R_{v_m} \end{pmatrix} = \frac{1}{jm} \begin{pmatrix} \rho_w \rho_w^T & \rho_w \rho_v^T \\ \rho_v \rho_w^T & \rho_v \rho_v^T \end{pmatrix}$$

where ρ_w and ρ_v are the residuals of the least squares problem defined in (54) and R_w, R_{wv} and R_v can be recovered by solving (43)-(45). Finally, the Kalman filter gain is given by

$$K = (A_d P C_d^T + R_{wv})(C_d P C_d^T + R_v)^{-1},$$

where P is the solution of the Ricatti equation

$$P = A_d P A_d^T + R_w - (A_d P C_d^T + R_{wv})(C_d P C_d^T + R_v)^{-1}(A_d P C_d^T + R_{wv})^T.$$

6. Case study

In order to test the algorithm, the LPV system

$$\begin{aligned} \dot{x}(t) &= A_0 x(t) + A_1 \left(p(t) \otimes x(t)\right) + B_0 u(t) + B_1 \left(p(t) \otimes u(t)\right), \\ y(t) &= C_0 x(t) + C_1 \left(p(t) \otimes x(t)\right) + D_0 u(t) + D_1 \left(p(t) \otimes u(t)\right) + v(t), \end{aligned} \tag{63}$$

with

$$\begin{aligned} A_0 &= \begin{pmatrix} -1 & -10 \\ 10 & -1 \end{pmatrix}, & A_1 &= \begin{pmatrix} 0 & -5 \\ 5 & 0 \end{pmatrix}, \\ B_0 &= \begin{pmatrix} 10 \\ 10 \end{pmatrix}, & B_1 &= \begin{pmatrix} 1 \\ -1 \end{pmatrix}, \\ C_0 &= \begin{pmatrix} 2 & 2 \end{pmatrix}, & C_1 &= \begin{pmatrix} 0.1 & -0.1 \end{pmatrix}, \\ D_0 &= \begin{pmatrix} 0 \end{pmatrix}, & D_1 &= \begin{pmatrix} 0 \end{pmatrix}. \end{aligned}$$

was simulated using a pseudo random binary sequence (PRBS) with 10 stages and a switching time $T_{sw} = 7T_s$ as input and a piecewise constant scheduling sequence, i.e., $p(t) = p_k$ for $kT_s \leq t < (k+1)T_s$. Two sets of simulations were performed. In the first set, p_k was a unit variance zero mean gaussian white noise sequence and, in the second, $p(t)$ was a zero mean square wave with period $T = 1$, duty cycle 50% and amplitude 1. The data was collected with a sampling period $T_s = 10^{-3}$ seconds.

6.1. *Monte Carlo simulations with a white noise scheduling signal*

First, the system was simulated with with no noise and the LPV identification algorithm was executed without downsampling (m=1). Figure 1, top, shows a logarithm representation ($\log \sigma_i - \log \sigma_{15}$) of the singular values

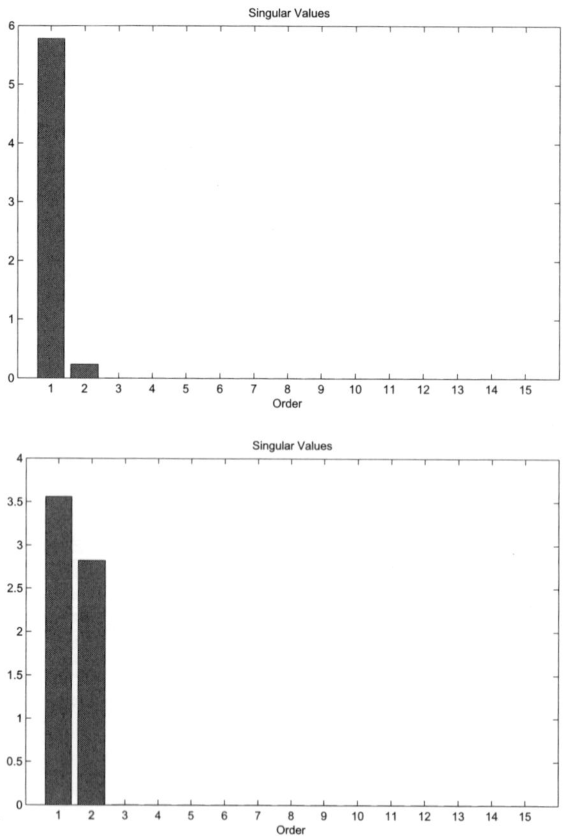

Fig. 1. Top: model order determination for $m = 1$. Bottom: model order determination for $m = 20$.

of $W_1 \mathcal{O}_i W_2$ in the first iteration. It can be seen that the second singular value is very small, indicating that a second order model might not be a robust choice for the identified LPV model. In fact, the algorithm crashed with m=1 due to the unobservability of the estimated model at the first iteration and it only converged with $m > 3$. The best results were obtained with downsampling factors around 20. Figure 1, bottom, shows the singular values for m=20, evincing a second order model as the obvious choice.

To evaluate the performance of the algorithm under noisy measurements, we run two Monte Carlo Simulations (MCS), each consisting in 100 experiments with SNR=20 dB and SNR=0 dB.

The quality of the models was assessed using the percentage simulation error (PSE), defined as:

$$PSE = \sqrt{\frac{\sum_{k=1}^{N} (y_{det}(k) - y_s(k))^2}{\sum_{k=1}^{N} y_{det}^2(k)}} \times 100, \tag{64}$$

where N is the length of the simulated input-output data record, y_{det} the output generated by the true model and y_s is the output generated by the estimated model.

The iterations were limited to 20 in every simulation in order to save simulation time. Figure 2 displays scatter plots of the eigenvalues of the estimated

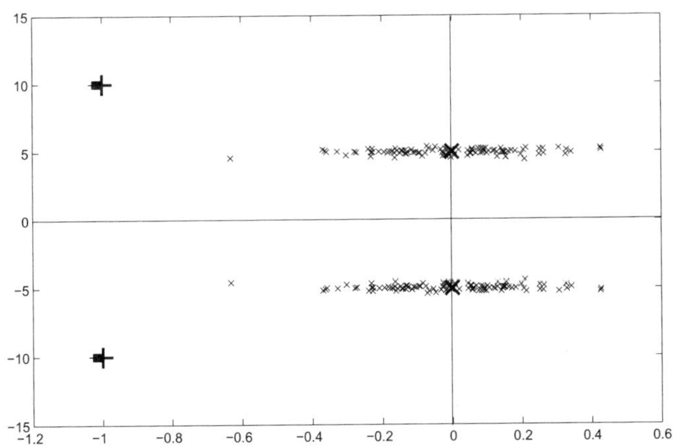

Fig. 2. White noise scheduling sequence and SNR=20 dB. + represents the estimated eigenvalues of A_0, × the estimated eigenvalues of A_1, $+$ the true eigenvalues of A_0 and $\times$ the true eigenvalues of A_1.

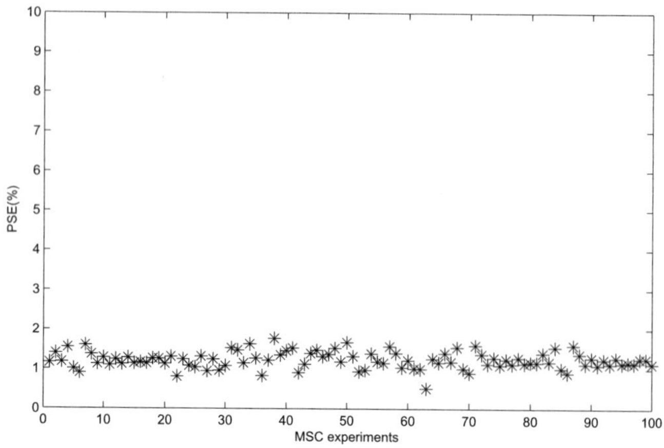

Fig. 3. PSE with a white noise scheduling sequence and SNR=20 dB.

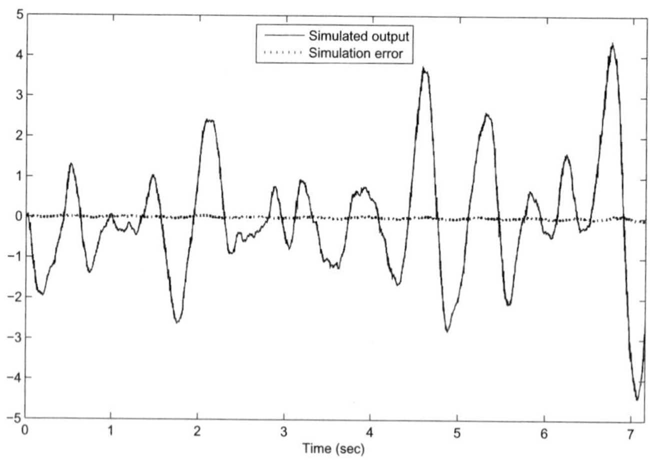

Fig. 4. Simulated output of a randomly chosen estimated models and the respective simulation error with a white noise scheduling sequence and SNR=20 dB.

A_0 and A_1 matrices for SNR=20dB. We can see that the estimated values concentrate around the true value. In Figure 3, we can see the PSE for all MCS. Notice that in every simulation it remains very small, indicating therefore a good accuracy of every estimated model. In Figure 4, the simulated output of a randomly chosen estimated model is compared with its simulation error.

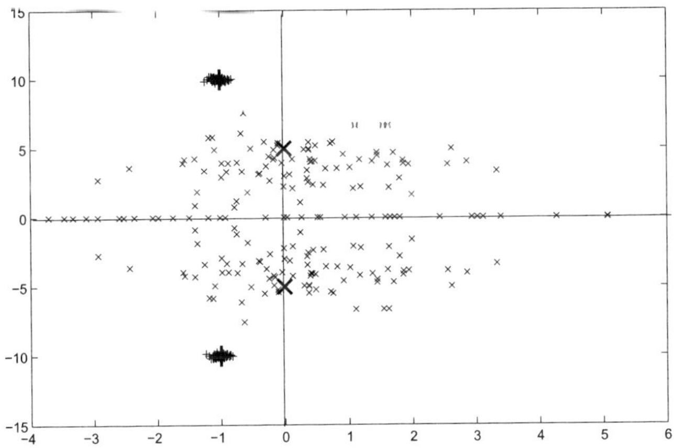

Fig. 5. White noise scheduling sequence and SNR=0 dB. + represents the estimated eigenvalues of A_0, $\times$ the estimated eigenvalues of A_1, $+$ the true eigenvalues of A_0 and $\times$ the true eigenvalues of A_1.

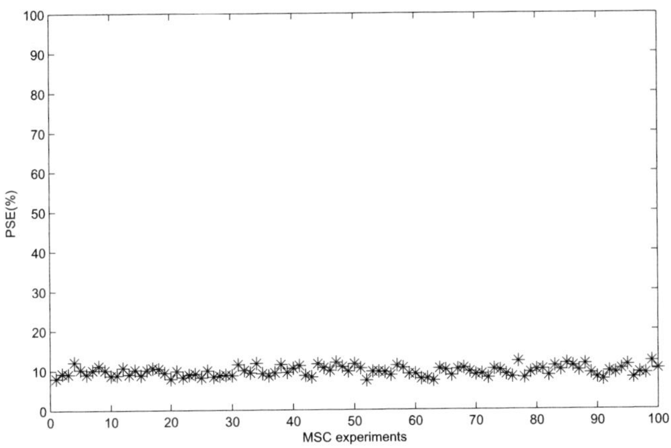

Fig. 6. PSE with a white noise scheduling sequence and SNR=0 dB.

Figure 5 exhibits the scatter plots of the eigenvalues of the estimated A_0 and A_1 matrices for SNR=0dB. As expected, the eigenvalues appear much more scattered, specially those of A_1. However, from the PSE plot, in Figure 6, we can see that the accuracy of the estimated models still remains within acceptable limits. This can be confirmed by Figure 7 where the simulated output of a randomly chosen estimated model is compared with its simulation error.

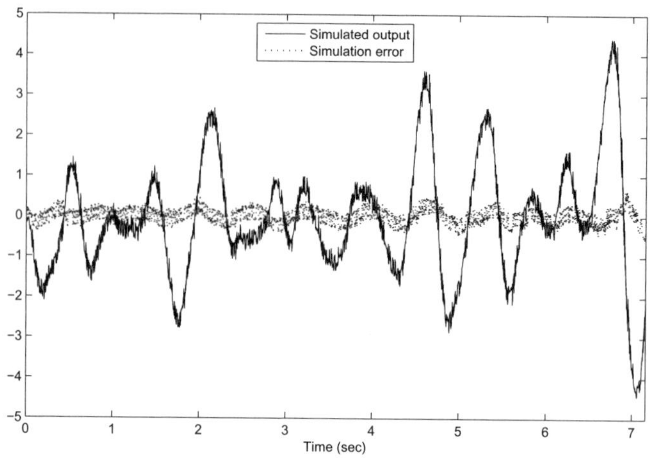

Fig. 7. Simulated output of a randomly chosen estimated models and the respective simulation error with a white noise scheduling sequence and SNR=0 dB.

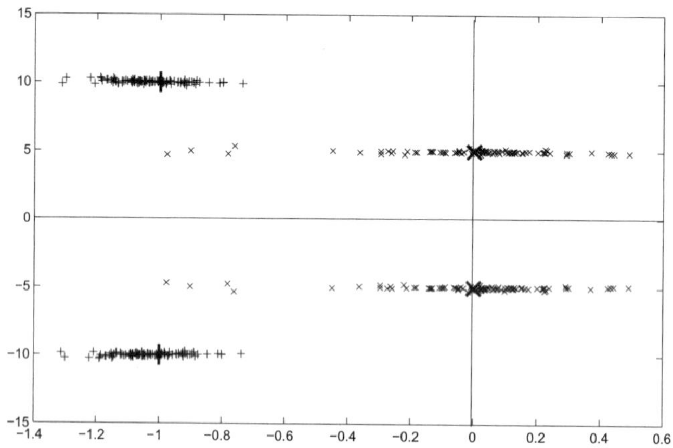

Fig. 8. Square wave scheduling sequence and SNR=20 dB. + represents the estimated eigenvalues of A_0, $\times$ the estimated eigenvalues of A_1, $+$ the true eigenvalues of A_0 and $\times$ the true eigenvalues of A_1.

6.2. *Monte Carlo simulations with a square wave scheduling signal*

The MCS were repeated, but with SNR=20dB and SNR=10dB[a].

[a]For SNR=0dB the algorithm did not converge due to instability of the estimated models.

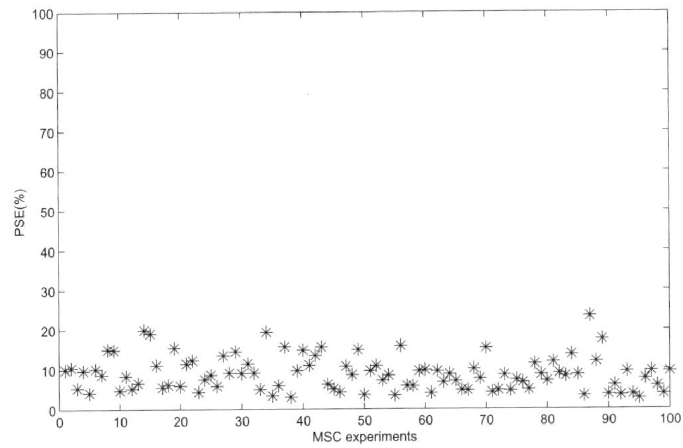

Fig. 9. PSE with a square wave scheduling sequence and SNR=20 dB.

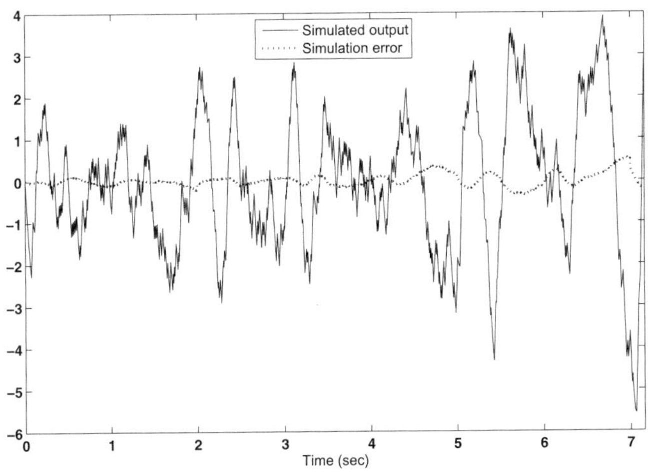

Fig. 10. Simulated output of a randomly chosen estimated models and the respective simulation error with a square wave scheduling sequence and SNR=20 dB.

In Figure 8, we can observe a higher dispersion of the estimated eigenvalues for SNR=20dB, accentuated for SNR=10dB as can be seen in Figure 11. This leads to a loss of precision, confirmed by PSE plots in Figures 9 and 12 and highlighted in Figures 10 and 13.

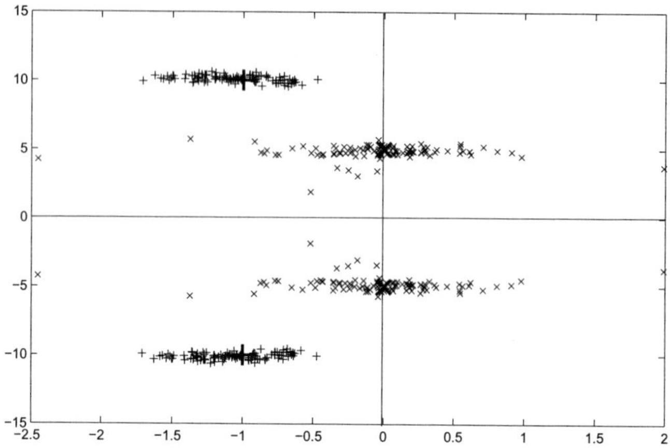

Fig. 11. Square wave scheduling sequence and SNR=10 dB. + represents the estimated eigenvalues of A_0, $\times$ the estimated eigenvalues of A_1, $+$ the true eigenvalues of A_0 and $\times$ the true eigenvalues of A_1.

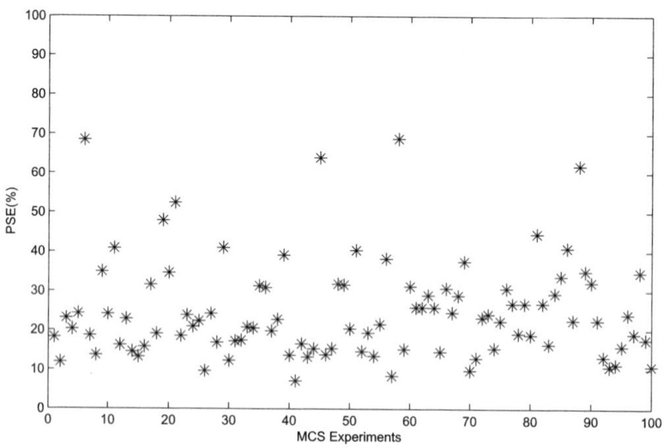

Fig. 12. PSE with a square wave scheduling sequence and SNR=10 dB.

7. Conclusion

A new subspace identification method for continuous-time LPV systems with affine parameter dependence has been proposed. This new approach is based on the successive approximations algorithm previously developed for discrete-time LPV systems. This algorithm identifies a discrete-time model which is next converted into an equivalent continuous one. In order to

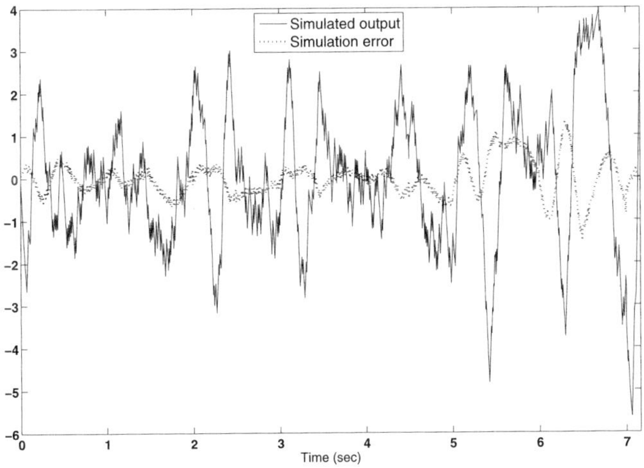

Fig. 13. Simulated output of a randomly chosen estimated models and the respective simulation error with a square wave scheduling sequence and SNR=10 dB.

preserve the affine parameter dependence in the LPV discrete-time model, a high sampling rate is required. But high sampling rates give rise to numerical problems due to the clustering of the zeros and poles around $1 + j0$. A trade-off solution was found through downsampling. This is done by partitioning the data set into m subsets, where m is the downsampling factor. Next, the data sets are combined into a single set and the resulting discrete model is identified using a subspace algorithm.

The proposed method is illustrated in the identification of a second order LPV system with two complex poles. For a white noise scheduling parameter, the estimated models where very close to the true ones even in very extreme noise conditions. For a square wave scheduling signal, the estimated models were also acceptable although with a significant accuracy loss.

References

Corno, M., Tanelli, M., Savaresi, S. M. and Fabbri, L. (2011). Design and validation of a gain scheduled controller for the electronic throttle body in ride-by-wire racing motorcycles, *IEEE Transactions on Control Systems Technology* **19**, 1, pp. 18–30.

De Caigny, J., Camino, J. F. and Swevers, J. (2008). Identification of MIMO LPV models based on interpolation, in *Proceedings of International Conference on Noise and Vibration Engineering* (Leuven, Belgium).

De Caigny, J., Camino, J. F. and Swevers, J. (2011). Interpolation-based modelling of MIMO LPV systems, *IEEE Transactions on Control Systems Technology* **19**, 1, pp. 46–63.

Ferreres, G. (2011). Computation of a flexible aircraft LPV/LFT model using interpolation, *IEEE Transactions on Control Systems Technology* **19**, 1, pp. 132–139.

Gaspar, P., Szabó, Z. and Bokor, J. (2006). Continuous-time parameter identification for LPV models using adaptive observers robust control design, in *Proceedings of the IFAC Symposium on Robust Control Design* (Toulouse, France).

Hu, Y. and Yurkovitch, S. (2011). Linear parameter varying battery model identification using subspace methods, *Journal of Power Sources* **196**, pp. 2913–2923.

Larimore, W. E. (1990). Canonical variate analysis in identification, filtering, and adaptive control, in *Proceedings of the IEEE Conference on Decision and Control*, Vol. 2 (Honolulu, HI, USA).

Lopes dos Santos, P., Azevedo Perdicoúlis, T.-P., Ramos, J. A., Jank, G. and Martins de Carvalho, J. L. (2011a). Indirect continuous-time system identification—a subspace downsampling approach, in *Submitted to the Conference on Decision and Control* (Orlando, FL, USA).

Lopes dos Santos, P., Azevedo Perdicoúlis, T.-P., Ramos, J. A., Martins de Carvalho, J. L., Jank, G. and Milhinhos, J. (2011b). An LPV modeling and identification approach to leakage detection in high pressure natural gas transportation networks, *IEEE Transactions on Control Systems Technology* **19**, 1, pp. 77–92.

Lopes dos Santos, P., Ramos, J. A. and Martins de Carvalho, J. L. (2007). Identification of linear parameter varying systems using an iterative deterministic-stochastic subspace approach, in *Proceedings of the European Control Conference* (Kos, Greece).

Lopes dos Santos, P., Ramos, J. A. and Martins de Carvalho, J. L. (2008a). Identification of LPV systems using successive approximations, in *Proceedings of the IEEE Conference on Decision and Control* (Cancún, Mexico).

Lopes dos Santos, P., Ramos, J. A. and Martins de Carvalho, J. L. (2008b). Subspace identification of linear parameter varying systems with innovation-type noise model driven by general inputs and a measurable white noise time varying parameter vector, *International Journal of Systems Science* **39**, 9, pp. 897–911.

Lopes dos Santos, P., Ramos, J. A. and Martins de Carvalho, J. L. (2009). Identification of bilinear systems with white noise inputs: an iterative deterministic-stochastic subspace approach, *IEEE Transactions on Control Systems Technology* **17**, 3.

Luspay, T., Kulcsár, B., van Wingerden, J., Verhaegen, M. and Bokor, J. (2011). Linear parameter varying identification of freeway traffic models, *IEEE Transactions on Control Systems Technology* **19**, 1, pp. 31–45.

Mercère, G., Palsson, H. and Poinot, T. (2011). Continuous-time linear parameter varying identification of a cross flow heat exchanger: a local approach, *IEEE Transactions on Control Systems Technology* **19**, 1, pp. 64–76.

Novara, C., Ruiz, F. and Milanese, M. (2011). Direct identification of optimal SM-LPV filters and application to vehicle yaw raw estimation, *IEEE Transactions on Control Systems Technology* **19**, 1, pp. 5–17.

Overschee, P. V. and Moor, B. D. (1996). *Subspace Identification for Linear Systems: Theory-Implementation-Applications* (Kluwer Academic Publishers, Dordrecht, The Netherlands).

Pfifer, H. and Hecker, S. (2011). Generation of optimal linear models for LFT-based robust stability analysis and control design, *IEEE Transactions on Control Systems Technology* **19**, 1, pp. 118–131.

Rugh, W. and Shamma, J. (2000). Research on gain scheduling, *Automatica* **36**, 10, pp. 1401–1425.

Scheerer, C. W. (1996). Mixed H_2/H_∞ control for time-varying and linear parametrically-varying systems, *International Journal of Robust and Nonlinear Control* **6**, 9–10, pp. 929–952.

Sung, S. W., Lee, S. Y., Kwak, H. J. and Lee, I. (2001). Continuous-time subspace system identification method, *Industrial and Engineering Chemistry Research* **40**, pp. 2886–2896.

Szabó, Z., Marcos, A., Mostaza, D., Kerr, M., Rödönyi, Bokor, J. and Bennani, S. (2011). Development of an integrated LPV/LFT framework: modeling and data-based validation tool, *IEEE Transactions on Control Systems Technology* **19**, 1, pp. 93–103.

Tanelli, M., Ardagan, D. and Lovera, M. (2011). Identification of LPV state space models for autonomic web service systems, *IEEE Transactions on Control Systems Technology* **19**, 1, pp. 93–103.

Tóth, R. (2008). *Modeling and identification of linear parameter-varying systems—an orthonormal basis Function Approcah*, Ph.D. thesis, Technische Universiteit Delft, The Netherlands.

Tóth, R. (2010). *Modeling and Identification of Linear Parameter-Varying Systems*, Notes in Control Systems (Springer-Verlag, Berlin, Germany).

Tóth, R., Felici, F., Heuberger, P. S. C. and den Hof, P. M. J. V. (2008). Crucial aspects of zero-order hold lpv state-space system discretization, in *Proceedings of the IFAC World Congress* (Seoul, Korea).

van Wingerden, J. W. and Verhaegen, M. (2009). Subspace identification of bilinear and LPV systems for open- and closed-loop data, *Automatica* **45**, pp. 372–381.

Verdult, V. (2002). *Nonlinear system identification: a state-space approach*, Ph.D. thesis, University of Twente, The Netherlands.

Verdult, V. and Verhaegen, M. (2002). Subspace identification of multivariable linear parameter-varying systems, *Automatica* **38**, pp. 805–814.

Verdult, V. and Verhaegen, M. (2005). Kernel methods for subspace identification of multivariable LPV and bilinear systems, *Automatica* **41**, pp. 805–814.

Verhaegen, M. and Dewilde, P. (1992a). Subspace model identification, Part 1—the output-error state-space model identification class of algorithms, *Inter-*

national Journal of Control **56**, pp. 1211–1241.

Verhaegen, M. and Dewilde, P. (1992b). Subspace model identification, Part 2—the output-error state-space model identification class of algorithms, *International Journal of Control* **56**, pp. 1187–1210.

Zhou, K. and Doyle, J. (1998). *Essentials of robust control* (Prentice-Hall, Upper Saddle River, NJ, USA).

Chapter 10

LPV system identification using series expansion models

R. Tóth, P. S. C. Heuberger & P. M. J. Van den Hof

Delft Center for Systems and Control
Delft University of Technology
Mekelweg 2, 2628 CD Delft, The Netherlands
r.toth@tudelft.nl
p.s.c.heuberger@tudelft.nl
p.m.j.vandenhof@tudelft.nl

In this chapter an Orthonormal Basis Functions (OBFs) model structure is proposed in the LPV context with advantageous properties in terms of estimation and realization. A solid system theoretic basis for the description of LPV systems in terms of these model structures is presented together with a general approach to LPV identification. Data-driven model structure selection is also discussed in this setting and the stochastic properties of the employed identification schemes are analyzed. The introduced approaches are demonstrated on an industrially relevant example.

1. Introduction

Describing nonlinear systems by *linear parameter-varying* (LPV) models has become an attractive approach to address control of complicated systems with regime-dependent (linear) behavior. However in LPV *data-driven modeling* or *identification* of such systems, it is a delicate issue to decide what kind of model structure will be used to capture the underlying dynamic behavior. As the LPV class is more like a modeling philosophy than an actual interpretation of first-principle relations, like the *linear time-invariant* (LTI) or *affine nonlinear*, etc. model classes, the actual possibilities are numerous with a lot of sensitive details. Beside the question of deciding what an adequate choice of the *scheduling variable* p is (see [Tóth

(2010); Kwiatkowski *et al.* (2006)]), the representation form of the model structure and the parametrization itself are also of crucial importance.

In the current literature, many successful approaches have been developed using model structures that are formulated in the form of *state-space* (SS) and *linear-fractional representations* (LFR), *input-output* (IO) representations or *series-expansion* forms. For a recent overview see [Tóth (2010); Casella and Lovera (2008)]. Despite the significant advances this field have recently seen, crucial questions about consequences of using an SS or an IO model, equivalence of representations and hence model parameterizations, and the degree of usefulness of the obtained models often remain undiscussed. However, these questions become important if the methods are applied in an engineering context.

By taking a fresh look on these problems, in this chapter, an *orthonormal basis functions* (OBFs) based LPV model structure using series expansion is introduced. This model structure appears to have advantageous properties, compared to other model structures, in terms of estimation and realization. A solid system theoretic basis for the description of LPV systems by LPV-OBF models is presented, together with a general approach to LPV identification both in the *local* (identification for constant p and interpolation) and the *global* (identification with varying p) setting. Data-driven model structure selection is also discussed and the stochastic properties of the employed identification schemes are investigated. Finally, the introduced approaches are demonstrated on an industrially relevant example.

2. Perspectives of series-expansion models

In order to shed more light on the specific problems of SS or IO models in the LPV context and why a different representation can be advantageous, we will take a closer look on these LPV representation forms. In specific, we will introduce SS, LFR and IO representations, discuss some important properties, and try to seek an answer to the question: does it really matter how the model equations are formulated or it is just a matter of personal choice? As we will see, such choices might have heavy consequences and it is often attractive to avoid the involved problems, perhaps by using alternative formulations like series-expansion forms.

A general dynamic description of a discrete-time LPV system $\mathscr{S}$ can be formalized as a convolution:

$$y(k) = \sum_{i=0}^{\infty} h_i(p,k)u(k-i), \tag{1}$$

in terms of the input signal $u : \mathbb{Z} \to \mathbb{R}^{n_\mathrm{u}}$ and the *scheduling variable* $p : \mathbb{Z} \to \mathbb{P}$, with range $\mathbb{P} \subseteq \mathbb{R}^{n_\mathrm{P}}$ often called the *scheduling "space"*. Here $y : \mathbb{Z} \to \mathbb{R}^{n_\mathrm{y}}$ denotes the output signal of $\mathscr{S}$ and $k \in \mathbb{Z}$ is the *discrete time* (DT). The coefficients h_i of Eq. (1) are *static* or *dynamic* matrix functions of p with arbitrary complexity ranging from simple linear to rational or real *meromorphic*[a] dependence bounded on $\mathbb{P}$. A static function (dependence) means that $h_i(p,k)$ depends only on $p(k)$. The situation where $h_i(p,k)$ depends on multiple but finite many time-shifted instances of p, like $\{p(k+\tau_1), \ldots, p(k-\tau_2)\}$ with $\tau_1, \tau_2 \geq 0$, is called *dynamic dependence.* To have a convenient notation to express both static and dynamic dependencies, we introduce the operator $\diamond : (\mathscr{R}, \mathbb{P}^{\mathbb{Z}}) \to \mathbb{R}^{\mathbb{Z}}_{\infty}$, where $\mathscr{R}$ is the set of all real meromorphic functions with finite dimensional domain, such that $(h_i \diamond p)(k) = h_i(p(k+\tau_1), \ldots, p(k-\tau_2))$.

In Eq. (1), the sequence $\{h_i\}_{i=0}^{\infty}$ defines the varying linear dynamical relation between u and y. This description is a series-expansion representation of $\mathscr{S}$ in terms of the so called *pulse basis* $\{q^{-i}\}_{i=0}^{\infty}$, where q is the time-shift operator, i.e. $q^{-i}u(k) = u(k-i)$. Thus Eq. (1) is also called the *impulse response representation* (IIR) of $\mathscr{S}$. It can be proven that for an asymptotically stable $\mathscr{S}$, the expansion (1) is convergent [Tóth (2010)].

An important property of LPV systems is that for a constant scheduling signal, i.e. $p(k) = \mathsf{p}$ for all $k \in \mathbb{Z}$, Eq. (1) is equal to a convolution describing an LTI system as each $h_i(p,k)$ is constant. Thus, LPV systems can be seen to be similar to LTI systems, but with a different signal behavior due to the variation of each h_i. Note that there are many formal definitions of LPV systems based on particular model structures and parameterizations. The convolution form (1) can be seen as their generalization.

Two important formulations are LPV *state-space* (SS) representations and LFRs, commonly used in the control literature. LPV-SS representations of a given LPV system $\mathscr{S}$, denoted as $\mathfrak{R}_{\mathrm{SS}}(\mathscr{S})$, are often defined under the assumption of static dependence in the form of

$$qx = A(p)x + B(p)u, \tag{2a}$$

$$y = C(p)x + D(p)u, \tag{2b}$$

[a] $h : \mathbb{R}^n \to \mathbb{R}$ is a real meromorphic function if $h = f/g$ with f, g analytic and $g \neq 0$.

where $x : \mathbb{Z} \to \mathbb{R}^{n_\mathrm{x}}$ is the state variable and A, B, C, D, with appropriate dimensions, are rational matrix functions of p, bounded on $\mathbb{P}$. The LFR of $\mathscr{S}$, denoted by $\mathfrak{R}_{\mathrm{LFR}}(\mathscr{S})$, is defined as

$$\begin{bmatrix} qx \\ z \\ y \end{bmatrix} = \begin{bmatrix} A & B_1 & B_2 \\ C_1 & D_{11} & D_{12} \\ C_2 & D_{21} & D_{22} \end{bmatrix} \begin{bmatrix} x \\ w \\ u \end{bmatrix}, \tag{3a}$$

where $\{A, \ldots, D_{22}\}$ are constant matrices with appropriate dimensions and

$$w(k) = \Delta(p(k))z(k), \tag{3b}$$

with $\Delta : \mathbb{P} \to \mathbb{R}^{n_\mathrm{w} \times n_\mathrm{z}}$ being the (linear) function of p. Commonly, Δ has a block diagonal structure and it is assumed to vary in a *polytope*. Additionally, x, w, z are latent (auxiliary) variables of $\mathfrak{R}_{\mathrm{LFR}}(\mathscr{S})$.

A particular drawback of the SS and the LFR forms is that the estimation of states or latent variables together with the underlying matrix coefficients under noisy measurements of y, like $y(k) + v(k)$ where $v(k)$, is stochastic noise process (not necessary white), is difficult, commonly requiring simplifying assumptions and approximations. Due to this complexity, stochastic implications of estimation are not well understood and the curse of dimensionality casts a constant shadow over these approaches. Usually the computational and parametrization simplicity can not be exploited yet as much as in the LTI case, thus often only simple dependencies like static and linear are considered. Furthermore, estimation is effected by non-uniqueness of the parametrization in both the global and local settings, which often causes interpolation to be unpredictable in the SS case [Tóth (2010); Tóth *et al.* (2011e)]. However a major advantage of the SS and LFR representation based approaches is that the delivered models are ready for control synthesis without further processing and a state-space representation, like in the LTI case, is efficient to describe MIMO relations.

An other important class of representations are IO representations in terms of polynomial forms, denoted by $\mathfrak{R}_{\mathrm{IO}}(\mathscr{S})$, which express the IO signal relations in their natural difference equation form:

$$y(k) = -\sum_{i=1}^{n_\mathrm{a}} (a_i \diamond p)(k) q^{-i} y(k) + \sum_{j=0}^{n_\mathrm{b}} (b_j \diamond p)(k) q^{-j} u(k). \tag{4}$$

The coefficients a_i and b_i are often assumed to be static or dynamic functions of p, representing from simple linear (affine) to rational or meromorphic dependence, as this setting is flexible to handle complicated dependences. Particularly attractive features of these structures are that their identification can be addressed via the extension of the LTI *prediction-error* (PE) framework [Tóth (2010); Tóth *et al.* (2011b)] enabling the stochastic analysis of the estimates, treatment of general noise models [Tóth *et al.* (2011b); Laurain *et al.* (2010)], experiment design [Dankers *et al.* (2011); Wei and Del Re (2006); Khalate *et al.* (2009)], model structure selection and direct identification of the involved dependencies [Tóth *et al.* (2011c, 2009b); Hsu *et al.* (2008)] often in a computationally attractive manner. However, such model structures have a serious disadvantage: the delivered IO model needs to be converted to an SS or an LFR form as the main stream LPV control-synthesis approaches are formulated in terms of these representations. Due to the fact that multiplication with q in Eq. (4) is not commutative over the p-dependent coefficients a_i and b_j, the involved realization theory is more complicated than in the LTI case and often introduces complicated rational dynamic dependence on p in the resulting SS forms [Tóth *et al.* (2011e)]. Even if there exist some strategies to avoid such complications in specific situations (see [Tóth *et al.* (2011a)]), the burden of the realization is likely to rise difficulties in applications.

With respect to the previously mentioned representation forms, a particularly attractive model structure in the LPV case follows by the truncation of Eq. (1) to a finite number of expansion terms:

$$y(k) = \sum_{i=0}^{n} (h_i \diamond p)(k) u(k-i), \tag{5}$$

which is the LPV form of the well-known LTI *finite impulse response* (FIR) models. Such models have attractive properties in terms of identification in opposite to the challenging identification problem of (2a-b) or (3a-b). In particular, they benefit from the advantages of IO models as their identification can be addressed via the PE framework. An important property of Eq. (5) is *linearity-in-the-coefficients* that allows to use *linear regression* for the estimation of $\{h_i\}_{i=1}^{n}$ if they are linearly parameterized:

$$(h_i \diamond p)(k) = \sum_{j=0}^{n_i} \theta_{i,j} (f_{i,j} \diamond p)(k), \tag{6}$$

where $\theta_{i,j} \in \mathbb{R}^{n_y \times n_u}$ are the unknown parameters and $f_{i,j}$ are prior selected functions. Furthermore, noise or disturbances in the system can be modeled in an *output error* (OE) sense with this model structure, which allows independent parametrization of the noise model. However, a well known disadvantage of FIR models, both in the LTI and the LPV cases, is that the expansion may have a slow convergence rate, meaning that it requires a relatively large number of parameters for an adequate approximation of the system. In order to benefit from the same properties, but achieve faster convergence rate of the expansion, it is attractive to use basis functions which, opposite to q^{-i}, have infinite impulse responses. A particular choice of such a basis follows through the use of *orthonormal basis functions* (OBF's), which are specific basis functions in $\mathscr{H}_2$ (Hardy space of square integrable complex functions) and have already proven their usefulness in LTI identification (see [Heuberger *et al.* (2005)]) . This is the idea of the representation we would like to use to formulate expansion-based model structures for LPV identification which are beneficial both from the estimation and the utilization point of view. As we will see, the proposed structures represent an attractive trade-off between SS and IO forms, combining the advantages of both representations based model structures.

3. Orthonormal basis function models

In this section we explore the possibilities for using series-expansion model structures for LPV systems, using the concept of OBF's. A major motivation is the linear-in-the-coefficients property of these structures, which is very beneficial in PE identification. A second merit of these structures is that they allow a relatively simple interpolation of local linear models with varying McMillan degree and they have a direct SS and LFR representation. Furthermore it was shown in [Boyd and Chua (1985)] that models composed from an OBF filter bank followed by a static nonlinearity are general approximators of nonlinear systems with fading memory.

3.1. *Series-expansion representations*

We will start to develop these expansions and the concepts of OBF's by following a local point of view. It is well known that an LPV system has an LTI behavior if the considered scheduling trajectory is constant, i.e. $p(k) \equiv \mathsf{p}$. Thus such a frozen aspect of the system can be represented by a transfer function $F_{\mathsf{p}}(z)$ with $z \in \mathbb{C}$ being the complex frequency. If $F_{\mathsf{p}} \in \mathscr{H}_2^{n_{\mathrm{y}} \times n_{\mathrm{u}}}$, then F_{p} can be written as

$$F_{\mathsf{p}}(z) = W_0 + \sum_{i=1}^{\infty} W_i \phi_i(z), \tag{7}$$

where $\{\phi_i\}_{i=1}^{\infty}$ is a basis for $\mathscr{H}_2$ and $W_i \in \mathbb{R}^{n_y \times n_u}$. In the theory of *generalized orthonormal basis functions* (GOBF's), the functions $\phi_i(z)$ are generated by applying *Gram-Schmidt orthonormalization* to the sequence of functions

$$\frac{1}{z-\xi_1}, \frac{1}{z-\xi_2}, \ldots, \frac{1}{z-\xi_{n_g}}, \frac{1}{(z-\xi_1)^2}, \frac{1}{(z-\xi_2)^2}, \cdots \tag{8}$$

with stable pole locations $\xi_1, \ldots, \xi_{n_g} \in \mathbb{D} = \{z \in \mathbb{C} \mid |z| < 1\}$. The choice of these *basis poles* determines the rate of convergence of the series expansion (7). Note that it is possible to also develop such an expansion using basis functions in $\mathscr{H}_2^{n_y \times n_u}$ but for the sake of simplicity we only consider here the so called scalar case. For more on multidimensional bases see [Heuberger *et al.* (2005)]. An alternative derivation of the basis functions is based on a balanced LTI realization $\{A_g, B_g, C_g, D_g\}$ of the *inner function*

$$G_g(z) = \prod_{i=1}^{n_g} \frac{1 - z\xi_i^*}{z - \xi_i}, \tag{9}$$

where $\{\phi_i(z)\}_{i=0}^{\infty}$ are the scalar elements of the vector functions

$$(zI - A_g)^{-1} B_g G_g^j(z), \quad j \in \mathbb{N}. \tag{10}$$

By using a truncated expansion in Eq. (7), an attractive OBF model structure for LTI identification results, with a well worked-out theory in terms of variance and bias expressions [Heuberger *et al.* (2005)]. The series expansion (7) can be extended to LPV systems (see [Tóth (2010)]), via the expansion of each q^{-i} in Eq. (1) in terms of $\{\phi_i\}_{i=1}^{\infty}$. Thus, an LPV system can be written as

$$y(k) = (W_0 \diamond p)(k)u(k) + \sum_{i=1}^{\infty} (W_i \diamond p)(k)\phi_i(q)u, \tag{11}$$

where W_i are matrix functions with dynamic dependence on p. An obvious choice of model structure is to use a truncated expansion, i.e. truncating Eq. (11) to a finite sum in terms of $\{\phi_i\}_{i=1}^{n}$:

$$y(k) \approx (W_0 \diamond p)(k)u(k) + \sum_{i=1}^{n} (W_i \diamond p)(k)\phi_i(q)u. \tag{12}$$

Note that these expansions are formulated in the time domain (using the shift operator q), as there exists no frequency-domain expression for LPV systems. Similar to the FIR case, this structure is linear in the coefficients $\{W_i\}_{i=1}^{n}$. Furthermore, it is proven that structures like (12), i.e. an OBF filter bank followed by a static nonlinearity, are general approximators of nonlinear systems with fading memory, i.e. nonlinear dynamic systems with convolution representation [Boyd and Chua (1985)]. An important question that arises is whether the basis functions ϕ_i can be chosen such that a fast convergence rate can be achieved for all possible trajectories of p, i.e. how $\{\phi_i(q)\}_{i=1}^n$ with minimal n should be chosen such that the approximation error is adequate for the problem at hand.

It is important to note that in case of an unstable LPV system, it is possible to factorize the system representation into a stable and unstable part by using co-prime factorization [Wood *et al.* (1996)]. A convergent impulse response representation of the unstable part can be characterized in terms of the basis $\{q^i\}_{i=1}^{\infty}$ and hence OBF models, like (12), of such systems can be formulated in a two-sided expansion.

3.2. *Basis selection*

In the previously introduced modeling concept, it has a prime importance to achieve an efficient selection of the basis $\{\phi_i\}_{i=1}^{\infty}$, which provides a fast convergence rate of the series-expansion (11). This makes it possible to capture the dynamics of the system with a small n in Eq. (12). By taking a closer look at Eq. (12), an important implication is that if p is constant, i.e. $p(k) \equiv \mathsf{p}$, then the error of the approximation for a given n depends on the expansion error of the frozen LTI transfer function F_p in terms of $\{\phi_i(q)\}_{i=1}^n$. This means that to achieve a fast convergence rate, i.e. small approximation error by Eq. (12), it is necessary to choose $\{\phi_i(q)\}_{i=1}^n$ such that the maximum expansion error of F_p for all $\mathsf{p} \in \mathbb{P}$ is minimal. Even if such a condition is not sufficient [Tóth *et al.* (2009a)], it gives an important tool for efficient basis selection in terms of the classical *Kolmogorov n-width* result of [Pinkus (1985)] extended to OBFs in [Oliveira e Silva (1996)]. This result states that for a given LTI *inner function* G_g with poles $\Xi_{n_\mathrm{g}} \subset \mathbb{D}$, the OBF's generated by G_g (see Section 3.1) are optimal in the n-width sense for the set of LTI systems having poles in the region

$$\Omega(\Xi_{n_\mathrm{g}}, \rho) = \{z \in \mathbb{D} \mid |G_\mathrm{g}(z^{-1})| \leq \rho\}. \tag{13}$$

Here ρ is the rate of convergence in the series expansion, and n should be a multiple of the number of basis poles n_g in Ξ_{n_g}. See Fig. 1, taken from [Tóth *et al.* (2009a)], for an example of these regions.

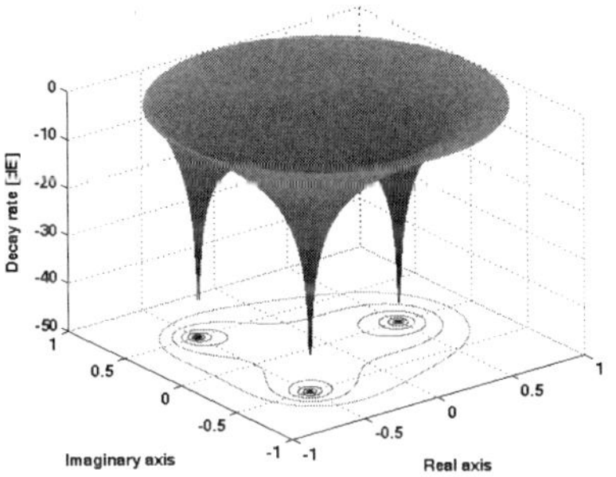

Fig. 1. Example of the function $|G_{\text{g}}(z^{-1})|$ and the region $\Omega(\Xi_{n_{\text{g}}}, \rho)$ for an inner function G_{g} with 3 poles and various values of ρ (decay rate). Note that if z_0 is a pole of G_{g}, then $G_{\text{g}}(z_0^{-1}) = 0$.

For the basis-selection problem we are dealing with the inverse problem, i.e. given a region of poles $\Omega_* = \{\xi \in \mathbb{C} \mid \exists \mathrm{p} \in \mathbb{P} \text{ s.t. } \xi \text{ is a pole of } F_{\mathrm{p}}(z)\}$, approximate this region as

$$\Omega_* \subseteq \Omega(\Xi_n, \rho) = \{z \in \mathbb{D} \mid |G_{\text{g}}(z^{-1})| \leq \rho\}. \tag{14}$$

The n optimal OBF poles $\Xi_n = \{\xi_1, \cdots, \xi_n\}$ are therefore obtained by solving the following *Kolmogorov n-width measure minimization* problem,

$$\min_{\Xi_n \subset \mathbb{D}} \rho = \min_{\Xi_n \subset \mathbb{D}} \max_{z \in \Omega_*} \left|G_{\text{g}}(z^{-1})\right| = \min_{\Xi_n \subset \mathbb{D}} \max_{z \in \Omega_*} \left| \prod_{k=1}^{n} \frac{1 - z\xi_k^*}{z - \xi_k} \right|. \tag{15}$$

For a given $\Xi_n = \{\xi_1, \cdots, \xi_n\}$ and Ω_*, the region $\Omega(\Xi_n, \rho_*)$ with ρ_* is the minimum of ρ such that (14) is satisfied, is called the *Kolmogorov bound* of Ξ_n. Smallness of this region w.r.t. Ω_* and ρ_* together indicate how well the basis functions are tuned w.r.t. Ω_*.

In order to select an efficient basis, it is obviously required to obtain knowledge about the system to be modeled. In the LPV case, this knowledge is Ω_*: the set of poles of all possible local linear models. In practice this knowledge is generally not available and one has to resort to limited prior-information resources, such as expert knowledge or preliminary identification experiments. A possible simple selection scheme, which delivers adequate results in practice, is given by the following steps:

(1) Identify a number of local linear models in several different operating regimes p_i, i.e. using data with a constant scheduling signal $p(k) \equiv \mathrm{p}_i$.

(2) Plot all poles of the identified models in the complex plane.

(3) Cluster the poles in groups and find optimal cluster centers (these centers will be used as basis poles) which minimize (15).

Alternatively, other basis optimization schemes based on recently developed sparse estimators can also be used via *orthogonal matching pursuit* or ℓ_1-optimization approaches [Tropp and Wright (2010)]. Basis selection is recently developing.

In the next section we present an efficient approach to obtain a simultaneous solution for the problems of reconstructing Ω_* from experimental data and the minimization of the Kolmogorov measure.

3.3. *A fuzzy clustering approach*

Objective-function-based fuzzy clustering algorithms, such as *fuzzy c-max* (FcM) clustering, have been used in a wide collection of applications [Bezdek (1981); Kaymak and Setnes (2002)]. Generally, FcM partitions the data into overlapping groups to capture the underlying structure [Jain and Dubes (1988)]. In this section we describe the extension of the classical FcM approach to the so-called *Fuzzy-Kolmogorov c-Max* (FKcM) algorithm, originally developed in [Tóth *et al.* (2009a)], which enables the determination of the region Ω_* on the basis of observed frozen poles with membership based, overlapping areas. We assume that we are given a set of observed/identified poles $Z = \{z_1, \ldots, z_N\} \subset \Omega_*$.

Let c be the number of clusters, that we wish to discern and let $v_i \in \mathbb{D}$ denote the *cluster center* of the i-th cluster. Denote $\mathbb{I}_{\tau_1}^{\tau_2} = \{n \in \mathbb{N} \mid \tau_1 \leq n \leq \tau_2\}$. Furthermore we define membership functions $\mu_i : \mathbb{D} \to [0,1]$, that determine for each $z \in \mathbb{D}$ the "degree of membership" to cluster i. By using a *threshold value* $0 < \varepsilon \leq 1$, we obtain a set

$$\Omega_\varepsilon = \{z \in \mathbb{D} \mid \exists i \in \mathbb{I}_1^c,\ \mu_i(z) \geq \varepsilon\}. \tag{16}$$

To measure dissimilarity of Z with respect to each cluster, we introduce distances $d_{i,k} = \kappa(v_i, z_k)$ between v_i and z_k, where κ, defined by

$$\kappa(x,y) = \left| \frac{x-y}{1-x^*y} \right|, \tag{17}$$

is a metric on $\mathbb{D}$, referred to as the *Kolmogorov metric.*

Analogously we define $\mu_{i,k} = \mu_i(z_k)$ and we regulate the membership functions by the so-called *fuzzy constraints*:

$$\sum_{i=1}^{c} \mu_{i,k} = 1 \quad \text{for } \forall k \in \mathbb{I}_1^N \quad \text{and} \quad 0 < \sum_{k=1}^{N} \mu_{i,k} \quad \text{for } \forall i \in \mathbb{I}_1^c.$$

With these preliminaries we can now formulate the problem we will consider.

Problem 10.1. *For a set of pole locations Z and for a given number of clusters c, find a set of cluster centers $\{v_i\}_{i=1}^{c}$, a set of membership functions $\{\mu_i\}_{i=1}^{c}$, and the maximum of ε, such that*

- *Ω_ε contains Z and it describes the underlying distribution of Z in terms of a chosen dissimilarity measure κ.*
- *With respect to Ω_ϵ, the OBF's, with poles Ξ_c in the cluster centers $\{v_i\}_{i=1}^{c}$, are optimal in the KnW sense, where $n = c$ and with the corresponding decay rate ρ as small as possible.*

Let $V = [\, v_1 \, \dots \, v_c \,]^\top$ and $U_k = [\, \mu_{1,k} \, \cdots \, \mu_{c,k} \,]^\top$ and denote by U the matrix with U_k 's as columns. Fuzzy clustering can be viewed as the minimization of the FcM-*functional* [Bezdek (1981)], $\mathscr{J}_m$, which in the FKcM case can be formulated as

$$\mathscr{J}_m(V,U) = \max_{k\in\mathbb{I}_1^N} \sum_{i=1}^{c} \mu_{i,k}^m d_{i,k}. \tag{18}$$

Here the design parameter $m \in (1,\infty)$ defines the *fuzziness* of the resulting partition in the sense that m determines how sharp the separation is between the clusters by the membership functions. For $1 \ll m$, each μ_i flattens till all poles in Z belong to all clusters with equal membership, while for $m = 1$, each $z \in Z$ belongs to only one cluster with a nonzero membership. It can be observed that Eq. (18) corresponds to a *worst-case (max) sum-of-error* criterion, contrary to the *mean-squared-error* criterion of the original FcM, see [Bezdek (1981)].

The crucial property of this functional is that it can be shown [Tóth *et al.* (2009a)] that for large values of m minimization of J_m is equivalent to the Kolmogorov measure minimization problem (15):

Theorem 1. *Given a data set $Z \subset \mathbb{D}$ with N elements, and a vector of cluster centers $V \in \mathbb{D}^c$, such that $d_{i,k} = \kappa(v_i, z_k) \neq 0$ for all $(i,k) \in \mathbb{I}_1^c \times \mathbb{I}_1^N$. Define U_m as a membership matrix of V minimizing Eq. (18) for $m > 1$. Then $\mathscr{J}_m(U_m, V) = c^{1-m} \max_{k\in\mathbb{I}_1^c} \left(\prod_{i=1}^{c} d_{i,k}\right)^{1/c} + \mathscr{O}(e^{-m})$. Furthermore, $\mathscr{J}_m(U_m, V)$ decreases monotonically with m, and $\lim_{m\to\infty} \mathscr{J}_m(U_m,V)=0$.*

This theorem gives that, by solving Problem 10.1 via the solution of Eq. (18), which can be obtained in terms of an alternating optimization (*Picard iteration*) [Tóth *et al.* (2009a)], one can simultaneously cluster the observed poles in such a way that the resulting cluster centers will approximate arbitrary well (governed by m) the n-width optimal OBF poles for the reconstructed frozen pole regions. See Fig. 2 for an example of the basis selection mechanism. For further details as well as a description of the optimization algorithm see [Tóth (2010)], where also the robust extension of the basis selection is discussed. This robust extension allows to solve the optimal basis selection problem when the local pole estimates are given up to an uncertainty region due to the effects of the measurement noise.

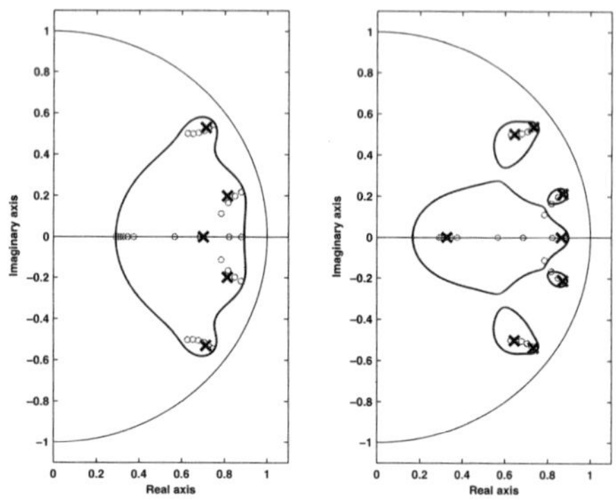

Fig. 2. Example of the basis selection procedure, using fuzzy clustering with fuzziness parameter $m = 8$. The 30 observed poles (i.e the set Z) are given with circles. The resulting cluster centers are depicted with a black x. The lines represent the Kolmogorov bound $\Omega(\Xi_c, \rho_*)$ w.r.t. Z. On the left hand side $c = 5$ clusters are determined, on the right hand side $c = 8$.

For the determination of the actual number of clusters in these algorithms, *adaptive cluster merging* can be applied (see [Kaymak and Setnes (2002); Tóth *et al.* (2009a)]). Starting from an initial number of clusters (typically around $N/2$), the adaptive merging steers the algorithm towards the natural number of groups that can be observed in the data.

3.4. *OBF's-based model structures*

Now we can define the OBF model structures we intend to use for addressing the identification of LPV systems. Assume that the basis selection step has been completed and we are given a set of n_{f} basis functions $\{\phi_i(z)\}_{i=1}^{n_{\mathrm{f}}}$ and the data-generating LPV system $\mathscr{S}_{\mathrm{o}}$ is affected by a stochastic disturbance $v(k)$ in an output additive sense. Note that v can represent a wide variety of noise processes from pure (white) measurement noise to process noise correlated with y, p and/or u. The input-output dynamics of a truncated

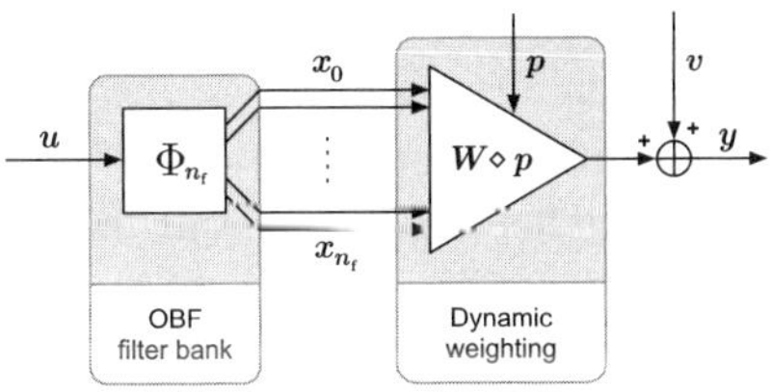

(a)

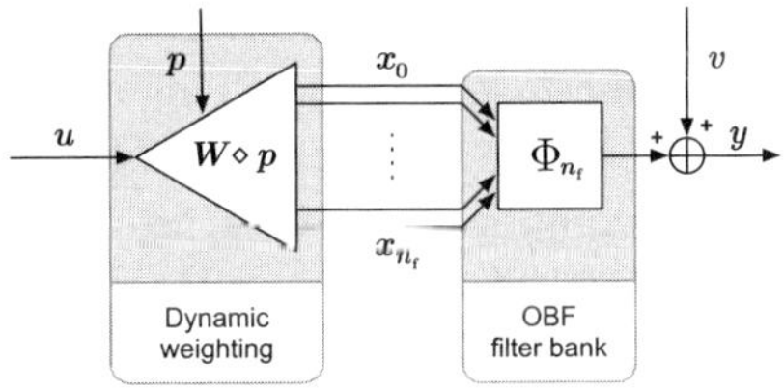

(b)

Fig. 3. IO signal flow graph of (a) the W-LPV OBF model described by (19) and (b) the H-LPV OBF model described by (20).

LPV series-expansion model in terms of $\{\phi_i(z)\}_{i=1}^{n_f}$ can now be written as

$$y(k) = \sum_{i=0}^{n_f} (W_i \diamond p)(k)\,\phi_i(q)u(t) + v(k), \tag{19}$$

with $\phi_0(\centerdot) = 1$. Equation (19) corresponds to a so-called *output error* (OE) model. Introduce $\Phi_{n_f} = \left[\phi_1 \ldots \phi_{n_f}\right]^\top$ and $W = \left[W_1 \ldots W_{n_f}\right]^\top$. Then the model structure (19) can be visualized as in Fig. 3a, where $x_i(k) = \phi_i(q)u(k)$. Because of the close resemblance of this structure to classical *Wiener models*, this model structure is referred to as a *Wiener* LPV OBF (W-LPV OBF) model. A closely related model structure, depicted in Fig. 3b, is the so-called *Hammerstein* LPV OBF model:

$$y(k) = \sum_{i=0}^{n_f} \phi_i(q)\,(W_i \diamond p)(k)\,u(k) + v(k). \tag{20}$$

The truncated expansion (20) can be obtained by deriving the series expansion (11) such that $\{W_i\}_{i=0}^{\infty}$ appear after the basis $\{\phi_i\}_{i=0}^{\infty}$. Such a reordering has no effect in the LTI case, but for LPV systems, due to the non-commutativity of multiplication of any p-dependent coefficients with q^{-1}, it results in a set of different expansion coefficients. Even if such expansions are equal in the asymptotic sense, in case of a finite truncation they have different approximation capabilities (see [Tóth (2010)]).

In the sequel we will restrict attention to the Wiener model structure. A particularly interesting feature of Eq. (19) is that it can be written in a state-space form

$$qx = Ax + Bu, \tag{21a}$$

$$y = (W \diamond p)x + (W_0 \diamond p)u + v, \tag{21b}$$

where the constant matrices A and B are completely determined by the basis functions $\{\phi_i\}_{i=1}^{n_f}$. This illustrates that the dependency on p is only present in the output equation (21b). The direct SS realization (21a-b) avoids complications that are present in the general IO case and at the same time allows to directly control the resulting dependency of the SS model by the chosen parametrization of each W_i. Similarly, direct LFR realization schemes of these models also exist in case the dependencies of each expansion coefficients are polynomial. This is a clear benefit over IO models. Moreover, as we will see, we can preserve all the advantages of the IO setting for the identification of (19) which is attractive compared to SS identification.

4. Identification via OBF models

In this section, the identification of the previously introduced OBF model structure (19) is discussed in a PE setting. With respect to the actual estimation of (19) we distinguish between two methods: a local and a global approach, which not only correspond to different estimation concepts but, as we will see, can also result in rather different model estimates. However, to derive these estimation schemes, first the question of parametrization of the coefficients is investigated.

4.1. *Parametrization of the coefficient dependence*

Beside the selection of basis functions in the considered LPV-OBF setting, model structure selection also contains an equally important part: the parametrization of the dependence of the coefficients W_i on p:

$$W_i \diamond p = \psi_i(\theta) \diamond p, \tag{22}$$

where ψ_i is function defined via some constant parameters $\theta \in \mathbb{R}^n$. The aim of the identification is to estimate θ based on a measured data record.

To simplify the estimation problem, in the LPV literature often a linear parametrization of the structural dependence is used. In fact, coefficients like W_i are considered to be a linear combination of fixed matrix functions $\psi_{i,j} : \mathbb{P} \to \mathbb{R}^{n_y \times n_u}$:

$$W_i = \sum_{j=0}^{n_\psi} \theta_{i,j} \odot \psi_{i,j}, \tag{23}$$

where $\psi_{i,0} = 1$, $\theta_{i,j} \in \mathbb{R}^{n_y \times n_u}$, and $\odot$ denotes the *Hadamard*, i.e. element-by-element, matrix product. A linear parametrization not only reduces the complexity of the associated estimation problem but also makes the problem of adequate selection of the underlaying structural dependence well-posed [Tóth (2010)]. In terms of Eq. (23), the selection problem of an adequate parametrization translates to a search for a set of functions $\{\psi_i\}_{i=0,j=0}^{n_f,n_\psi}$ such that the true expansion coefficients W_i^o of the system with respect to the used basis functions $\{\phi(q)\}_{i=1}^{n_\psi}$ satisfy $W_i^o \in \mathrm{Span}(\{\psi_{i,j}\}_{j=0}^{n_\psi})$. In case of a black-box scenario, the choice of $\{\psi_{i,j}\}$ can be arbitrary. One can consider all $\{\psi_{i,j}\}$ to be rational functions or polynomials with a fixed degree and a fixed order of dynamic dependence. However the number of possible choices is enormous. Including a too large set of functions $\{\psi_{i,j}\}$ can easily lead to over-parametrization, while restriction of $\{\psi_{i,j}\}$ to only a few basic functions can lead to serious structural bias. In order to assist the selection of an efficient set of functional dependencies in the parametrization of linear regression models, recently practically applicable approaches have been proposed in [Tóth *et al.* (2011c, 2009b)] and [Hsu *et al.* (2008)]. In [Hsu *et al.* (2008)] a *dispersion functions* based method while in [Tóth *et al.* (2011c)] a *support vector machine* approach, both originating from the *machine learning* field, has been developed to basically learn the underlying static or dynamic nonlinear dependence of the coefficients with great efficiency. In [Tóth *et al.* (2009b)] a coefficient shrinkage method, the so-called *non-negative garrote* (NNG) approach originating from statistics, has been introduced for this purpose. The NNG uses regularization in terms of weights to penalize individual elements of the parameter vector θ. In this way, the approach starts with a relatively large set of possible functional dependencies from which those functions that do not contribute significantly to the validity of the estimated model are eliminated by decreasing their weights.

4.2. *Prediction error concept*

Based on the considered parametrization of $\{W_i\}_{i=0}^{n_f}$ given by Eq. (23), the deterministic part of (19) can be written in the operator form

$$G(q,\theta) = \sum_{i=0}^{n_f} \sum_{j=0}^{n_\psi} (\theta_{i,j} \odot \psi_{i,j}) \phi_i(q), \tag{24}$$

with an overall parameter vector $\theta \in \mathbb{R}^{n_\theta}$ containing the elements of $\theta_{i,j}$.

Similarly, we can introduce a parametrized noise model w.r.t. v such as

$$v(k) = \big(H(q,\theta) \diamond p\big)e(k), \tag{25}$$

where e is a zero mean white noise process,

$$H(q,\theta) = \sum_{i=0}^{\infty} (h_i(\theta) \diamond p) q^{-i}, \tag{26}$$

is the IRR of a parametrized asymptotically stable LPV filter and θ also contains the parameters associated with this noise model. Then, the overall model $\mathscr{M}_\theta = (G(q,\theta), H(q,\theta))$ is represented as

$$y_\theta(k) = \big(G(q,\theta) \diamond p\big)(k)u(k) + \big(H(q,\theta) \diamond p\big)(k)e(k). \tag{27}$$

If there is a θ_o such that the data-generating system $\mathscr{S}_\mathrm{o}$ satisfy that $\mathscr{S}_\mathrm{o} = \mathscr{M}_{\theta_\mathrm{o}}$, then the aim of identification is to estimate θ_o based on measured data $\mathcal{D}_N$. In any other case, an $\mathscr{M}_\theta$ is searched for that describes the behavior of $\mathscr{S}_\mathrm{o}$, the best in terms of a given criterion function. To simplify the following discussion, in the sequel we treat only the case of $H(q,\theta) = I$ and e being a vector of independent zero white noise processes. However, we will briefly return to the concept of general noise models later.

It is possible to show (see [Tóth (2010); Tóth *et al.* (2011b)]) that w.r.t. (19) with $H(q,\theta) = I$, the conditional expectation of $y_\theta(k)$ in the ℓ_2 sense under the information set of $\{u(\tau)\}_{\tau=1}^{k}$ and $\{p(\tau)\}_{\tau=1}^{k}$ is equal to

$$\hat{y}_\theta(k\,|\,k-1) := \big(G(q,\theta) \diamond p\big)(k)u(k). \tag{28}$$

Then the basic philosophy of PE based identification is that w.r.t. a given model set $\mathcal{M} = \{\mathscr{M}_\theta \mid \theta \in \mathbb{R}^{n_\theta}\}$ with parameter space $\Theta \subseteq \mathbb{R}^{n_\theta}$ and a data set $\mathcal{D}_N$, to find θ such that the *one-step-ahead predictor* (28) associated with $\mathscr{M}_\theta$ provides a *prediction error*

$$e_\theta(k) = y(k) - \hat{y}_\theta(k|k-1), \tag{29}$$

which resembles a zero mean white noise "as much as possible".

Based on the predictor form (28), many different classical identification criteria can be applied to meet the above goal. A particularly interesting choice is the *least-squared* (LS) prediction-error criterion

$$\mathscr{V}(\theta,\mathcal{D}_N) = \frac{1}{N}\sum_{k=1}^{N} e_\theta^\top(k)e_\theta(k) = \frac{1}{N}\|e_\theta\|_{\ell_2}^2 \tag{30}$$

where the residual $e_\theta(k)$ is given by (29). As we will see, linear parametrization of the expansion coefficients (see Eq. (23)) yields that w.r.t. (30), the estimation of θ reduces to a linear regression problem. In other cases, when the parametrization of the coefficients is nonlinear, the estimation corresponds to a nonlinear optimization problem.

To guarantee a unique solution of (30), one condition is that the set of functions $\{\psi_{i,j}\}$ are chosen such that $\mathscr{M}_\theta$ is globally identifiable:

Definition 10.1. The model structure $(G(q,\theta), H(q,\theta))$, defined by Eq. (27), with a parameter domain $\Theta \subseteq \mathbb{R}^{n_\theta}$ and with $H(q,\theta) = I$ is globally identifiable, if for any $\theta_1, \theta_2 \in \Theta$, the corresponding one-step-ahead predictors (see Eq. (28)) are distinguishable:

$$G(q,\theta_1) = G(q,\theta_2) \;\Rightarrow\; \theta_1 = \theta_2.$$

In terms of the 1-step ahead predictor (28) of the considered OBF model, a necessary and sufficient condition to guarantee identifiability is that in the parameterization, each function set $\{\psi_{i,j}\}_{j=0}^{n_\psi}$ is a set of linearly independent functions. Another condition for the unique solution of (30), is the *informativity* of the data set $\mathcal{D}_N$:

Definition 10.2. For a model structure $(G(q,\theta), H(q,\theta))$, defined by Eq. (27), with a parameter domain $\Theta \subseteq \mathbb{R}^{n_\theta}$ and with $H(q,\theta) = I$, a data set $\mathcal{D}_N = \{u(k), y(k), p(k)\}_{k=1}^N$ is called informative, if the following holds for the one-step-ahead predictors (see Eq. (28)):

$$\big(G(q,\theta_1) \diamond p\big)u = \big(G(q,\theta_2) \diamond p\big)u \;\Rightarrow\; G(q,\theta_1) = G(q,\theta_2).$$

This means that if $\mathcal{D}_N$ is informative w.r.t. a globally identifiable $\mathscr{M}_\theta$, then the global optimum of (30) is unique. Now, having the model structure and the required concepts of identification established, we can introduce identification schemes of W-LPV-OBF models.

4.3. *Local estimation approach*

As a first approach, we aim at the identification of (19) based on the frozen aspects of the underlying data-generating system. This so-called local approach uses a number N_{loc} of "local" experiments, i.e. data collection with a constant scheduling signal $p(k) \equiv \mathsf{p}_\tau \in \mathbb{P}$, resulting in data sequences

$\mathcal{D}_{N,\mathrm{p}_\tau} = \{u(k), y(k), \mathrm{p}_\tau\}_{k=1}^{N}$ for $\tau \in \mathbb{I}_1^{N_{\mathrm{loc}}}$. Based on these data, N_{loc} LTI-OBF models defined as

$$G_\tau(q, \eta_\tau) = \sum_{i=0}^{n_\mathrm{f}} W_{i,\tau}\phi_i(q), \quad H(q, \eta_\tau) = I, \tag{31}$$

with $\eta_\tau = [\, W_{0,\tau} \ldots W_{n_\mathrm{f},\tau} \,] \in \mathbb{R}^{n_\mathrm{y} \times (n_\mathrm{u} \cdot (n_\mathrm{f}+1))}$, are estimated using the LS criterion. Note that – under the condition that the data sets are informative – there exist unique analytic solutions to these estimation problems. The estimated coefficients can now be considered as "samples" of the function $W_i \diamond p$, in the sense that $W_i \diamond p$ with $p(k) \equiv \mathrm{p}_\tau$ is equal to $W_{i,\tau}$. As a second step, we use interpolation to obtain estimates of the function $W_i \diamond p$, for instance by assuming a polynomial dependency of $(W_i \diamond p)(k)$ on $p(k)$, or by making use of splines etc. Note that as the only information available about the system is in the form of $\mathcal{D}_{N,\mathrm{p}_\tau}$ where p is constant, the estimation assumes that each W_i has only static dependence on p.

Let Vec(•) denote *vectorization* of a matrix in the sense that

$$M = \begin{bmatrix} \alpha_{1,1} & \alpha_{1,2} \\ \alpha_{2,1} & \alpha_{2,2} \end{bmatrix} \quad \Rightarrow \quad \mathrm{Vec}\,(M) = \begin{bmatrix} \alpha_{1,1} & \alpha_{1,2} & \alpha_{2,1} & \alpha_{2,2} \end{bmatrix}^\top .$$

Furthermore, introduce the diagonal matrix row construction as

$$\mathrm{Diag}_{\mathrm{row}}(M) = \begin{bmatrix} \alpha_{1,1} & \alpha_{1,2} & 0 & 0 \\ 0 & 0 & \alpha_{2,1} & \alpha_{2,2} \end{bmatrix}.$$

Then the local concept of estimation is formalized in terms of Algorithm 10.1.

4.4. *Global estimation approach*

Opposite to the local approach, in the global case we aim at the estimation of (19) in terms of the parametrized model structure (27) with $H(q, \theta) = I$ using a data set $\mathcal{D}_N$ where p is varying. This data set is assumed to be informative w.r.t. (27). Then, in terms of the LS criterion, a unique analytic solution – under the condition that Eq. (27) is globally identifiable – can be obtained for θ via Algorithm 10.2.

A similar algorithm can be introduced for the identification of the Hammerstein LPV-OBF model structure (20), including the estimation of initial conditions, see [Tóth (2010)] for a detailed discussion. In case the noise model is parameterized in terms of an LPV filter, i.e. $H(q, \theta) \neq I$, the

Algorithm 10.1 OBF based LPV identification, local method

Input: an OBF set $\Phi_{n_\mathrm{f}} = \{\phi_i\}_{i=0}^{n_\mathrm{f}}$ with $\phi_0(\centerdot) = 1$, scheduling points $\mathcal{P} = \{\mathrm{p}_\tau\}_{\tau=1}^{N_\mathrm{loc}} \subset \mathbb{P}$, data records $\mathcal{D}_{N,\mathrm{p}_\tau} = \{u(k), y(k), \mathrm{p}_\tau\}_{k=1}^{N}$ of $\mathscr{S}$, an identification criterion $\mathscr{V}$, and the local OBF model structure (31) with $\eta_\tau = [\,W_{0,\tau} \ldots W_{n_\mathrm{f},\tau}\,]$. Assume that each $\mathcal{D}_{N,\mathrm{p}_\tau}$ is informative w.r.t. (31).

1: for each $\tau \in \mathbb{I}_1^{N_\mathrm{loc}}$, calculate $\hat{\eta}_\tau = \arg\min_{\eta_\tau} \mathscr{V}(\eta_\tau, \mathcal{D}_{N,\mathrm{p}_\tau})$.

2: choose a set of matrix functions $\{\psi_{i,j} : \mathbb{P} \to \mathbb{R}^{n_\mathrm{y} \times n_\mathrm{u}}\}_{i=0,j=0}^{n_\mathrm{f},n_\psi}$ where the scalar functions $\{[\psi_{i,j}]_{k,l}\}_{j=0}^{n_\psi}$ are linearly independent for each (i,k,l) and all $\hat{\theta}_{i,j} = \arg\min_{\theta_{i,j} \in \mathbb{R}^{n_\mathrm{y} \times n_\mathrm{u}}} \|\hat{\Gamma}_i - \sum_{j=0}^{n_\psi} \Psi_{i,j} \mathrm{Vec}(\theta_{i,j})\|$ achieves the least interpolation error in terms of the considered norm $\|\centerdot\|$, where $\hat{\Gamma}_i = [\,\hat{W}_{i,1}^\top \ldots \hat{W}_{i,N_\mathrm{loc}}^\top\,]^\top$ and $\Psi_{i,j} = [\,\mathrm{Diag}_\mathrm{row}(\psi_{i,j}(\mathrm{p}_1))^\top \ldots \mathrm{Diag}_\mathrm{row}(\psi_{i,j}(\mathrm{p}_{N_\mathrm{loc}}))^\top\,]^\top$.

3: **return** estimated model (27).

estimation problem becomes non-linear in θ. If $v(k)$ is not white or correlated with u, then Algorithm 10.2 results in a biased estimate. To handle such cases it is possible to use a recently developed *instrumental variable* approach [Laurain *et al.* (2010)], which can provide statistically efficient estimation for linear regression models such as (32).

Algorithm 10.2 OBF based LPV identification, global method

Input: an OBF set $\Phi_{n_\mathrm{f}} = \{\phi_i\}_{i=0}^{n_\mathrm{f}}$ with $\phi_0(\centerdot) = 1$, matrix functions $\{\psi_{i,j} : \mathbb{P} \to \mathbb{R}^{n_\mathrm{y} \times n_\mathrm{u}}\}_{i=0,j=0}^{n_\mathrm{f},n_\psi}$ where the scalar functions $\{[\psi_{i,j}]_{k,l}\}_{j=0}^{n_\psi}$ are linearly independent for each (i,k,l), a data record $\mathcal{D}_N = \{u(k), y(k), p(k)\}_{k=1}^{N}$ of $\mathscr{S}$, an identification criterion $\mathscr{V}$, and the OBF model structure:

$$G(q,\theta) = \sum_{i=0}^{n_\mathrm{f}} \sum_{j=1}^{n_\psi} (\theta_{i,j} \odot \psi_{ij}) \phi_i(q), \quad H(q,\theta_\tau) = I, \tag{32}$$

with $\theta = [\,\theta_{0,0} \ldots \theta_{n_\mathrm{f},n_\psi}\,]$. Assume that $\mathcal{D}_N$ is informative.

1: calculate the signals $x_{i,j} = \psi_{i,j} \odot (1_{n_\mathrm{y}} \otimes \phi_i(q) u^\top)$ and let $\Gamma = \mathrm{Diag}_\mathrm{row}([\,x_{0,0} \ldots x_{n_\mathrm{g},n_\psi}\,])$ giving that $y(k) = \Gamma(k)\mathrm{Vec}(\theta) + \varepsilon_\theta(k)$.

2: estimate θ in terms of $\hat{\theta} = \arg\min_\theta \mathscr{V}(\theta, \mathcal{D}_N)$. In case of (30), $\hat{\theta} = \left(\frac{1}{N}\Gamma_\mathrm{d}^\top \Gamma_\mathrm{d}\right)^{-1} \left(\frac{1}{N}\Gamma_\mathrm{d}^\top Y\right)$ with $Y = [\,y^\top(1) \ldots y^\top(N)\,]^\top$ and $\Gamma_\mathrm{d} = [\,\Gamma^\top(1) \ldots \Gamma^\top(N)\,]^\top$.

3: **return** estimated model (27).

4.5. *Properties of estimation*

Similar to the classical LTI identification framework, it is possible to show that under minor conditions, the parameter estimates of local and global W-LPV approaches are convergent and consistent. Convergence means that for $N \to \infty$ the parameter estimate $\hat{\theta}$ converges, i.e. $\hat{\theta} \to \theta_*$, with probability 1, while consistency means that the convergence point θ_* is equal to θ_o, the parameters of the data generating system $\mathscr{S}_o$. Obviously, the latter property requires that the data generating system is in the model class $\mathcal{M}$. Furthermore, asymptotic bias and variance expressions can be derived, see [Tóth (2010)] for details.

It is important to mention that all the above presented results and algorithms are implicitly based on the assumption that the sequence of p is measurable/observable in the system without any measurement noise or disturbance. In the LPV literature, such an assumption is generally taken as a technical necessity and the resulting methods based on it are almost exclusively applied in practical situations where measurements of p are polluted by noise with various stochastic properties. The reason for this theoretical gap lies in the difficulty to establish a conditional expectation of $y(k)$ in the situation when instead of $p(k)$ only its noisy observations are available. Recently it has been proved that using estimated moments, a one-step-ahead predictor of $y(k)$ can be formulated if p is observed up to an additive white noise independent from v and the resulting formulation still allows linear-regression based estimation under an LS criterion [Tóth *et al.* (2011b)]. However, consistency and convergence properties of estimation in that case are currently not well-understood.

4.6. *Global* versus *local approach*

As demonstrated, both the global and the local approaches provide attractive ways of identifying an LPV system. An obvious question is when to use which approach. In most situations the global approach is considered to be more attractive as it provides estimation of the system with a varying trajectory of p, giving a better possibility to approximate the global dynamic behavior of the system instead of just the frozen aspects. As shown, estimation in the global case can be formulated in a simple least-squares setting and cumbersome problems of interpolation are avoided due to the fixed functional dependencies of the parametrization. The better understood behavior of the stochastic nature of estimation in the global setting also suggests that it is a theoretically more sound approach than the local method if informativity of $\mathcal{D}_N$ can be guaranteed.

However, practical use of LPV identification on industrial problems often turns out to favor different properties. In most practical applications, identification must be accomplished in a closed-loop setting due to instability of the plant or because the current production can not be disturbed in the favor of identification. In terms of the local approach, the well worked out methods of the LTI framework can be fully used to solve the identification problem in a divide-and-conquer manner. The use of frequency-based identification is also supported in the local setting. The latter is important in mechatronic applications where the often tight modeling specifications with respect to the frozen behaviors are only available in the frequency domain. Usually, such specifications can not be addressed in the global setting. On the other hand, interpolation can result in unexpected global behavior as the local identification approach only focuses on the frozen aspects of the system. However, such a drawback can be avoided by using data with varying p to assist the interpolation. As a general recipe, the use of the global approach is advised whenever there is enough possibility to perturb the system for an informative data record and if the model specifications are not given in the frequency domain. In other situations, the use of a local approach is advised due to its higher capability to meet the target performance under the given information content of available data sets.

There is an important aspect of the proposed identification methods if in the data-generating system p is not an external (free) signal but depends on internal signals like inputs, outputs or state-variables. Such systems are called *quasi*-LPV. For many quasi LPV systems, p can generally be not held constant. In such cases, only the global method is applicable, as the local approach needs identification of the system w.r.t. constant scheduling trajectories. Violation of the freedom of p and how this affects the previous results are generally not well understood and these questions are subject of current research.

5. Identification of a high-performance positioning device

In the sequel, the benefits of the proposed LPV-OBF identification approach are demonstrated on the data-driven modeling of an industrially relevant application: an xy-positioning table.

The conventional design of high-performance positioning devices used commonly in the production of *integrated circuits* (ICs) – with usual servo error requirements in the order of $[1, 50]$ [nm] – involves a *long stroke*, called the xy-positioning table, moved by two linear motors on parallel rails, see Fig. 4. This table is controlled in the x, y-translational and the z-rotational motion *degrees of freedom* (DOF's). The linearized (or so called local) dynamics of this device varies with its actual position, often manifesting in terms of position-dependent resonant dynamics. Based on [Tóth *et al.* (2011d)], we present in the following an LPV identification study of an xy-table aiming at a highly accurate fit of the resulting model w.r.t. the frozen frequency responses of the original plant. To generate-data, we will use a first-principle model which makes it possible to compare the results of the modeling analytically with the original system.

5.1. *First-principle modeling*

The first-principle modeling concept of a conventional xy-positioning table is described in Fig. 4. In this modeling concept, it is assumed that the long stroke has no displacement in the x-direction, i.e. $x_2 = 0$. Under this assumption, the dynamical behavior of this multiple mass-damper-spring system $\mathscr{S}$ can be described via the following differential equation:

$$r_{\mathrm{M}}\ddot{w} + r_{\mathrm{B}}(x_1)\dot{w} + r_{\mathrm{K}}(x_1)w = r_{\mathrm{F}}u \tag{33}$$

where $w = [\ x_1 \quad y_1 \quad R_{\mathrm{z}1} \quad y_2 \quad R_{\mathrm{z}2}\ ]^\top$, $u = [\ F_{\mathrm{x}} \quad F_{\mathrm{left}} \quad F_{\mathrm{right}}\ ]^\top$ and r_{M} and r_{F} are full-rank block diagonal matrices with appropriate dimensions and r_{B} and r_{K} are linear matrix functions of x_1. By taking $p = x_1$ as

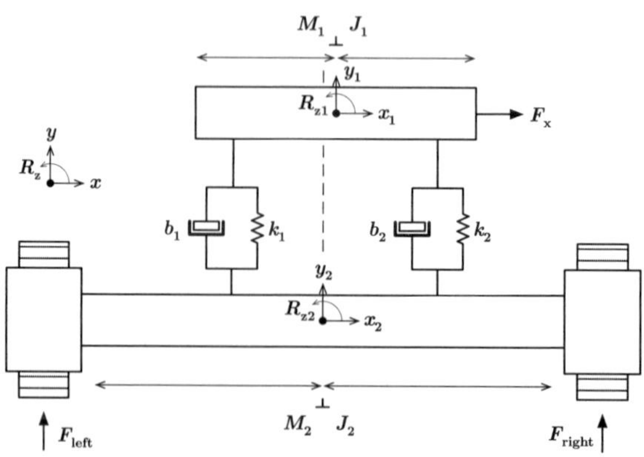

Fig. 4. First-principle modeling concept of xy-positioning tables.

a scheduling variable with $\mathbb{P} = [\mathrm{x}_{\min}, \mathrm{x}_{\max}] \subset \mathbb{R}$, the differential equation Eq. (33) corresponds to an LPV-IO representation $\mathfrak{R}_{\mathrm{IO}}(\mathscr{S})$, with input-output partition $(u, [x_1\ y_1\ R_{\mathrm{z1}}])$.

In Eq. (33), forces in one direction have influence on the movements in other directions (see [Tóth *et al.* (2011d)]). Thus, to enable the design of SISO controllers, the plant dynamics are commonly decoupled in practice by using pre- and post-transformation matrices T_{u} and T_{y} implemented directly into the hardware. As in the LPV case full decoupling of the IO channels is currently not a well-understood concept, the decoupling of the plant is developed by using a rigid-body formulation of Eq. (33), providing approximately decoupled dynamics (i.e. approximately diagonal) in the low-frequency region.

Based on the above given considerations, the rigid-body decoupled plant can be written as:

$$\mathfrak{R}'_{\mathrm{IO}}(\mathscr{S}) = T_{\mathrm{y}}(p) * \mathfrak{R}_{\mathrm{IO}}(\mathscr{S}) * T_{\mathrm{u}}(p). \tag{34}$$

The matrix T_{y} is defined by the variables to be controlled: $y' = [\ y_1 - R_{\mathrm{z1}}x_1 \quad R_{\mathrm{z1}}]^\top$ which are the actual measurements available from xy-positioning tables (besides the measurement of x_1). T_{u} is developed by assuming arbitrary slow variation of x_1 and aiming for $T_{\mathrm{u}}P_0T_{\mathrm{y}} = I$ where P_0 is the static gain of the system, giving a set of new input variables u' satisfying $[\ F_{\mathrm{x}} \quad F_{\mathrm{left}} \quad F_{\mathrm{right}}\ ]^\top = T_{\mathrm{u}}(p)u'$.

The frozen FRF's of the first-principle model of a real-life xy-positioning table w.r.t. (u', y'), are depicted at different x_1-positions in terms of Bode magnitude plots in Fig. 5. To protect the interest of the manufacturer, frequency and time have been scaled throughout this example. The following observations are crucial:

- The system dynamics can clearly be separated into an unstable rigid body part dominant in the low frequency band (below 1) and a x_1-position dependent stable flexible part dominant in the frequency band $[1, 3]$ which is symmetric in magnitude to the $x_1 = 0$ position (phase has a 180° drop at $x_1 = 0$ due to sign change).
- In the diagonal channels, rigid body dynamics correspond to a 2nd order integrator, while in the off-diagonal channels, due to the decoupling, only a small proportional term can be observed.
- At $x_1 = 0$, the off-diagonal transfer functions become approximately

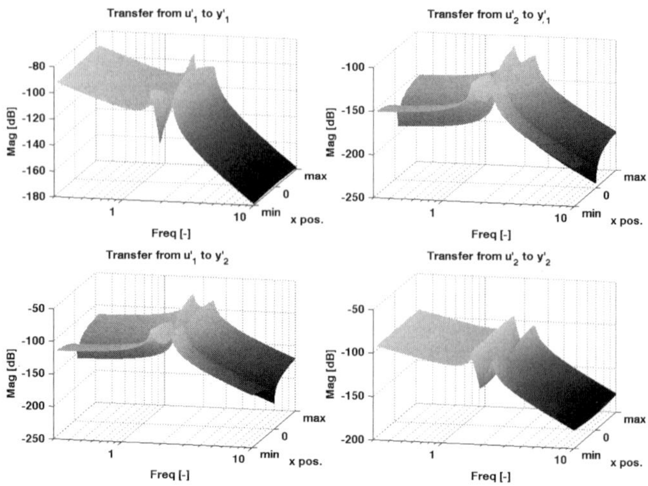

Fig. 5. Bode magnitude plot of the 2 × 2 MIMO xy-table model at different x-positions.

zero, indicating perfect decoupling. The order of the transfer functions of the diagonal channels (each with order 6) also drops by 2 at $p = 0$.

5.2. *Simulation conditions*

As the underlying system is unstable, meaningful simulations or measurements can only be obtained under closed-loop conditions. For this purpose robust continuous-time LTI single-loop controllers $K_{\mathrm{y}_1}(s)$ and $K_{\mathrm{R}_{\mathrm{z}1}}(s)$ have been designed for the model satisfying moderate specs. in terms of performance. The complete closed control loop of the system is given in Fig. 6, which corresponds to a simplified control architecture used in practice. Additionally, to record DT data for identification purposes, the inputs and outputs of the xy-positioning table in Fig. 6 are sampled with a sampling frequency of 20 (i.e. 10× the highest interesting frequency point: 2).

To give a realistic setting for identification, noise affecting both the closed loop control and the data acquisition is also considered in the form:

$$\hat{y}_1(k) = y_1'(k) + v_1(k), \quad \hat{y}_2(k) = y_2'(k) + v_2(k), \tag{35}$$

with v_1 and v_2 independent white noise processes with normal distributions: $v_1(k) \in \mathscr{N}(0, \frac{1}{3} \cdot 10^{-7})$ and $v_2(k) \in \mathscr{N}(0, \frac{5}{3} \cdot 10^{-6})$. Such noise levels are typical for the considered laser-interferometers based high-accuracy position measurements. Note that these noise conditions seem to be not so significant, but due to the relatively small range of movement and the tight error specifications they are challenging enough.

5.3. *OBFs based LPV identification*

In modeling of xy-tables it is important to achieve accuracy of the model w.r.t. both constant and varying trajectories of $p(t) = x_1(t)$. For constant values of p, typical specifications require that the magnitude of the error in terms of the frozen *frequency response function* (FRF) of the system, i.e. $F_{\mathrm{p}}(\mathrm{i}\omega)$ needs to be less than -40dB. It is generally true that it is very difficult to include identification constraints into global methods which would guarantee a specified upper bound of the model accuracy for constant p (local fit). As LPV systems do not have a transfer function representation, it is only possible to include frequency domain constraints for local type of approaches. Thus in the following we will study local identification of the xy-table using the introduced OBF approach.

5.3.1. *Choice of model structure*

Based on the observations in Section 5.1, it is attractive to separate the system dynamics into an additive "rigid-body part," which is not dependent on p, and a remaining "flexible part" contains the varying-poles related dynamical aspects of the system. Identifying the flexible part with a fixed DT rigid body filter, provides the means to enforce the well-known fact that the low-frequency behavior of the system is governed by decoupled 2nd-order integrators with an additional zero at -1 for each diagonal IO channel:

$$\phi_{\mathrm{R}}(z) = \frac{z+1}{(z-1)^2}. \tag{36}$$

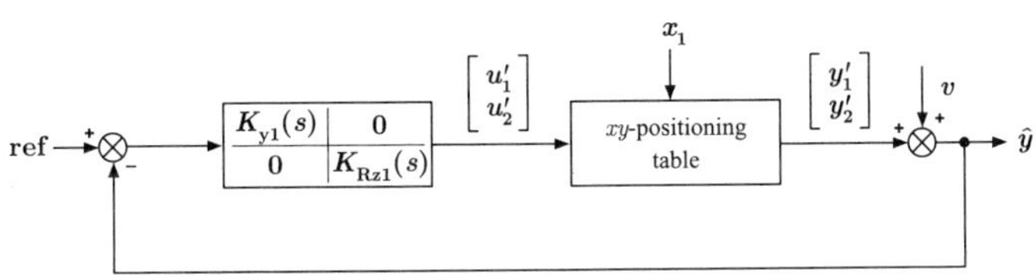

Fig. 6. Simplified closed-loop control scheme of the xy-table mechanism with measurement error v.

It can also be observed in Fig. 5 that the "moving" pole locations of the underlying IO channels of the system are the same. This implies that the optimal set of OBFs, which provide the fastest convergence rate, is the same for each channel. Furthermore, local approaches can only identify coefficients, like W_i in Eq. (12), with static dependence. Thus the overall model structure can be chosen as

$$\hat{y}_\theta = \begin{bmatrix} c_1\phi_{\mathrm{R}}(q) & 0 \\ 0 & c_2\phi_{\mathrm{R}}(q) \end{bmatrix} u + \sum_{i=1}^{n_{\mathrm{f}}} W_i(p)\phi_i(q)u \tag{37}$$

where $\{\phi_i(q)\}_1^{n_{\mathrm{f}}}$ is a set of SISO OBF's and $c_1, c_2 \in \mathbb{R}$ with $W_i : \mathbb{P} \to \mathbb{R}^{2\times 2}$ are the unknown coefficients to be estimated. As a next step, it is important to design our experiments which will give the information upon which adequate selection of the basis functions and the estimation of the expansion coefficients will be accomplished.

5.3.2. *Experiment design and data generation*

The first step of experiment design for local identification is the gridding of $\mathbb{P}$. This refers to designing the points on the x_1-axis around which local LTI identification of the setup will be performed. It is important that the gridding must be dense enough to capture important dynamic changes of the plant for different x_1-positions. By analyzing the rate of change of the frozen poles and zeros of the system w.r.t. $\mathbb{P} = [\mathrm{x}_{\min}, \mathrm{x}_{\max}]$, a grid of 21 equidistant points is chosen.

In order to generate informative data for frequency-domain identification at the designated x_1-positions, orthogonal multisines with normalized amplitude are generated based on 2^{14} equidistant frequency points $\mathcal{W} = \{\omega_k\}_{k=1}^{2^{14}}$ in the range $[10^{-4}, 10]$. This frequency range has been chosen to contain the relevant dynamical aspect of the plant in terms of rigid body and flexible modes. The orthogonality of the generated multisine signals $r_{11}, r_{12}, r_{21}, r_{22}$ can be understood in the following manner: the *discrete-time Fourier transforms* $R_{11}(\omega), \ldots, R_{22}(\omega)$ of these signals, satisfy that

$$R(\omega_k)R^{\mathrm{H}}(\omega_k) = \begin{bmatrix} \lambda_1(\omega_k) & 0 \\ 0 & \lambda_2(\omega_k) \end{bmatrix} \prec I, \quad \omega_k \in \mathcal{W},$$

$$\text{where} \quad R(\omega_k) = \begin{bmatrix} R_{11}(\omega_k) & R_{12}(\omega_k) \\ R_{21}(\omega_k) & R_{22}(\omega_k) \end{bmatrix},$$

where $\bullet^{\mathrm{H}}$ denotes the *Hermitian conjugate*. This property ensures high accuracy frequency domain estimates in closed loop even under heavy measurement noise. In the experiments, first signals r_{11} and r_{21} are used as references for y_1' and y_2' (see Fig. 6) and with constant x_1 equal to a grid point. Then the whole experiment is repeated by using r_{12} and r_{22}. The two set of responses for y_1' and y_2' are required to uniquely estimate the 2x2 MIMO FRF of the plant at the considered x_1 position. Note that the normalized reference signals are multiplied with 10^{-4} to remain in the operating range of the setup.

With the designed multisine sequence, data is generated based on the closed loop model starting from zero initial conditions. To generate an appropriately long data record for the attenuation of both the transient and noise effects, the designed multisines are repeated 25 times. For validation purposes, noise free data records are also generated.

5.3.3. *FRF estimate of the local behaviors*

The data records that are collected in the previous step can now be used to deliver estimates of the FRF of the system at the chosen x_1-positions. Consider the data sets $\mathcal{D}_{\mathrm{p},1} = \{y'_{\mathrm{ref1}}(k), u'_{\mathrm{ref1}}(k), r_{\mathrm{ref1}}(k)\}_{k=1}^{N_\mathrm{d}}$, $\mathcal{D}_{\mathrm{p},2} = \{y'_{\mathrm{ref2}}(k), u'_{\mathrm{ref2}}(k), r_{\mathrm{ref2}}(k)\}_{k=1}^{N_\mathrm{d}}$ collected from the model with $x_1 = \mathsf{p}$ and reference signals $r_{\mathrm{ref}i} = [r_{1i}\ r_{2i}]^\top$. Denote the *fast Fourier transform* (FFT) of these signals taken on one period of the time-domain data as $R_{\mathrm{ref1}}(\omega)$, $U'_{\mathrm{ref1}}(\omega)$, $Y'_{\mathrm{ref1}}(\omega)$ and $R_{\mathrm{ref2}}(\omega)$, $U'_{\mathrm{ref2}}(\omega)$, $Y'_{\mathrm{ref2}}(\omega)$ respectively. Due to the periodic nature of the excitation, it is true that after the transients have died out, the FFTs of each period of the measured data records only differ from each other in terms of the additive noise. Therefore, by chopping off the transient part of the data records (first 5-10 periods) and averaging the results of the FFT on the remaining periods, the effect of the noise can be averaged out. Thus in the sequel consider these spectra as the averaged FFT of the non-transient periods. Let

$$\begin{aligned}
\bar{U}(\omega) &= \left[U'_{\mathrm{ref1}}(\omega)\ U'_{\mathrm{ref2}}(\omega)\right], \\
\bar{Y}(\omega) &= \left[Y'_{\mathrm{ref1}}(\omega)\ Y'_{\mathrm{ref2}}(\omega)\right], \\
\bar{R}(\omega) &= \left[R_{\mathrm{ref1}}(\omega)\ R_{\mathrm{ref2}}(\omega)\right].
\end{aligned}$$

The classical way to estimate the FRF of the plant for a given frequency point $\omega_k \in \mathcal{W}$ is $\hat{F}(\omega_k) = \bar{Y}(\omega_k) \cdot \bar{U}^{-1}(\omega_k)$. However, it is well known that such an *empirical transfer function estimate* is biased in case of closed-loop

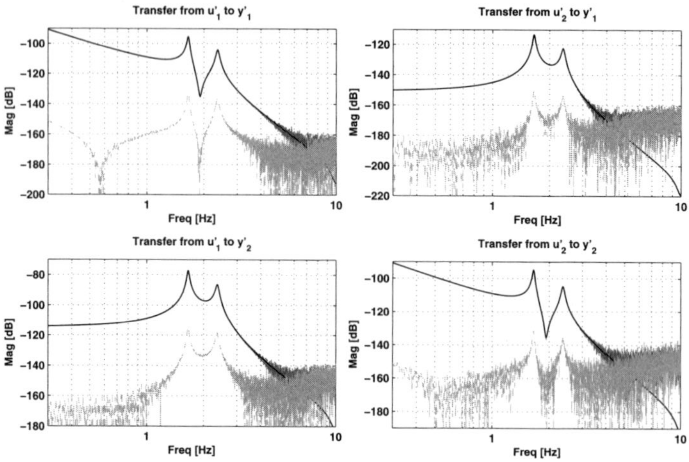

Fig. 7. Bode magnitude plot of the estimated FRF of the plant at position $\mathrm{x}_{\min}$. Original plant (black), estimated FRF (grey), error (light grey).

data. To have an unbiased estimate it is better to consider

$$\hat{F}(\omega_k) = \left(\bar{Y}(\omega_k)\bar{R}^{\mathrm{H}}(\omega_k)\right) \cdot \left(\bar{U}(\omega_k)\bar{R}^{\mathrm{H}}(\omega_k)\right)^{-1}. \tag{38}$$

Among many choices of unbiased closed loop estimators, Eq. (38) has also been observed in the literature to deliver good results under heavy noise settings [Wernholt and Gunnarsson (2007)].

By using the data records and the estimation approach Eq. (38), FRF estimates of the plant at the considered scheduling points have been calculated. During the calculation the first 10 periods in the records have been removed to attenuate the effect of initial conditions. The results at position $x_{\min}$ are depicted in Fig. 7. From this figure it is obvious that the method delivers almost perfect estimates of the frozen FRFs on each IO channel. Furthermore the considered noise only significantly affects the high-frequency band beyond the flexible modes, which shows that accurate frequency-domain information is available to recover the most important dynamical aspects of the plant from measured data.

5.3.4. *Selection of the OBF filter banks*

To arrive at an adequate selection of the OBF functions in Eq. (12) the FK*c*M approach introduced in Section 3.3 is used. To obtain an estimate of the frozen pole locations of the xy-positioning table model at the considered

Table 1. Achieved Kolmogorov n-width cost by the FKcM provided OBF's for fuzziness $m = 35$ and different number of OBFs n_g.

n_g	estimated frozen poles	true frozen poles
4	-61.061 dB	-1.651 dB
8	-125.884 dB	-7.084 dB
12	-197.604 dB	-12.031 dB
16	-257.905 dB	-16.845 dB

x_1-positions, a general *curve fitting* method can be applied on the previously obtained FRF estimates. Here the approach of the FREQID toolbox has been used [de Callafon and Van den Hof (1996)]. To arrive at the correct number of poles a MIMO common denominator model with 8th order has been estimated with curve fitting. The worst-case absolute error of the resulting pole estimates w.r.t. the true pole locations of the system at the given x-positions has been 0.07%.

Next the FKcM approach is applied on the obtained highly-accurate pole estimates. By analyzing the results of the algorithm based on the estimated pole locations, it has been observed that nearly optimal basis selection can be achieved if the fuzziness m is set to 35. Using this fuzziness value the algorithm has been executed on the estimated pole locations. The algorithm has been used with different number of optimized basis functions n_g and the results are summarized in Table 1 and given in Fig. 8. In this table, as a performance measure, the Kolmogorov n-width cost (see Eq. (15)) with $n = n_g$ has been computed in dB both for the obtained pole estimates and also for all true frozen pole locations Ω_* of the xy-positioning table in the considered x_1-region. From Fig. 8 and Table 1 it follows that the obtained OBF poles achieve very small representation error w.r.t the estimated poles (small Kolmogorov cost). However, w.r.t. the true pole locations a dramatic difference can be observed between the set of 4 or 16 basis functions. Further analysis of the results shows that at least 12 basis functions are needed to adequately represent the varying system dynamics on $\mathbb{P}$. After posteriori assessment of the estimation with 12 and 16 basis functions it is concluded that 16 basis functions are required to meet with the specifications. It can also be shown that beyond 16 basis functions the improvement in model accuracy is not significant. Thus in the sequel, we will consider the OBF set with 16 basis functions selected by the FKcM approach to formulate Eq. (37).

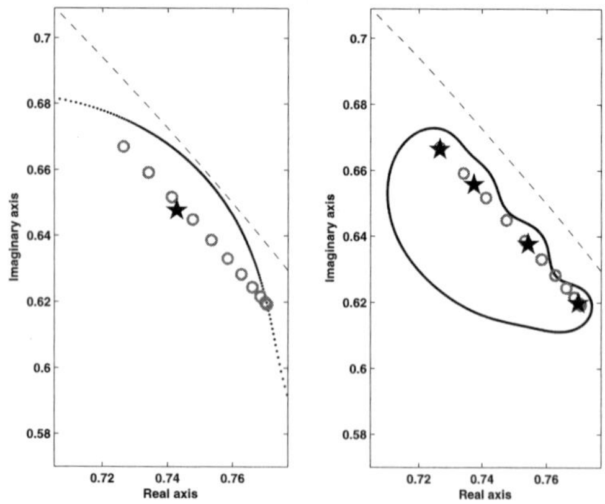

Fig. 8. Zoomed in figures depicting the results of the basis selection w.r.t. the estimated frozen poles (i.e the set Z) given with circles. The cluster centers (basis poles) are given with black $\star$. The black lines represent the Kolmogorov bound $\Omega(\Xi_c, \rho_*)$ w.r.t. the frozen pole set Ω_* of the system, while the dashed lines are the perimeter of the unit disc. On the left hand side $c = 4$ clusters are determined, on the right hand side $c = 16$.

5.3.5. *Estimation of the expansion coefficient dependence*

By having the OBF filter banks chosen, the last remaining step of identification is to estimate the constants c_1, c_2 and the expansion coefficients $W_i(p)$ in Eq. (37). For this purpose the already calculated FRF estimates of the system are used. Note that the frequency response of the OBF filters and the first-principles suggested rigid body filters can be computed w.r.t. the frequency points of the FRF estimate and in terms of the model structure these frequency responses should approximate the estimated FRFs by linear combination. Thus, estimation of the samples of the expansion coefficients W_i at each considered grid point $\mathsf{p} \in \mathcal{P}$ reduces to a simple linear regression. After solving the linear regression, the resulting samples of each W_i can be interpolated using any approach like polynomial, spline, Chebysev, etc. After investigation of the obtained results with each method, it has been concluded that a polynomial interpolation provides the most efficient solution in terms of the complexity/accuracy trade off. Regarding polynomial interpolation it has been concluded that for the case of 16 OBF functions a polynomial order of 15 is minimally required to achieve a good approximation of the frozen dynamics. By using the FRF estimates, the samples of the expansion coefficients of the OBF filter banks obtained in the previous section with $n_\mathrm{g} = 16$ have been estimated and these samples have been interpolated with 17th order polynomials. The results are depicted in Fig. 9. These figures show that using only a few estimated samples of the coefficient functions, a close approximation of the polynomial dependencies can be obtained. This concludes the identification as the delivered model now can be explicitly realized in an LFR form.

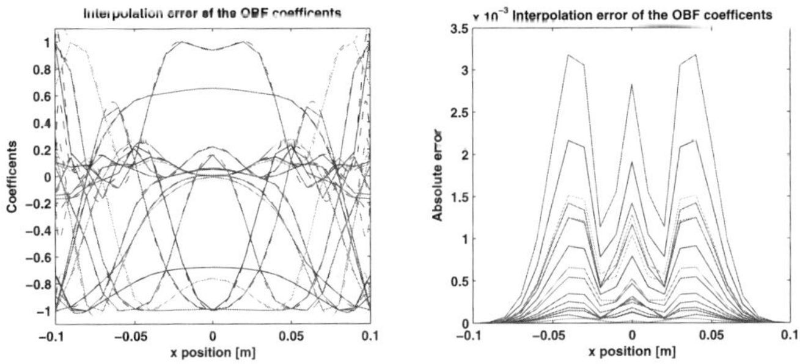

(a) (b)

Fig. 9. Optimal coefficient functions W_i of the OBF's $n_g = 16$ (solid lines) w.r.t. the frozen transfer functions between y_1' and u_1' at the grid positions x_1 together with their polynomial approximation (dashed lines).

5.4. *Validation of the model*

As a final step it remains to validate the obtained model in both the frequency and the time domain.

5.4.1. *Frequency-domain validation*

The obtained LPV-OBF model can be compared in terms of its frozen frequency responses to the behavior of the first-principle model. For the OBF model with 16 basis functions and 17th-order polynomial dependence, the frozen frequency responses for each IO channel have been computed on a fine grid $\mathcal{P} \subset \mathbb{P}$ (1000 points) together with the response of the true system and visualized in terms of Bode plots. The results at $\mathrm{x}_{\min}$ are given in Fig. 10 which corresponds to the worst-case model fit. By analyzing these results the following observations can be made:

- The overall difference between the magnitude error and magnitude of the transfer functions is aprox. 40 dB.
- However, the error increases with an aprox. 20 dB around the anti-resonance mode. This results as a side effect of linear regression. Using better tuned weights this error can be decreased if necessary.

This means that specifications in terms of frequency-domain accuracy could be achieved with the investigated identification approach.

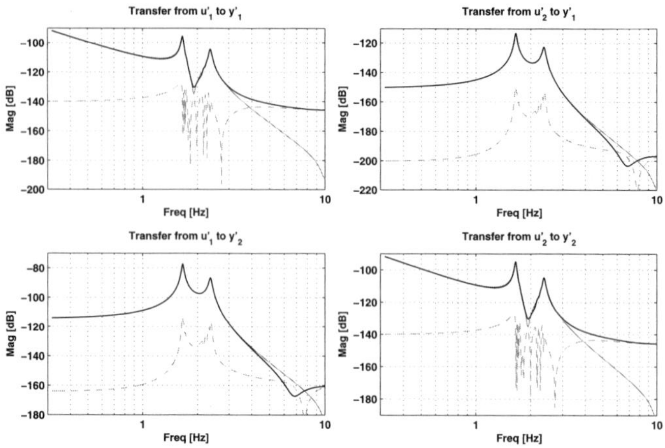

Fig. 10. Bode magnitude plot of the frozen frequency response at position x_{min} of the estimated LPV-OBF model obtained with 16 basis functions and 17th-order polynomial coefficient dependence. Original plant (black), estimated OBF model (grey), approximation error (dashed grey).

5.4.2. *Time-domain validation*

As a next step we investigate the time-domain behavior of the identified LPV-OBF model. First the open-loop response of the model is computed by using recorded u and p signals from a closed-loop simulation of the original xy-positioning table model for a monotone increasing p which corresponds to a fast sweep over $\mathbb{P}$. The used reference signals here are a zero set-point for R_{z1} and a typical step-like pattern for y_1, designed in terms of optimal speed, acceleration and jerk profile. The resulting responses of the LPV-OBF model (after re-transformation with $T_y(p)$) are given in gray in Fig. 11 while the response of the original plant is given in black. The error is dominated by a small difference that looks like the step response of an integrator. This yields that the identified LPV-OBF model is capable to reproduce the response of the system with high accuracy and the main source of the error is related to small differences between the unstable part of system and the LPV-OBF model. This hypothesis is also validated by the closed-loop response of the LPV-OBF model given with dashed gray in Fig. 11. It is important to note that we considered validation of the model with varying p, while the LPV-OBF model was obtained purely on the basis of the frozen behavior of the system. By achieving an acceptable error which meets the aimed specs. we conclude that the proposed identification approach can deliver high-quality model estimates.

5.4.3. *Economical size*

As we could see, a high number of OBFs and a high-order polynomial coefficient dependence were needed to capture the dynamics of the xy-table with the desired accuracy. This means that the final LFR form of the identified model (37) is relatively large with $\dim(x) = 4 + 2 \cdot 16 = 36$ and $\dim(z) = 2 \cdot 17 = 34$. However, by applying recent methods in LPV model reduction, like the approach of [Petersson and Löfberg (2009)], this LFR form can be reduced to state dimension 8 and with $\dim(z) = 5$, without a significant loss of accuracy. The explanation lays in the fact that in the considered model structure Eq. (37) all dependencies on p are at the output-side. Therefore in terms of realization, there is a certain freedom to consider states and input contributions which also depend on p and hence the total dimension of the model can be reduced.

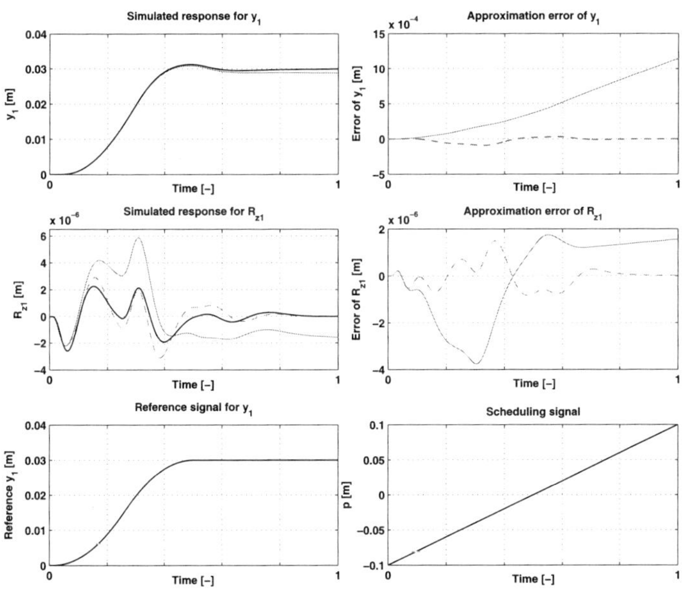

Fig. 11. Time-domain validation of the estimated LPV-OBF model using 16-basis functions with 17th-order polynomial coefficient dependence. Simulated closed-loop response of the original plant is given with black for the reference and scheduling signals depicted in the lower two figures. The simulated response of the OBF model based on the closed loop input signals of the plant is given with grey (open-loop validation) while the closed-loop response of the model for the given reference signal is depicted with dashed grey (closed-loop validation).

6. Conclusion

Based on a series-expansion representation of LPV systems, model structures using orthonormal basis functions have been developed which allow system identification in an attractive manner. These models represent an interesting trade-off between state-space and input-output models by having a direct state-space realization and allowing the efficient use of linear regression for the estimation of the underlying parameters with well-understood stochastic properties. Identification schemes of these models have been proposed both in the local and global setting and attractive properties of these approaches have been demonstrated on the identification of an industrially relevant application.

References

Bezdek, J. C. (1981). *Pattern recognition with fuzzy objective function algorithms* (Plenum Press, New York, NY, USA).

Boyd, S. and Chua, L. O. (1985). Fading memory and the problem of approximating nonlinear operators with Volterra series, *IEEE Transactions on Circuits and Systems* **32**, 11, pp. 1150–1161.

Casella, F. and Lovera, M. (2008). LPV/LFT modelling and identification: overview, synergies and a case study, in *Proceedings of the IEEE Multiconference on Systems and Control* (San Antonio, TX, USA).

Dankers, A. G., Tóth, R., Heuberger, P. S. C., Bombois, X. and den Hof, P. M. J. V. (2011). Identifiability and the informativity of data sets for LPV-ARX identification, in *Proceedings of the IEEE Conference on Decision and Control* (Orlando, FL, USA).

de Callafon, R. A. and Van den Hof, P. M. J. (1996). FREQID - frequency domain identification toolbox for use with Matlab, *Selected Topics in Identification, Modeling and Control* **9**, pp. 129–134.

Heuberger, P. S. C., Van den Hof, P. M. J. and Bo Wahlberg (2005). *Modeling and identification with rational orthonormal basis functions* (Springer-Verlag, London, UK).

Hsu, K., Vincent, T. L. and Poolla, K. (2008). Nonparametric methods for the identification of linear parameter varying systems, in *Proceedings of the International Symposium on Computer-Aided Control System Design* (San Antonio, TX, USA).

Jain, A. K. and Dubes, R. C. (1988). *Algorithms for clustering data* (Prentice Hall, Englewood Cliffs, NJ, USA).

Kaymak, U. and Setnes, M. (2002). Fuzzy clustering with volume prototypes and adaptive cluster merging, *IEEE Transactions on Fuzzy Systems* **10**, 6, pp. 705–711.

Khalate, A. A., Bombois, X., Tóth, R. and Babuška, R. (2009). Optimal experimental design for LPV identification using a local approach, in *Proceedings of the IFAC Symposium on System Identification* (Saint-Malo, France).

Kwiatkowski, A., Werner, H. and Boll, M. T. (2006). Automated generation and assessment of affine LPV models, in *Proceedings of the IEEE Conference on Decision and Control* (San Diego, CA, USA).

Laurain, V., Gilson, M., Tóth, R. and Garnier, H. (2010). Refined instrumental variable methods for identification of LPV Box-Jenkins models, *Automatica* **46**, 6, pp. 959–967.

Oliveira e Silva, T. (1996). A n-width result for the generalized orthonormal basis function model, in *Proceedings of the IFAC World Congress* (Sydney, Australia).

Petersson, D. and Löfberg, J. (2009). Optimization based LPV-approximation of multiple model systems, in *Proceedings of the European Control Conference* (Budapest, Hungary).

Pinkus, A. (1985). *n-widths in approximation theory* (Springer-Verlag, New York, NY, USA).

Tóth, R. (2010). *Modeling and identification of linear parameter-varying systems*, Lecture Notes in Control and Information Sciences (Springer-Verlag).

Tóth, R., Abbas, H. and Werner, W. (2011a). On the state-space realization of LPV input-output models: practical approaches, *in print, IEEE Transactions on Control Systems Technology* .

Tóth, R., Bitar, E., Heuberger, P. S. C., Van den Hof, P. M. J. and Poolla, K. (2011b). A prediction-error identification framework for linear parameter-varying systems, *Submitted to Automatica* .

Tóth, R., Heuberger, P. S. C. and Van den Hof, P. M. J. (2009a). Asymptotically optimal orthonormal basis functions for LPV system identification, *Automatica* **45**, 6, pp. 1359–1370.

Tóth, R., Laurain, V., Zheng, W., and Poolla, K. (2011c). A support vector machine approach for LPV linear-regression models, in *Proceedings of the IEEE Conference on Decision and Control* (Orlando, FL, USA).

Tóth, R., Lyzell, C., Enqvist, M., Heuberger, P. S. C. and Van den Hof, P. M. J. (2009b). Order and structural dependence selection of LPV-ARX models using a nonnegative garrote approach, in *Proceedings of the IEEE Conference on Decision and Control* (Shanghai, China).

Tóth, R., van de Wal, M., Heuberger, P. S. C. and Van den Hof, P. M. J. (2011d). LPV identification of high performance positioning devices, in *Proceedings of the American Control Conference* (San Francisco, CA, USA).

Tóth, R., Willems, J. C., Heuberger, P. S. C. and Van den Hof, P. M. J. (2011e). The behavioral approach to linear parameter-varying systems, *in print, IEEE Transactions on Automatic Control.*

Tropp, J. A. and Wright, S. J. (2010). Computational methods for sparse solution of linear inverse problems, *Proceedings of the IEEE* **98**, 6, pp. 948–958.

Wei, X. and Del Re, L. (2006). On persistent excitation for parameter estimation of quasi-LPV systems and its application in modeling of diesel engine torque, in *Proceedings of the IFAC Symposium on System Identification* (Newcastle, Australia).

Wernholt, E. and Gunnarsson, S. (2007). Analysis of methods for multivariable frequency response function estimation in closed loop, in *Proceedings of the Conference on Decision and Control* (New Orleans, LA, USA).

Wood, G. D., Goddard, P. J. and Glover, I. (1996). Approximation of linear parameter-varying systems, in *Proceedings of the IEEE Conference on Decision and Control* (Kobe, Japan).

Chapter 11

System identification of linear parameter varying state-space models

A. Wills & B. Ninness

School of Electrical Engineering and Computer Science
University of Newcastle
Callaghan, NSW 2308, Australia
Adrian.Wills@newcastle.edu.au
Brett.Ninness@newcastle.edu.au

This chapter examines the estimation of multivariable linear models for which the parameters vary in a time-varying manner that depends in an affine fashion on a known or otherwise measured signal. These locally linear models which depend on a measurable operating point are known as linear parameter varying (LPV) models. The contribution here relative to previous work on the topic is that in the Gaussian case, an expectation-maximisation (EM) algorithm-based solution is derived and profiled.

1. Introduction

This chapter considers the problem of estimating the parameters in linear parameter varying (LPV) model structures on the basis of observed input-output data records. LPV models are linear time varying structures wherein the time dependence is affinely related to a known "scheduling" signal. They have proven to be of importance in aerospace [Marcos and Balas (2004)], engine control [Balas (2002)], and compressor control applications [Giarré *et al.* (2006)] amongst others [Rugh and Shamma (2000)].

Due to their relevance, the problem of estimating LPV models has attracted significant research attention. The contribution [Lee and Poolla (1999)] was the first to not require state measurements, and work only with input-output measurements. It examined least squares prediction error (PE) estimation of multiple-input multiple-output (MIMO) state space structures with possible noise corruptions on the output measurements via a gradient-based search technique. This approach has been further examined and applied in [Verdult *et al.* (2003, 2004); Fujimori and Ljung (2006)].

The estimation of single-input single-output (SISO) transfer function LPV model structure has also been addressed. The work in [Bamieh and Giarré (2002)] has considered estimation of ARX-type LPV models via least mean square (LMS) and recursive least squares (RLS) algorithms. The recent paper [Laurain *et al.* (2010)] addresses more general Box–Jenkins type LPV models via an instrumental variable method.

A final key strand of research has been the development of subspace-based methods, again for the estimation of MIMO state-space LPV models [van Wingerden and Verhaegen (2009); Felici *et al.* (2007); Verdult and Verhaegen (2005, 2002)]. A main advantage of these methods is that they are non-iterative in nature, and their effectiveness has been demonstrated empirically.

This chapter addresses the estimation of MIMO state-space LPV systems that are affected by both state and measurement noise. Among that many possibilities, here a maximum-likelihood (ML) criteria is used. A key rationale for the consideration is the availability of underpinning theory assuring their general consistency, even in the presence of quite arbitrary stable closed loop control [Ljung (1978)]. The particular importance of this for LPV systems has been emphasised and explained in [van Wingerden and Verhaegen (2009)].

There are several possible choices of algorithm for maximising the likelihood function. This chapter will develop and illustrate a new estimation approach using the expectation-maximisation (EM) algorithm. The performance of this technique for other linear time invariant, bilinear, frequency domain and other identification problems has been previously demonstrated [Isaksson (1993); Goodwin and Feuer (1999); Gopaluni (2008); Gibson and Ninness (2005); Gibson *et al.* (2005); Wills *et al.* (2009)]. As will be illustrated here, it also delivers an effective and numerically robust solution for LPV estimation. Furthermore, it has reasonable computational requirements which scale modestly with increasing system state and input dimension.

Having developed the EM method for MIMO LPV systems, the chapter profiles it empirically via a simulation study.

2. Problem formulation

This chapter considers the following linear time-varying state-space model structure

$$x_{t+1} = A_t x_t + B_t u_t + w_t, \tag{1a}$$

$$y_t = C_t x_t + D_t u_t + e_t. \tag{1b}$$

Here $x_t \in \mathbf{R}^n$ is the system state, $u_t \in \mathbf{R}^m$ is the system input, and $y_t \in \mathbf{R}^\ell$ is the system output. The noise terms $w_t \in \mathbf{R}^n$ and $e_t \in \mathbf{R}^\ell$ are assumed to be zero mean i.i.d. process for which

$$\text{Cov}\left\{\begin{bmatrix} w_t \\ e_t \end{bmatrix}\right\} = \begin{bmatrix} Q & S \\ S^T & R \end{bmatrix}$$

$$\triangleq \Pi \succ 0. \tag{1c}$$

The time variations in the system matrices in (1a), (1b) are assumed to be of the following linear forms

$$A_t = \sum_{i=0}^{n_\alpha} \alpha_t(i) A(i), \quad A(i) \in \mathbf{R}^{n\times n}\ \forall i, \tag{1d}$$

$$B_t = \sum_{i=0}^{n_\beta} \beta_t(i) B(i), \quad B(i) \in \mathbf{R}^{n\times m}\ \forall i \tag{1e}$$

$$C_t = \sum_{i=0}^{n_\alpha} \alpha_t(i) C(i), \quad C(i) \in \mathbf{R}^{\ell\times n}\ \forall i, \tag{1f}$$

$$D_t = \sum_{i=0}^{n_\beta} \beta_t(i) D(i), \quad D(i) \in \mathbf{R}^{\ell\times m}\ \forall i, \tag{1g}$$

where the signals

$$\alpha_t = \begin{bmatrix} \alpha_t(0) \\ \vdots \\ \alpha_t(n_\alpha) \end{bmatrix} \in \mathbf{R}^{n_\alpha}, \tag{1h}$$

$$\beta_t = \begin{bmatrix} \beta_t(0) \\ \vdots \\ \beta_t(n_\beta) \end{bmatrix} \in \mathbf{R}^{n_\beta} \tag{1i}$$

are termed "*scheduling parameters*" that are assumed known or measurable. A model of this form (1a)-(1h) is an example of a *linear parameter varying* (LPV) system, and in practice the values taken by the scheduling parameter specify a given system operating point.

In what follows, it will prove important to note the equivalent formulation

$$\begin{bmatrix} x_{t+1} \\ y_t \end{bmatrix} = \begin{bmatrix} \mathscr{A} & \mathscr{B} \\ \mathscr{C} & \mathscr{D} \end{bmatrix} \begin{bmatrix} (\alpha_t \otimes x_t) \\ (\beta_t \otimes u_t) \end{bmatrix} + \begin{bmatrix} w_t \\ e_t \end{bmatrix}, \tag{2}$$

where

$$\mathscr{A} = \begin{bmatrix} A(0) \dots A(n_\alpha) \end{bmatrix}, \tag{3}$$
$$\mathscr{B} = \begin{bmatrix} B(0) \dots B(n_\beta) \end{bmatrix}, \tag{4}$$
$$\mathscr{C} = \begin{bmatrix} C(0) \dots C(n_\alpha) \end{bmatrix}, \tag{5}$$
$$\mathscr{D} = \begin{bmatrix} D(0) \dots D(n_\beta) \end{bmatrix} \tag{6}$$

and $\otimes$ denotes the Kronecker tensor product [Bernstein (2005)].

Assume that the system matrices $\mathscr{A}, \mathscr{B}, \mathscr{C}, \mathscr{D}$, the disturbance covariance Π and the initial state x_1 are appropriately parametrized by the elements of a vector $\theta \in \mathbf{R}^{n_\theta}$. Details on how this may be achieved are provided in what follows.

Then the problem considered in this chapter is the estimation of a value of θ that best explains (in a manner soon to be made precise) an observed length N data record

$$U = [u_1, \cdots, u_N], \tag{7a}$$
$$Y = [y_1, \cdots, y_N] \tag{7b}$$

of the input-output response.

3. Maximum-likelihood estimation

One approach taken in this chapter to address this estimation problem is the employment of the classical Maximum-Likelihood (ML) technique. This requires the formulation of the joint density $p_\theta(Y)$ of the observation, which may be expressed using Bayes' rule as

$$p_\theta(Y) = p_\theta(y_0) \prod_{t=1}^{N} p_\theta(y_t \mid y_{t-1}, \cdots, y_1). \tag{8}$$

A maximum likelihood estimate $\widehat{\theta}_{\mathrm{ML}}$ of the parameter vector θ is then defined as any value $\widehat{\theta}_{\mathrm{ML}}$ satisfying

$$\widehat{\theta}_{\mathrm{ML}} \in \left\{\theta \in \Theta : L(\theta) \leq L(\theta'), \forall \theta' \subset \Theta\right\}, \tag{9}$$

where

$$L(\theta) \triangleq -\log p_\theta(Y) \tag{10}$$

and $\Theta \subseteq \mathbf{R}^{n_\theta}$ is a user chosen compact subset. In the above, the common notation p_θ is used to refer to a range of different densities, with the arguments to p_θ indicating what is intended. Furthermore, note that while the form of p_θ will depend on U, it is not formally an argument since it is not a random variable.

In the case where stochastic elements w_t, e_t are Gaussian distributed

$$\begin{bmatrix} w_t \\ e_t \end{bmatrix} \sim \mathscr{N}(0, \Pi) \tag{11}$$

then the density $p_\theta(y_t \mid Y_{t-1})$ will also have a Gaussian density and lead to (ignoring constant terms that do not affect the minimisation (9)) [Åström (1980)]

$$L(\theta) = \sum_{t=1}^{N} \log\det\{\Lambda_t(\theta)\} + \varepsilon_t^T(\theta)\Lambda_t^{-1}\varepsilon_t(\theta), \tag{12}$$

where

$$\varepsilon_t(\theta) = y_t - \widehat{y}_{t|t-1}(\theta), \tag{13a}$$
$$\Lambda_t(\theta) = \mathrm{Cov}\{\widehat{y}_{t|t-1}(\theta)\} \tag{13b}$$

with $\widehat{y}_{t|t-1}(\theta)$ being the mean square optimal one step ahead predictor of y_t. This can be computed using a standard Kalman filter.

$$\widehat{y}_{t|t-1} = C_t\widehat{x}_{t|t-1} + D_t u_t, \tag{14}$$
$$\widehat{x}_{t+1|t} = A_t\widehat{x}_{t|t-1} + B_t u_k + K_t e_t, \tag{15}$$
$$K_t = \left(A_t P_{t|t-1} C_t^T + S\right)\Lambda_t^{-1}, \tag{16}$$
$$P_{t+1|t} = A_t P_{t|t-1} A_t^T + Q - K_t \Lambda_t K_t^T, \tag{17}$$
$$\Lambda_t = C_t P_{t|t-1} C_t^T + R, \tag{18}$$

where $\widehat{x}_{t|t-1}$ is the mean square optimal one step ahead state prediction which has associated covariance $P_{t|t-1}$.

4. The expectation-maximisation (EM) algorithm for ML estimation

Consider for a moment that in addition to U, Y a record

$$X = [x_1, x_2, \cdots, x_{N+1}] \tag{19}$$

of the state sequence for a system modeled by (1a), (1b) is also available. Via the definition of conditional probability, this would permit the negative log-likelihood (10) to be expressed as

$$\begin{aligned} L(\theta) &\triangleq \log p_\theta(Y) & (20)\\ &= \log p_\theta(X,Y) - \log p_\theta(X|Y). & (21)\end{aligned}$$

The reason for considering this idea is that, as will be seen presently, under the LPV model (1a), (1b) the associated log-likelihood $p_\theta(X,Y)$ can be maximised in closed form. However, it is unreasonable to assume that a record of the state sequence X is available. To address this, the log-likelihood $\log p_\theta(X,Y)$ can be approximated by its expected value

$$\mathscr{Q}(\theta, \widehat{\theta}_k) \triangleq \mathbf{E}_{\widehat{\theta}_k}\{\log p_\theta(X,Y) \mid Y\} \tag{22}$$

conditional on the data Y observed, and an initial estimate $\widehat{\theta}_k$. Due to the expectation operator $\mathbf{E}_{\widehat{\theta}_k}\{\cdot \mid Y\}$ being linear, this approximation $\mathscr{Q}(\theta, \widehat{\theta}_k)$ can also be maximised in closed form.

Furthermore, via (20), this approximation is related to $L(\theta)$ according to

$$L(\theta) = \mathscr{Q}(\theta, \widehat{\theta}_k) - \mathscr{V}(\theta, \widehat{\theta}_k), \tag{23}$$

where

$$\mathscr{V}(\theta, \widehat{\theta}_k) \triangleq \mathbf{E}_{\widehat{\theta}_k}\{\log p_\theta(X \mid Y) \mid Y\}. \tag{24}$$

Therefore

$$\begin{aligned} L(\theta) - L(\widehat{\theta}_k) = &\left[\mathscr{Q}(\theta, \widehat{\theta}_k) - \mathscr{Q}(\widehat{\theta}_k, \widehat{\theta}_k)\right] + \\ &\left[\mathscr{V}(\widehat{\theta}_k, \widehat{\theta}_k) - \mathscr{V}(\theta, \widehat{\theta}_k)\right]. \end{aligned} \tag{25}$$

The second term

$$\mathscr{V}(\widehat{\theta}_k, \widehat{\theta}_k) - \mathscr{V}(\theta, \widehat{\theta}_k) = \int \log \left[\frac{p_{\widehat{\theta}_k}(X \mid Y)}{p_\theta(X \mid Y)} \right] p_{\widehat{\theta}_k}(X \mid Y)\, \mathrm{d}X \tag{26}$$

is the Kullback-Leibler divergence metric between $p_\theta(X|Y)$ and $p_{\widehat{\theta}_k}(X \mid Y)$, and hence is non-negative. As a result, any value of θ for which $\mathscr{Q}(\theta, \widehat{\theta}_k) > \mathscr{Q}(\widehat{\theta}_k, \widehat{\theta}_k)$ implies that $L(\theta) > L(\widehat{\theta}_k)$.

This suggests a strategy of maximising $\mathscr{Q}(\theta, \widehat{\theta}_k)$, which must increase $L(\theta)$ via (25), and then setting $\widehat{\theta}_{k+1}$ equal to this maximiser and repeating the process. This procedure is known as the expectation-maximisation (EM) algorithm, which can be stated in in abstract form as follows.

Algorithm 4.1. EM Algorithm

(1) **E-Step**

$$\text{Calculate:} \qquad \mathscr{Q}(\theta, \widehat{\theta}_k); \tag{27}$$

(2) **M-Step**

$$\text{Compute:} \qquad \widehat{\theta}_{k+1} = \arg\max_{\theta \in \Theta} Q(\theta, \widehat{\theta}_k). \tag{28}$$

---

5. EM for LPV models

Application of the EM Algorithm to the LPV model (1a), (1b) requires development of how the E-Step and M-Step may be computed.

5.1. *E-Step*

The E-Step may be calculated by the following two lemmas.

Lemma 1. *With regard to the LPV model structure* (1a)-(1h) *and the following assumption on the initial state*

$$x_1 \sim \mathscr{N}(\mu, P_1) \tag{29}$$

the joint log-likelihood approximation $\mathscr{Q}(\theta,\widehat{\theta}_k)$ defined via (19), (22) is given by

$$\begin{aligned}-2\mathscr{Q}(\theta,\widehat{\theta}_k) = &\log\det P_1 + N\log\det\Pi +\\ &\mathrm{Tr}\left\{P_1^{-1}\mathbf{E}_{\widehat{\theta}_k}\{(x_1-\mu)(x_1-\mu)^T \mid Y\}\right\}+\\ &\mathrm{Tr}\left\{\Pi^{-1}\left[\Phi - \Psi\Gamma^T - \Gamma\Psi^T + \Gamma\Sigma\Gamma^T\right]\right\},\end{aligned} \tag{30}$$

where

$$\Gamma \triangleq \begin{bmatrix}\mathscr{A} & \mathscr{B}\\ \mathscr{C} & \mathscr{D}\end{bmatrix} \tag{31}$$

$$\Phi \triangleq \sum_{t=1}^{N}\begin{bmatrix}\mathbf{E}_{\widehat{\theta}_k}\{x_{t+1}x_{t+1}^T \mid Y\} & \widehat{x}_{t+1|N}y_t^T\\ y_t\widehat{x}_{t+1|N}^T & y_ty_t^T\end{bmatrix} \tag{32}$$

$$\Psi \triangleq \sum_{t=1}^{N}\begin{bmatrix}\alpha_t^T\otimes\mathbf{E}_{\widehat{\theta}_k}\{x_{t+1}x_t^T \mid Y\} & \beta_t^T\otimes\widehat{x}_{t+1|N}u_t^T\\ \alpha_t^T\otimes y_t\widehat{x}_{t|N}^T & \beta_t^T\otimes y_tu_t^T\end{bmatrix} \tag{33}$$

$$\Sigma \triangleq \sum_{t=1}^{N}\begin{bmatrix}\alpha_t\alpha_t^T\otimes\mathbf{E}_{\widehat{\theta}_k}\{x_tx_t^T \mid Y\} & \alpha_t\beta_t^T\otimes\widehat{x}_{t|N}u_t^T\\ \beta_t\alpha_t^T\otimes u_t\widehat{x}_{t|N}^T & \beta_t\beta_t^T\otimes u_tu_t^T\end{bmatrix} \tag{34}$$

with

$$\widehat{x}_{t|N} \triangleq \mathbf{E}_{\widehat{\theta}_k}\{x_t \mid Y\}. \tag{35}$$

Proof. The Markov property of the LPV model structure (2) together with Bayes' rule yields

$$p_\theta(Y, X_{N+1}) = p_\theta(x_1)\prod_{t=1}^{N}p_\theta(x_{t+1}, y_t \mid x_t). \tag{36}$$

Furthermore, via (1c),(2)

$$p_\theta\left(\begin{bmatrix}x_{t+1}\\ y_t\end{bmatrix}\middle| x_t\right) \sim \mathscr{N}(\Gamma z_t, \Pi), \tag{37}$$

where

$$\xi_t \triangleq \begin{bmatrix}x_{t+1}\\ y_t\end{bmatrix}, \quad z_t \triangleq \begin{bmatrix}\alpha_t\otimes x_t\\ \beta_t\otimes u_t\end{bmatrix} \tag{38}$$

and by the assumption (29) $p_\theta(x_1) = \mathcal{N}(\mu, P_1)$. Therefore, using (36) and excluding constant terms

$$\begin{aligned} -2\log p_\theta(Y, X_{N+1}) = &\log\det P_1 + N\log\det\Pi + \\ &(x_1-\mu)^T P_1^{-1}(x_1-\mu) + \\ &\sum_{t=1}^{N}(\xi_t - \Gamma z_t)^T\Pi^{-1}(\xi_t - \Gamma z_t). \end{aligned} \tag{39}$$

Applying the conditional expectation operator $\mathbf{E}_{\widehat{\theta}_k}\{\cdot \mid Y\}$ to both sides of equation (39) yields (30) with

$$\Phi = \mathbf{E}_{\widehat{\theta}_k}\{\xi_t\xi_t^T \mid Y\}, \tag{40}$$

$$\Psi = \mathbf{E}_{\widehat{\theta}_k}\{\xi_t z_t^T \mid Y\}, \tag{41}$$

$$\Sigma = \mathbf{E}_{\widehat{\theta}_k}\{z_t z_t^T \mid Y\}. \tag{42}$$

Using the definitions (35),(38), the fact that α_t, β_t are deterministic, and elementary properties of the Kronecker product then completes the proof.

□

This reduces much of the computation of $\mathcal{Q}(\theta, \widehat{\theta}_k)$ to that of calculating the Kalman smoothed state estimate $\widehat{x}_{t|N}$ together with its covariance $P_{t|N}$, which then delivers $\mathbf{E}_{\widehat{\theta}_k}\{x_t x_t^T \mid Y\} = P_{t|N} + \widehat{x}_{t|N}\widehat{x}_{t|N}^T$. What remains is the additional computation of $\mathbf{E}_{\widehat{\theta}_k}\{x_{t+1}x_t^T \mid Y\}$, which is not obtainable by a standard Kalman smoother.

The following lemma establishes how all these quantities my be obtained via numerically robust "square root" recursions.

Lemma 2. *The components*

$$\mathbf{E}_{\widehat{\theta}_k}\{x_t^T \mid Y\} \triangleq \widehat{x}_{t|N}^T, \tag{43}$$

$$\mathbf{E}_{\widehat{\theta}_k}\{x_t x_t^T \mid Y\} = \widehat{x}_{t|N}\widehat{x}_{t|N}^T + P_{t|N}^{1/2}P_{t|N}^{T/2}, \tag{44}$$

$$\mathbf{E}_{\widehat{\theta}_k}\{x_t x_{t-1}^T \mid Y\} = \widehat{x}_{t|N}\widehat{x}_{t-1|N}^T + M_{t|N} \tag{45}$$

required for the computation of (30) *via (32)-* (35) *may be robustly computed as follows. The smoothed state estimate* $\{\widehat{x}_{t|N}\}$ *is calculated via the reverse-time recursion*

$$\widehat{x}_{t|N} = \widehat{x}_{t|t} + J_t\left[\widehat{x}_{t+1|N} - \overline{A}_t\widehat{x}_{t|t} - \overline{B}_t u_t - SR^{-1}y_t\right], \tag{46}$$

$$J_t \triangleq P_{t|t}\overline{A}_t^T P_{t+1|t}^{-1}, \tag{47}$$

where

$$\overline{A}_t \triangleq A_t - SR^{-1}C_t, \qquad \overline{B}_t \triangleq B_t - SR^{-1}D_t. \tag{48}$$

The associated state covariance matrices are computed from their square roots, for example, $P_{t|t} = P_{t|t}^{1/2} P_{t|t}^{T/2}$*, via recursions involving the following* $\mathscr{Q}R$*-decompositions*

$$\begin{bmatrix} P_{t|t}^{T/2}\overline{A}_t^T & P_{t|t}^{T/2} \\ \overline{Q}^{T/2} & 0 \\ 0 & P_{t+1|N}^{T/2} J_t^T \end{bmatrix} = \mathscr{Q}^1 \begin{bmatrix} \mathscr{R}_{11}^1 & \mathscr{R}_{12}^1 \\ 0 & \mathscr{R}_{22}^1 \\ 0 & 0 \end{bmatrix}, \tag{49}$$

$$\begin{bmatrix} P_{t-1|t-1}^{T/2}\overline{A}_{t-1}^T \\ \overline{Q}^{T/2} \end{bmatrix} = \mathscr{Q}^2 \begin{bmatrix} \mathscr{R}_1^2 \\ 0 \end{bmatrix}, \tag{50}$$

$$\begin{bmatrix} R^{T/2} & 0 \\ P_{t|t-1}^{T/2} C_t^T & P_{t|t-1}^{T/2} \end{bmatrix} = \mathscr{Q}^3 \begin{bmatrix} \mathscr{R}_{11}^3 & \mathscr{R}_{12}^3 \\ 0 & \mathscr{R}_{22}^3 \end{bmatrix}, \tag{51}$$

where $\overline{Q} \triangleq Q - SR^{-1}S^T$ *and then setting*

$$P_{t|N}^{T/2} = \mathscr{R}_{22}^1, \tag{52}$$

$$P_{t|t-1}^{T/2} = \mathscr{R}_1^2, \tag{53}$$

$$P_{t|t}^{T/2} = \mathscr{R}_{22}^3. \tag{54}$$

The matrices $M_{N|N}$ *and* $M_{N+1|N}$ *are calculated via the initialisation*

$$M_{N|N} = (I - K_N C_N)\overline{A}_{N-1} P_{N-1|N-1}, \tag{55}$$

$$M_{N+1|N} = \overline{A}_N P_{N|N}, \tag{56}$$

followed by the the backwards recursion $\{M_{t|N}\}_{t=2}^{N-1}$ *given by*

$$M_{t|N} = P_{t|t} J_{t-1}^T + J_t (M_{t+1|N} - \overline{A}_t P_{t|t}) J_{t-1}^T. \tag{57}$$

Finally, the reverse time recursion (46) is initialised by running to $t = N$ *the (robust) Kalman filter recursions*

$$K_t = P_{t|t-1} C_t^T \left(C_t P_{t|t-1} C_t^T + R\right)^{-1} = \mathscr{R}_{12}^3, \tag{58}$$

$$\widehat{x}_{t|t-1} = \overline{A}_{t-1}\widehat{x}_{t-1|t-1} + \overline{B}_t u_{t-1} + SR^{-1} y_{t-1}, \tag{59}$$

$$\widehat{x}_{t|t} = \widehat{x}_{t|t-1} + K_t \left(y_t - C_t \widehat{x}_{t|t-1} - D_t u_t \right), \tag{60}$$

for $t = 1, \ldots, N$.

Proof. See Section 4.1 in [Gibson and Ninness (2005)]. □

5.2. *M-Step*

With the completion of the E-Step delivering $\mathscr{Q}(\theta, \widehat{\theta}_k)$, attention turns to maximising it with respect to θ. At this point, the details of how θ parametrizes the LPV model (1a)-(2) need to be established. For this purpose, we assume that θ is defined in a partitioned manner according to

$$\theta^T \triangleq [\beta^T, \eta^T]^T, \tag{61}$$

where recalling (1c), (29)

$$\beta^T \triangleq \left[\text{vec}\left\{\Gamma\right\}^T, \mu^T \right], \tag{62}$$

$$\eta^T \triangleq \left[\text{vec}\left\{\Pi\right\}^T, \text{vec}\left\{P_1\right\}^T \right], \tag{63}$$

and the vec$\{\cdot\}$ operator creates a vector from a matrix by stacking its columns on top of one another. We likewise partition the set Θ of candidate parameter vectors as

$$\Theta = \Theta_1 \times \Theta_2, \tag{64a}$$

$$\beta \in \Theta_1, \tag{64b}$$

$$\eta \in \Theta_2, \tag{64c}$$

where $\Theta_1 \subseteq \mathbf{R}^{n_\beta}$, $\Theta_2 \subseteq \mathbf{R}^{n_\eta}$ are both compact, and for which all $\eta \in \Theta_2$ imply symmetric positive definite Π, P_1.

With these definitions in place, local maximisers of $\mathscr{Q}(\theta, \widehat{\theta}_k)$ may be directly computed via the results of the following lemma.

Lemma 3. *Let* Σ *defined in* (34) *satisfy* $\Sigma > 0$ *and be used to define* $\widehat{\beta}$ *according to (*Ψ *is defined in* (33)*)*

$$\widehat{\beta} \triangleq \left[\text{vec}\left\{\widehat{\Gamma}\right\}^T, \widehat{\mu}^T \right]^T, \tag{65a}$$

$$\widehat{\Gamma} = \Psi \Sigma^{-1}, \tag{65b}$$

$$\widehat{\mu} \triangleq \widehat{x}_{1|N}. \tag{65c}$$

If Θ defined by (64) is such that $\widehat{\beta}$ lies within Θ_1, then for any fixed $\eta \in \Theta_2$, the point (65) is the unique maximiser

$$\widehat{\beta} = \arg\max_{\theta\in\Theta_1} \mathcal{Q}\left(\begin{bmatrix}\beta \\ \eta\end{bmatrix}, \widehat{\theta}_k\right). \tag{66}$$

Furthermore, $\widehat{\eta}$ given by

$$\widehat{\eta} \triangleq \left[\text{vec}\left\{\widehat{\Pi}\right\}^T, \text{vec}\left\{\widehat{P}_1\right\}^T\right]^T, \tag{67}$$

$$\widehat{\Pi} = \mathscr{R}_{22}^{T/2}\mathscr{R}_{22}^{1/2}, \tag{68}$$

$$\widehat{P}_1 \triangleq P_{1|N}^{1/2}P_{1|N}^{T/2}, \tag{69}$$

forms a stationary point of $\mathcal{Q}(\cdot, \widehat{\theta}_k)$ with respect to η. Here, $P_{1|N}^{1/2}$ is defined by (49) and $\mathscr{R}_{22}^{1/2}$ is defined by the Cholesky factorisation (see Algorithm 4.2.4 of [Golub and van Loan (1989)])

$$\begin{bmatrix}\Sigma & \Psi^T \\ \Psi & \Phi\end{bmatrix} = \begin{bmatrix}\mathscr{R}_{11} & \mathscr{R}_{12} \\ \varnothing & \mathscr{R}_{22}\end{bmatrix}^T \begin{bmatrix}\mathscr{R}_{11} & \mathscr{R}_{12} \\ \varnothing & \mathscr{R}_{22}\end{bmatrix}. \tag{70}$$

Note that the right hand side of the expression for Π in (68) is

$$\Pi = \Phi - \Psi\Sigma^{-1}\Psi^T \tag{71}$$

realised in a numerically robust fashion that ensures essential properties of symmetry and non negative-definiteness of the result.

Proof. This follows by an argument identical to that used to establish Lemma 4.3 of [Gibson *et al.* (2005)]. □

5.3. *A summary of the algorithm*

The preceding derivations are now summarised in the interests of clearly defining the new algorithm proposed here.

Algorithm 5.1 (EM Algorithm for LPV Systems).

(1) *Initialise the algorithm by choosing a parameter vector $\widehat{\theta}_0$.*

(2) **(E-Step)** *Employ the square-root implementation of the modified Kalman smoother presented in Lemma 2 in conjunction with the parameter estimate $\widehat{\theta}_k$ to calculate the matrices Φ, Ψ and Σ as shown in equations (32) through (34).*

(3) **(M-Step)** *In order to choose an updated parameter estimate* $\widehat{\theta}_{k+1}$, *select* Π, Γ, μ *and* P_1 *according to equations* (65), (68) *and* (69).

(4) If the algorithm has converged, terminate, otherwise return to step 2.

Regarding step 4, obvious strategies for gauging convergence involve copying those developed for gradient based search [Dennis and Schnabel (1983); Nocedal and Wright (1999)]. In particular, this chapter suggests a strategy of termination when relative likelihood increase on an iteration drops below a predetermined threshold. This approach is supported by empirical evidence. In the authors experience, once the rate of increase in the likelihood drops it rarely returns to higher levels.

6. Simulation study

In this section we demonstrate the utility of the EM method detailed above for the estimation of LPV systems from input-output measurements.

6.1. *SISO first order example*

We begin with a simple first order SISO LPV system in the interests of clearly profiling parameter estimation accuracy. For this purpose, consider an LPV model in the form of (1) where $n = 1$, $m = 1$, $p = 1$ and with time variation involving $n_\alpha = n_\beta = 2$ as follows

$$x_{t+1} = \alpha_t(1)A(1)x_t + \alpha_t(2)A(2)x_t + \beta_t(1)B(1)u_t + \beta_t(2)B(2)u_t + w_t \tag{72}$$

$$y_t = \alpha_t(1)C(1)x_t + \alpha_t(2)C(2)x_t + \beta_t(1)D(1)u_t + \beta_t(2)D(2)u_t + e_t \tag{73}$$

with parameter values

$$\Gamma = \begin{bmatrix} A(1) & A(2) & B(1) & B(2) \\ C(1) & C(2) & D(1) & D(2) \end{bmatrix} = \begin{bmatrix} 0.9 & 0.6 & 1 & 1 \\ 0.5 & 0.7 & 0 & 0 \end{bmatrix}. \tag{74}$$

The noise signals w_t, e_t are taken as i.i.d. normally distributed signals with covariance structure defined by

$$\begin{bmatrix} w_t \\ e_t \end{bmatrix} \sim \mathcal{N}\left(0, \begin{bmatrix} Q & S \\ S & R \end{bmatrix}\right) \tag{75}$$

with

$$[Q, S, R] = [0.01, 0, 0.01]. \tag{76}$$

Estimates of these parameters are collected in a vector θ according to

$$\theta^T = [A(1),\ A(2),\ B(1),\ B(2),\ C(1),\ C(2),\ D(1),\ D(2),\ Q,\ S,\ R].$$

For the purposes of estimation, $N = 1000$ samples of the output y_t of this model with input u_t being ± 1 amplitude square wave with period 100 samples was generated, and are illustrated as (respectively) the bottom and top plots in Figure 1. The middle plot illustrates the scheduling signals $\alpha_t(1)$ (solid) and $\alpha_t(2)$ (dashed) as unit amplitude non-negative sinusoidal waves π-radians out of phase. Finally, the scheduling signal β_t was set equal to this choice for α_t.

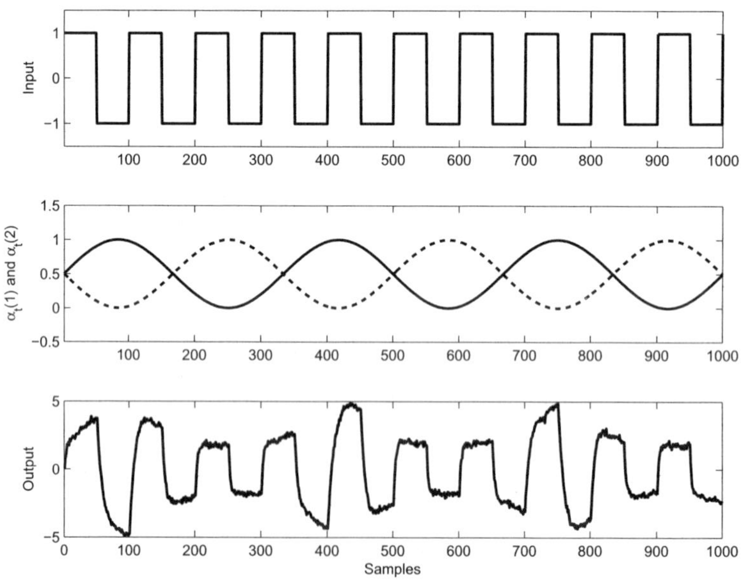

Fig. 1. Top: input signal; Middle: $\alpha_t(1)$ (solid) and $\alpha_t(2)$ (dashed) signals; Bottom: output signal.

In fact, $M = 1000$ realisations of this sort were generated, and for each one an estimate $\widehat{\theta}$ of the parameters θ were obtained via the EM Algorithm 5.1.

For this purpose, the EM method was initialised as

$$\widehat{\theta}_0^T = \left[10^{-5}[0.9,\ 0.6,\ 1,\ 1,\ 0.5,\ 0.7],\ 1,\ 0,\ 1\right], \tag{77}$$

so that the system matrices are almost zero and the noise covariance is 100 times larger than the true values.

Recall that the state-space model (1) is not uniquely parametrized. In general this presents a difficulty when comparing true system values, like those in (74), with estimated ones, since the estimates are likely to correspond to a different state coordinate system. However, for the first order example considered here, any similarity transformation of the state will cancel for the $A(1,2)$ terms and will also cancel when considering the product of $B(1)$ with $C(1)$ and $B(2)$ with $C(2)$.

This facilitates studying the accuracy of the EM-based estimates via the histograms of estimated $\widehat{A}(1,2)$, $\widehat{B}(1)\widehat{C}(1)$, $\widehat{B}(2)\widehat{C}(2)$ and $\widehat{D}(1,2)$ illustrated in Figure 2 together with true values as vertical lines. This illustrates the estimates to have empirical mean values close to the true values, and with

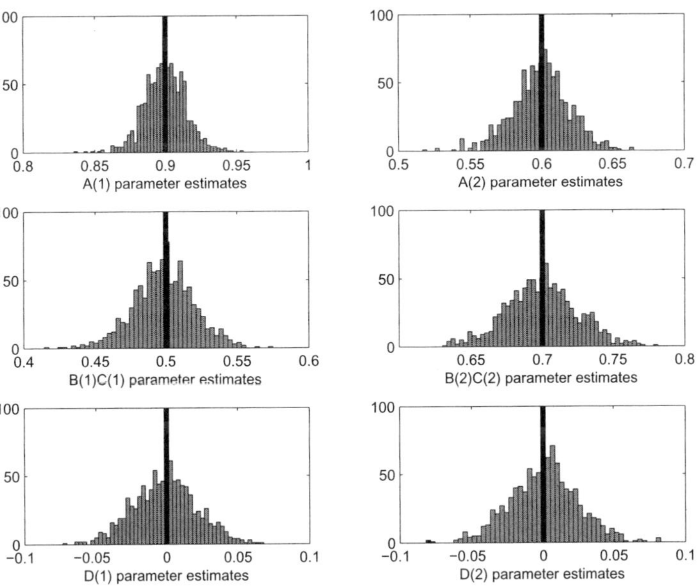

Fig. 2. Histograms of estimated parameter values.

relatively small dispersion about them, despite the initial estimates being quite far from the true values.

These encouraging results on a very simple system justify further study on more complex systems.

6.2. *MIMO* 10*'th order example*

We now profile a more challenging situation where the underlying LPV system was formed by creating two random $n = 10$'th order $m = 2$ input, $p = 2$ output LTI-systems $[A(1), B(1), C(1), D(1)]$ and $[A(2), B(2), C(2), D(2)]$ that were both ensured to be stable. These were employed in the LPV model structure (1) with $n_\alpha = n_\beta = 2$. The scheduling signal α_t was taken as the out of phase square waves shown in the middle plot of Figure 3, and the scheduling signal β_t was chosen as the out of phase positive sinusoidal signals shown in the bottom plot of Figure 3.

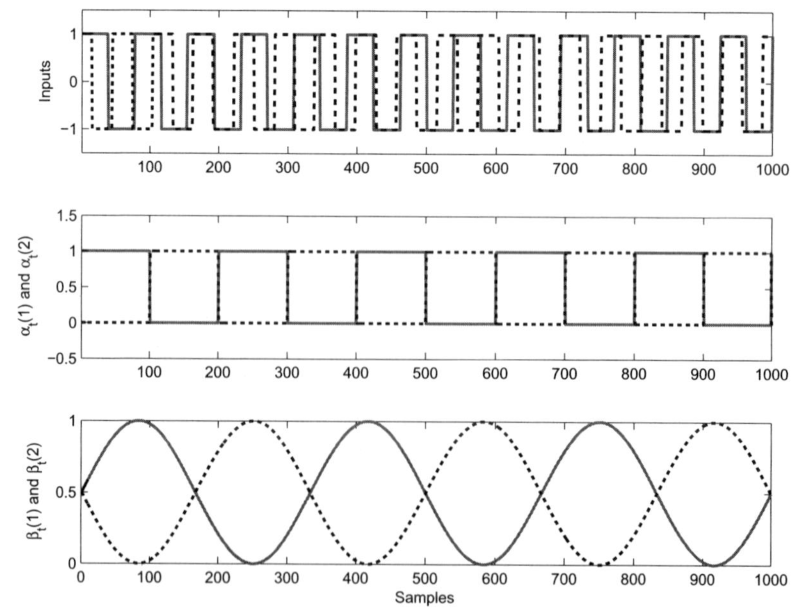

Fig. 3. Top: input signals; Middle: $\alpha_t(1)$ (solid) and $\alpha_t(2)$ (dashed) signals; Bottom: $\beta_t(1)$ (solid) and $\beta_t(2)$ (dashed) signals.

The state and measurement noises w_t, e_t were taken as i.i.d. processes according to

$$\begin{bmatrix} w_t \\ e_t \end{bmatrix} \sim \mathcal{N}(0,\ \Pi), \quad \Pi = \begin{bmatrix} 10^{-2} I_n & 0 \\ 0 & 10^{-2} I_p \end{bmatrix}. \tag{78}$$

The system was simulated using $N = 1000$ samples of the square wave input shown in the top plot of Figure 3 in order to provide an $N = 1000$ ouput response y_t.

Based on this data, the EM Algorithm 5.1 was used to compute a parameter estimate $\widehat{\theta}$ with deliberately poor initial estimate $\widehat{\theta}_0$ chosen as

$$\widehat{\theta}_0^T = [\text{vec}\left\{10^{-5}\Gamma\right\}^T,\ \text{vec}\left\{100\Pi\right\}^T]. \tag{79}$$

Unlike the previous example, it is more difficult to present a comparison of the true and estimated parameter values in the current case. Instead we adopt a common approach of profiling the measured system output y_t versus the predictor $\widehat{y}_{t|t-1}$ implied by the estimated model $\widehat{\theta}$ according to the Kalman filter (14).

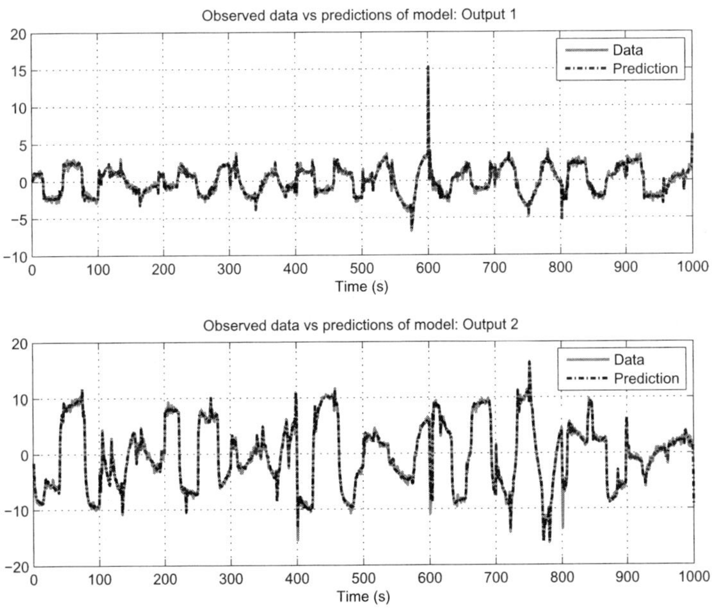

Fig. 4. Predicted (dashed) and actual (solid) output responses.

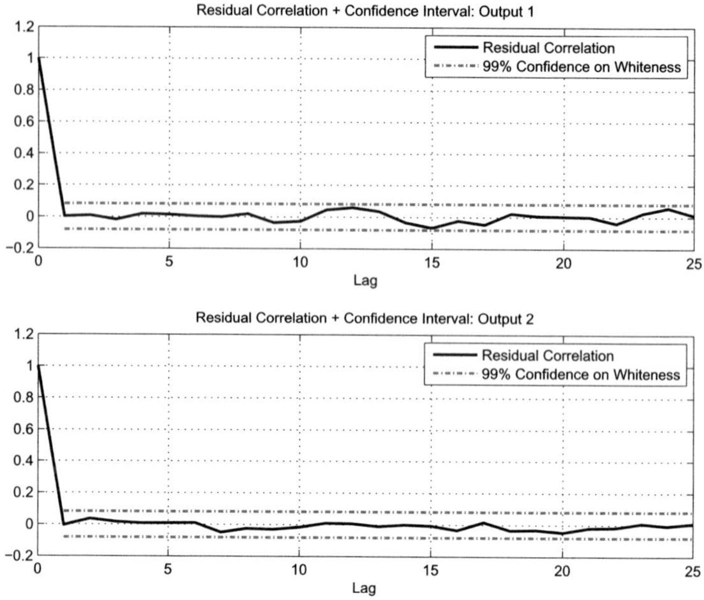

Fig. 5. Autocorrelation of residual errors.

This is shown in Figure 4, and illustrates a close match, consistent with an accurate estimate being obtained. To further study this, the sample correlation of the prediction residual $\varepsilon_t = y_t - \widehat{y}_{t|t-1}$ with itself, and with the input u_t together with 99% confidence bounds was computed and is shown in Figures 5 and 6. Again, these results indicate that an accurate estimate has been obtained.

While these results pertain to just one system and data realisation, the authors can report that they have trialed hundreds of realisations with identical results.

7. Conclusion

In this chapter the expectation-maximisation (EM) algorithm has been presented and examined for the purpose of finding Maximum-Likelihood estimates of state-space linear parameter varying models.

This has illustrated advantages of relatively straightforward implementation, apparently robust to initial parameter estimates, computational load

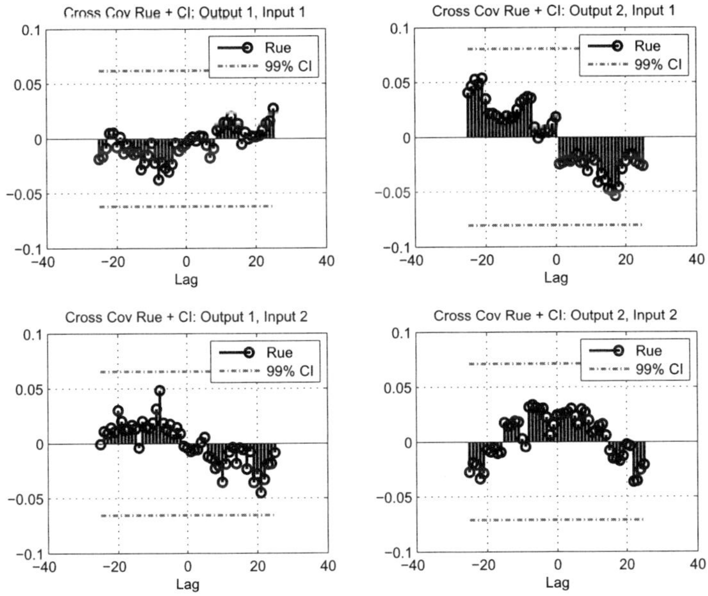

Fig. 6. Cross-correlation of residual error with input.

that scales linearly with data length N, and straighforward handling of high-order MIMO LPV systems. Balancing is a disadvantage of not being able to straightforwardly handle structured LPV systems.

Matlab-based code implementing and illustrating the EM approach described here is freely available from the authors upon request.

References

Åström, K. (1980). Maximum likelihood and prediction error methods, *Automatica* **16**, pp. 551–574.

Balas, G. J. (2002). Linear parameter-varying control and its application to a turbofan engine, *International Journal of Robust and Nonlinear Control* **12**, 9, pp. 763–796.

Bamieh, B. and Giarré, L. (2002). Identification of linear parameter varying models, *International Journal of Robust and Nonlinear Control* **12**, pp. 841–853.

Bernstein, D. S. (2005). *Matrix Mathematics: Theory, Facts and Formulas with Application to Linear Systems Theory* (Princeton University Press, Princeton, NJ, USA).

Dennis, J. and Schnabel, R. B. (1983). *Numerical methods for unconstrained optimization and nonlinear equations* (Prentice-Hall, Englewood Cliffs, NJ, USA).

Felici, F., van Wingerden, J. W. and Verhaegen, M. (2007). Subspace identification of MIMO LPV systems using a periodic scheduling sequence, *Automatica* **43**, 10, pp. 1684–1697.

Fujimori, A. and Ljung, L. (2006). Model identification of linear parameter varying aircraft systems, *Proceedings of the Institute of Mechanical Engineers, Part G Journal of Aerospace Engineering* **220**, G4, pp. 337–346.

Giarré, L., Bauso, D., Falugi, P. and Bamieh, B. (2006). LPV model identification for gain scheduling control:An application to rotating stall and surge control problems, *Control Engineering Practice* **14**, 4, pp. 351–361.

Gibson, S. and Ninness, B. (2005). Robust maximum-likelihood estimation of multivariable dynamic systems, *Automatica* **41**, 10, pp. 1667–1682.

Gibson, S., Wills, A. and Ninness, B. (2005). Maximum-likelihood parameter estimation of bilinear systems, *IEEE Transactions on Automatic Control* **50**, 10, pp. 1581–1596.

Golub, G. and van Loan, C. (1989). *Matrix Computations* (Johns Hopkins University Press, Baltimore, MD, USA).

Goodwin, G. and Feuer, A. (1999). Estimation with missing data, *Mathematical and Computer Modelling of Dynamical Systems* **5**, 3, pp. 220–244.

Gopaluni, R. B. (2008). A particle filter approach to identification of nonlinear processes under missing observations, *The Canadian Journal of Chemical Engineering* **86**, 6, pp. 1081–1092.

Isaksson, A. J. (1993). Identification of ARX-models subject to missing data, *IEEE Transactions on Automatic Control* **38**, 5, pp. 813–819.

Laurain, V., Gilson, M., Tth, R. and Garnier, H. (2010). Refined instrumental variable methods for identification of LPV box–jenkins models, *Automatica*, pp. 959–967.

Lee, L. H. and Poolla, K. (1999). Identification of linear parameter-varying systems using nonlinear programming, *Journal of Dynamic Systems, Management, and Control* **121**, pp. 71–78.

Ljung, L. (1978). Convergence analysis of parametric identification methods, *IEEE Transactions on Automatic Control* **AC-23**, 5, pp. 770–783.

Marcos, A. and Balas, G. (2004). Development of linear parameter varying models for aircraft, *Journal of Guidance, Control and Dynamics* **27**, 2, pp. 218–228.

Nocedal, J. and Wright, S. (1999). *Numerical Optimization* (Springer-Verlag, New York, NY, USA).

Rugh, W. and Shamma, J. (2000). Research on gain scheduling, *Automatica* **36**, 10, pp. 1401–1425.

van Wingerden, J. W. and Verhaegen, M. (2009). Subspace identification of bilinear and LPV systems for open- and closed-loop data, *Automatica* **45**, pp. 372–381.

Verdult, V., Bergboer, N. and Verhaegen, M. (2003). Identification of fully parametrized linear and nonlinear state-space systems by projected gradient search, in *Proceedings of the IFAC Symposium on System Identification* (Rotterdam, Netherlands).

Verdult, V., Lovera, M. and Verhaegen, M. (2004). Identification of linear parameter varying state space models with application to helicopter rotor dynamics, *International Journal of Control* **77**, 13, pp. 1149–1159.

Verdult, V. and Verhaegen, M. (2002). Subspace identification of multivariable linear parameter-varying systems, *Automatica* **38**, 5, pp. 805–814.

Verdult, V. and Verhaegen, M. (2005). Kernel methods for subspace identification of multivariable LPV and bilinear systems, *Automatica* **41**, 9, pp. 1557–1565.

Wills, A., Ninness, B. and Gibson, S. (2009). Maximum likelihood estimation of state space models from frequency domain data, *IEEE Transactions on Automatic Control* **54**, 1, pp. 19–33.

Chapter 12

Piecewise affine identification of interconnected systems with LFR structure

S. Paoletti & A. Garulli

Dipartimento di Ingegneria dell'Informazione
Università degli Studi di Siena
via Roma 56, 53100 Siena, Italy
paoletti@dii.unisi.it
garulli@ing.unisi.it

In this chapter, identification of systems composed by interconnected LTI systems and static nonlinearities is addressed. The interconnection is described by a linear fractional representation in which the nonlinear block is modeled by a piecewise affine map. A two-step iterative identification procedure, with non-increasing cost at each iteration, is proposed. Features and effectiveness of the presented identification procedure are illustrated through a numerical example and an application to the *silver-box* problem (a popular benchmark in nonlinear system identification).

1. Introduction

Identification of *linear parameter-varying* (LPV) systems is a broad research area which has attracted an increasing number of researchers in recent years (see, e.g., [Tóth (2008)], Chapter 2 of this book, and references therein). In particular, within the general framework of LPV systems, several model classes have been studied. Notable examples are *switched systems*, in which the scheduling parameters are discrete external signals that do not depend on the system internal variables, and *piecewise affine (PWA) systems*, where the scheduling parameters are a piecewise constant function of the system states and inputs.

The universal approximation properties of PWA maps [Lin and Unbehauen (1992); Breiman (1993)] are one of the main reasons for the steadily growing interest in PWA system identification. A variety of techniques have been proposed in the literature for this model class (see [Roll (2003)] and [Paoletti *et al.* (2007)] for an extensive overview). Most of these approaches deal with black-box identification and do not exploit any prior knowledge on the structure of the system generating the data. However, in real world systems some information on the system structure is often available. A typical example is that of Hammerstein or Wiener systems, in which a static nonlinearity is cascaded with a linear dynamical system. More complex interconnections can be represented by adopting the *linear fractional representation* (LFR) formalism (see, e.g., [Zhou *et al.* (1996); Hsu *et al.* (2005a)]).

While Hammerstein and Wiener systems are quite popular in the identification literature, much less work has been done on identification of general LFR systems. One of the first contributions in this respect can be found in [Previdi and Lovera (2003)], where parameter estimation is performed for an LFR model in which the forward part is represented by a classical linear regression and the feedback part is given by a nonlinear dynamic map parameterized by a neural network. Identification of LFR interconnections of linear dynamic systems and static nonlinear maps has been addressed in [Hsu *et al.* (2005b, 2008)], where the linear part is assumed to be known and a nonparametric approach for the estimation of the static nonlinear maps is proposed. In [Pepona *et al.* (2011)], a method has been proposed for identification of LFR models composed by the interconnection of a linear time-invariant block and a static PWA nonlinearity, both unknown. The method relies on an iterative scheme that alternates the identification of the linear and the nonlinear part, and can be seen as the extension to LFR models of classical iterative identification techniques such as [Narendra and Gallman (1966); Voros (2003); Bai and Li (2004)]. However, convergence of the proposed iterative procedure is not guaranteed. Moreover, the method employs a smoothing procedure for estimating the interconnection internal signals, which requires stability of the zeros of the estimated linear block.

In this paper, the same model class as in [Pepona *et al.* (2011)] is considered, and a new iterative procedure is proposed. The new procedure guarantees that the identification cost (namely, the squared norm of the residuals) does not increase at each step. The smoothing step is not required anymore and this removes the constraint on the stability of the zeros of the linear part. The main feature of the proposed approach is that it allows one to successfully exploit the prior knowledge on the system interconnection

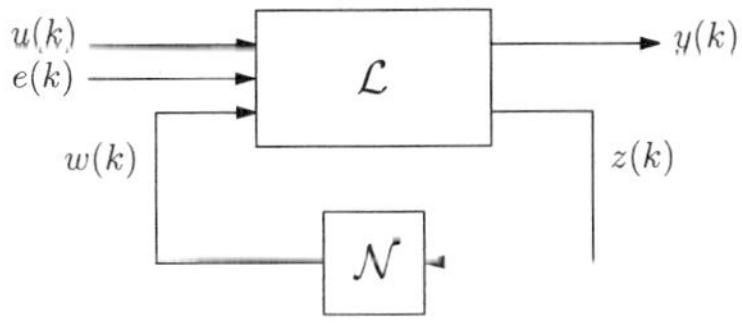

Fig. 1. LFR model structure.

structure. This is demonstrated by comparing the identified LFR models to those obtained by applying black-box PWA identification techniques directly to the overall interconnected system. Results are presented for two case studies: a simple academic example and the *silverbox* problem, a benchmark application in nonlinear system identification [Pintelon and Schoukens (2001); Nolcos (2004)].

The paper is organized as follows. The considered model class and identification problem are introduced in Section 2. Section 3 provides a basic background on black-box PWA system identification. The iterative identification algorithm is presented in Section 4, and applied in Section 5 to a numerical example and to the *silverbox* problem. Conclusions and future research directions are summarized in Section 6.

2. PWA-LFR models

The model class considered in the paper is firstly introduced. A discrete-time dynamical system formed by the interconnection of linear and nonlinear components can always be cast into the LFR form in Fig. 1 (see, e.g., [Hsu *et al.* (2008)]), where the block

$$\mathscr{L} = \begin{bmatrix} \mathscr{L}_{yu} & \mathscr{L}_{ye} & \mathscr{L}_{yw} \\ \mathscr{L}_{zu} & \mathscr{L}_{ze} & \mathscr{L}_{zw} \end{bmatrix} \tag{1}$$

is a linear time-invariant dynamical system (partitioned as usual in control theory), while the block $\mathscr{N}$ takes into account the system nonlinearities. Signals $u(k)$, $y(k)$ and $e(k)$ are the system input, output and noise at time $k \in \mathbb{Z}$, while $z(k)$ and $w(k)$ are internal signals representing respectively the input and the output of the nonlinear block. In this paper, attention is restricted to the case in which all the involved signals are scalar, and the nonlinearity is static.

The considered class of *PWA-LFR models* is defined by the equations

$$A(q)y(k) = B(q)u(k - n_{k_u}) + G(q)w(k - n_{k_w}) + \varepsilon(k) \tag{2a}$$

$$w(k) = f(z(k)), \tag{2b}$$

where the ARX model (2a) describes the linear block $\mathscr{L}$ in Fig. 1, and the static map $f(\cdot)$ in (2b) is a PWA approximation of the static nonlinearity $\mathscr{N}$. In (2a), $\varepsilon(k)$ is the error process, the integers n_{k_u} and n_{k_w} are fixed delays of the u and w signals, and $A(q), B(q), G(q)$ are polynomials of fixed degree n_a, n_b-1 and n_g-1, respectively, in the delay operator q^{-1}, namely

$$A(q) = 1 + a_1 q^{-1} + \ldots + a_{n_a} q^{-n_a} \tag{3a}$$

$$B(q) = b_1 + b_2 q^{-1} + \ldots + b_{n_b} q^{-n_b+1} \tag{3b}$$

$$G(q) = g_1 + g_2 q^{-1} + \ldots + g_{n_g} q^{-n_g+1}, \tag{3c}$$

with $b_1 \neq 0$ and $g_1 \neq 0$. In the following, it will be useful to rewrite (2a) in regression form as

$$y(k) = \vartheta^T \varphi(k) + \varepsilon(k), \tag{4}$$

where

$$\vartheta = [\ \vartheta_\phi^T\ \vartheta_\omega^T\ ]^T \tag{5}$$

is the unknown parameter vector, partitioned according to

$$\vartheta_\phi = [\ a_1\ \ldots\ a_{n_a}\ b_1\ \ldots\ b_{n_b}\ ]^T \tag{6a}$$

$$\vartheta_\omega = [\ g_1\ \ldots\ g_{n_g}\ ]^T, \tag{6b}$$

and

$$\varphi(k) = [\ \phi(k)^T\ \omega(k)^T\ ]^T \tag{7}$$

is the (partially unknown) regression vector, with

$$\phi(k) = [\ -y(k-1)\ \ldots\ -y(k-n_a)\ u(k-n_{k_u})\ \ldots\ u(k-n_b-n_{k_u}+1)\ ]^T \tag{8a}$$

$$\omega(k) = [\ w(k-n_{k_w})\ \ldots\ w(k-n_g-n_{k_w}+1)\ ]^T. \tag{8b}$$

The PWA map $f(\cdot)$ in (2b) has the form:

$$f(z) = \alpha_i z + \beta_i \quad \text{if } z \in \mathscr{Z}_i, \quad i = 1, \ldots, s, \tag{9}$$

where s is the number of modes, $\alpha_i, \beta_i \in \mathbb{R}$, $i = 1, ..., s$, are the parameters of each mode, and $\{\mathscr{Z}_i\}_{i=1}^s$ is a set of disjoint intervals covering the domain $\mathscr{Z} \subseteq \mathbb{R}$ where the map $f(\cdot)$ is defined, i.e. $\mathscr{Z} = \bigcup_{i=1}^s \mathscr{Z}_i$ and $\mathscr{Z}_i \bigcap \mathscr{Z}_j = \emptyset$ for any $i \neq j$.

Having in mind the objective to identify a PWA-LFR model (2) of the LFR system in Fig. 1, the following main assumption is considered.

Assumption 12.1. The signal $z(k)$ is either measured or can be inferred from past inputs $u(k)$ and outputs $y(k)$, i.e. there exists a known stable linear system $\mathscr{M}$ such that $z(k) = \mathscr{M} \begin{bmatrix} y(k) \\ u(k) \end{bmatrix}$.

Assumption 12.1 is similar to the "measurability" assumption on $z(k)$ adopted in [Hsu *et al.* (2008)], but in a context where $\mathscr{L}$ is assumed to be known and only $\mathscr{N}$ has to be identified. Such an assumption is motivated by the fact that the situation when $z(k)$ is known, while $w(k)$ is unknown, appears to be the most general situation worth of investigation if one wants to identify both $\mathscr{L}$ and $\mathscr{N}$. Indeed, if both $z(k)$ and $w(k)$ are known, the identification problem becomes trivially the separate identification of the linear system $\mathscr{L}$ and of the nonlinear system $\mathscr{N}$, which are well-studied problems, while if $w(k)$ is known and $z(k)$ is unknown, only the part $[\ \mathscr{L}_{yu}\ \mathscr{L}_{ye}\ \mathscr{L}_{yw}\ ]$ of $\mathscr{L}$ can be identified. Note that stability of the known linear system $\mathscr{M}$ in Assumption 12.1 is required to cope with the uncertainty on its initial condition. Under Assumption 12.1, the considered identification problem is the following:

Problem 12.1. Given the data set $\{u(k), y(k), z(k)\}_{k=1}^N$, identify a PWA-LFR model (2) of the system in Fig. 1, such that the following cost (squared norm of the residuals) is minimized:

$$J(\vartheta, f(\cdot)) = \sum_{k=\bar{n}+1}^{N} |y(k) - \vartheta^T \varphi(k)|^2, \tag{10}$$

where $\bar{n} = \max\{n_a, n_b + n_{k_u} - 1, n_g + n_{k_w} - 1\}$, while ϑ and $\varphi(k)$ are defined according to (5)-(8).

3. Black-box PWA system identification

By suitably rewriting equations (2) and (9), the PWA-LFR model can be equivalently described by a *piecewise affine ARX* (PWARX) model of the type

$$y(k) = g(\boldsymbol{r}(k)) + \varepsilon(k), \tag{11}$$

where $\boldsymbol{r}(k)$ is a finite-length regression vector composed by past inputs and outputs, and $g(\cdot)$ is a PWA map. The following simple example clarifies how this can be done.

Example 1. Consider the ARX model

$$y(k) + a_1 y(k-1) = b_1 u(k-1) + g_1 w(k-2) + g_2 w(k-3) + \varepsilon(k), \tag{12}$$

where the input $w(k)$ is a PWA function of $z(k) = y(k)$ with 2 modes, i.e.

$$w(k) = \begin{cases} \alpha_1 y(k) + \beta_1 & \text{if } y(k) < 0 \\ \alpha_2 y(k) + \beta_2 & \text{if } y(k) \geq 0. \end{cases} \tag{13}$$

This is a PWA-LFR model with $n_a = n_b = 1$, $n_g = 2$, $n_{k_u} = 1$, $n_{k_w} = 2$ and $s = 2$. By substituting (13) into (12), one obtains

$$\begin{aligned} y(k) = &- a_1 y(k-1) + b_1 u(k-1) + g_1 \alpha_{i(k-2)} y(k-2) + g_1 \beta_{i(k-2)} \\ &+ g_2 \alpha_{i(k-3)} y(k-3) + g_2 \beta_{i(k-3)} + \varepsilon(k), \end{aligned} \tag{14}$$

where $i(k)$ is the scheduling parameter, defined as

$$i(k) = \begin{cases} 1 & \text{if } y(k) < 0 \\ 2 & \text{if } y(k) \geq 0. \end{cases} \tag{15}$$

Equation (14) can be rewritten in PWARX form (11), by setting $\boldsymbol{r}(k) = [\ -y(k-1)\ y(k-2)\ y(k-3)\ u(k-1)\ ]^T$ and choosing the PWA map $g(\cdot)$ such that

$$g(\boldsymbol{r}) = \Theta_i^T \begin{bmatrix} \boldsymbol{r} \\ 1 \end{bmatrix} \tag{16}$$

with

$$\Theta_i = \begin{cases} [\ a_1\ \ g_1\alpha_1\ \ g_2\alpha_1\ \ b_1\ \ g_1\beta_1 + g_2\beta_1\ ]^T & \text{if } \boldsymbol{r}_2 < 0, \boldsymbol{r}_3 < 0 \\ [\ a_1\ \ g_1\alpha_1\ \ g_2\alpha_2\ \ b_1\ \ g_1\beta_1 + g_2\beta_2\ ]^T & \text{if } \boldsymbol{r}_2 < 0, \boldsymbol{r}_3 \geq 0 \\ [\ a_1\ \ g_1\alpha_2\ \ g_2\alpha_1\ \ b_1\ \ g_1\beta_2 + g_2\beta_1\ ]^T & \text{if } \boldsymbol{r}_2 \geq 0, \boldsymbol{r}_3 < 0 \\ [\ a_1\ \ g_1\alpha_2\ \ g_2\alpha_2\ \ b_1\ \ g_1\beta_2 + g_2\beta_2\ ]^T & \text{if } \boldsymbol{r}_2 \geq 0, \boldsymbol{r}_3 \geq 0. \end{cases} \tag{17}$$

Note that the resulting PWARX model has 4 modes and the PWA map $g(\cdot)$ is defined in a regressor space of dimension 4, while the PWA map of the original PWA-LFR model has only 2 modes and is defined on the scalar variable $z(k) = y(k)$.

Clearly, a possible way to estimate a PWA model of the system in Fig. 1 from input-output data $\{u(k), y(k)\}$, is to apply black-box PWA identification techniques to obtain a PWARX model of the type in (11). However, this approach has several drawbacks. As highlighted in Example 1, the number of modes of the resulting PWA map $g(\cdot)$ is larger (typically, much larger) than s in (9). The same holds for the dimension of the regression vector $\boldsymbol{r}(k)$ with respect to $z(k)$. This implies that, in general, PWA-LFR models are more parsimonious in terms of number of parameters than their corresponding PWARX representations. This suggests that exploiting the internal structure of the system should lead to a remarkable reduction of the computational burden required by the overall identification procedure, and give benefits in terms of accuracy of the identified model.

One of the aims of the present paper is to assess the benefits of exploiting structure in identification of interconnected systems, as compared to classical black-box approaches. For this reason, a quick review of black-box identification techniques for PWA models is given below.

3.1. *Review of PWA system identification*

The most straightforward way to estimate a PWA model (11) is to formulate directly the optimization problem:

$$\hat{g}(\cdot) = \arg \min_{g(\cdot)\in \mathrm{PWA}_s} \sum_{k=\bar{n}+1}^{N} |y(k) - g(\boldsymbol{r}(k))|^2, \tag{18}$$

where PWA_s denotes the set of PWA functions with s modes. Note that in this formulation the minimization is restricted to PWA_s functions $g(\cdot)$ with fixed number of modes. Indeed, if the number of modes is left free, the minimum cost in (18) can be made zero by associating one submodel to each data point (overfitting).

In general, the solution of (18) is a very hard problem. In [Roll *et al.* (2004)], the minimization problem is reformulated as a mixed-integer quadratic program (MIQP) for the class of Hinging Hyperplane models (a particular class of PWA models for which the PWA map is continuous). The MIQP problem can be solved for the global optimum, but this comes at the cost of a theoretically very high worst-case computational complexity, which means that the approach is mainly suitable for small-scale problems. Due to the difficulty of solving (18), researchers typically focused on heuristic and suboptimal approaches. In some cases, the parameters of the affine submodels and the coefficients of the hyperplanes defining the partition of the regressor domain are estimated simultaneously by minimizing directly the criterion function through numerical methods such as Gauss-Newton search (see, e.g., [Julián *et al.* (1998); Pucar and Sjöberg (1998)]). This way of tackling the identification problem is straightforward, but has the drawback that the optimization algorithm may get trapped in a local minimum. Techniques for reducing the risk of getting stuck in a local minimum can be used, at the cost of increased computational complexity. In other approaches, all parameters are identified simultaneously for a model with a very simple partition (e.g., with only two modes). If the resulting model is not satisfactory, new submodels are added, in order to improve the value of the criterion function. In other words, instead of solving it at once, the overall identification problem is divided into several steps, each consisting in an easier problem to solve. The algorithm proposed in [Breiman (1993)] is the most famous in this category, and is analyzed in [Pucar and Sjöberg (1998)]. In [Julián *et al.* (1998)] an iterative method is decribed for introducing new partitions in the regressor domain, when the error obtained is not satisfactory. As for the first category of approaches, there is still a risk of getting stuck in a local minimum. When adding new submodels, one should also take into consideration the risk of overfit.

The problem becomes easy if the number of modes s is fixed, and the partitions are either known or fixed *a priori*. In this case each regression vector $\boldsymbol{r}(k)$ can be associated to the submodel corresponding to the region $\mathscr{R}_i$ it belongs to. Hence, by introducing the quantities

$$\chi_i(k) = \begin{cases} 1 \text{ if } \boldsymbol{r}(k) \in \mathscr{R}_i \\ 0 \text{ otherwise,} \end{cases} \tag{19}$$

the identification problem reduces to the following optimization problem:

$$\min_{\Theta_i} \sum_{k=\bar{n}+1}^{N} \sum_{i=1}^{s} \chi_i(k) \, |y(k) - [\ \boldsymbol{r}(k)^T \ 1 \ ]^T \Theta_i|^2, \tag{20}$$

which is an ordinary least-squares problem in the unknowns Θ_i.

4. Structured identification of PWA-LFR models

This section describes an iterative approach to tackle Problem 12.1, thus enabling for the simultaneous identification of both the linear part $\mathscr{L}$ and the nonlinear part $\mathscr{N}$ of the system in Fig. 1.

In order to explain the proposed iterative approach, consider the cost (10) to be minimized. By recalling (5) and (7), (10) can be rewritten as

$$J(\vartheta, f(\cdot)) = \sum_{k=\bar{n}+1}^{N} |y(k) - \vartheta_\phi^T \phi(k) - \vartheta_\omega^T \omega(k)|^2. \tag{21}$$

On the other hand, by extending the PWA map (9) to vectors, i.e.

$$\tilde{f}(z_1, \ldots, z_d) = [\ f(z_1)\ \ldots\ f(z_d)\ ]^T, \tag{22}$$

and defining

$$\zeta(k) = [\ z(k - n_{k_w})\ \ldots\ z(k - n_{k_w} - n_g + 1)\ ]^T, \tag{23}$$

it turns out from (2b) and (8b) that:

$$\omega(k) = \tilde{f}(\zeta(k)). \tag{24}$$

By substituting (24) in (21), one finally obtains

$$J(\vartheta, f(\cdot)) = \sum_{k=\bar{n}+1}^{N} |y(k) - \vartheta_\phi^T \phi(k) - \vartheta_\omega^T \tilde{f}(\zeta(k))|^2. \tag{25}$$

The cost (25) should be minimized with respect to both the parameter vector ϑ and the PWA map $f(\cdot)$. Note however that the minimization of (25) with respect to the PWA map $f(\cdot)$ is not well-posed if one does not set an upper bound to the number of modes s of $f(\cdot)$. When s is not bounded above, one can always make the cost (25) equal to zero by suitably choosing $f(\cdot)$ with as many modes as data points $z(k)$. Therefore, in the following, it is assumed that s is fixed a priori, and (25) is minimized with respect to ϑ and $f(\cdot) \in PWA_s$, with PWA_s denoting the set of PWA functions of one variable with s modes.

The proposed iterative identification approach is to alternate the minimization of (25) with respect to $f(\cdot)$ and ϑ. In this way, one generates a sequence of values for $J(\vartheta, f(\cdot))$ which is non-increasing and bounded below. This guarantees that the sequence converges, and one can easily devise a stopping criterion which makes the procedure always terminate in finite time.

It is worthwhile to note that the presented approach, alternating between the estimation of the linear and the nonlinear part, stems from the classical iterative procedure for identification of Hammerstein systems originally proposed in [Narendra and Gallman (1966)]. This is motivated by the fact that, under Assumption 12.1, Problem 12.1 is an extension of the multi-input Hammerstein identification problem, in which one of the two input signals, namely $u(k)$, enters directly the linear block. The main difference with respect to the classical Hammerstein case is that the input $z(k)$, entering the nonlinear block, cannot be arbitrarily chosen for identification purposes, since it is an internal signal of the LFR system.

4.1. *Algorithm description*

The iterative algorithm for the identification of PWA-LFR models proposed in this paper is described next. After an initialization step, each algorithm iteration consists of two stages, namely the PWA approximation of the nonlinear part and the identification of the linear part.

Let $j = 1, 2, \ldots$ be the iteration counter, and $\vartheta^j = [\ (\vartheta^j_\phi)^T\ (\vartheta^j_\omega)^T\ ]^T$ denote the estimate of ϑ computed at the jth iteration.

Initialization

An estimate ϑ^0 of the parameter vector ϑ in (4) must be provided for initialization. This can be done in two steps:

i) The first part ϑ^0_ϕ, defined in (6a), is estimated by fitting an ARX model to the input-output data $\{u(k), y(k)\}$.

ii) The second part ϑ^0_ω, defined in (6b), could be chosen according to the available information on the interconnected system (see Section 5.2 for LFR models with Lur'e structure). If no prior knowledge is available, one may set the first entry of ϑ^0_ω to one[a] and the others equal to zero.

[a]This choice is without loss of generality, since it is always possible to scale the nonlinear map $f(\cdot)$.

PWA approximation of the nonlinear part

In this step, a PWA approximation $f^j(\cdot)$ of the static nonlinearity $\mathscr{N}$ is obtained by solving the following problem, derived from (25):

$$f^j(\cdot) = \arg \min_{f(\cdot)\in PWA_s} \sum_{k=\bar{n}+1}^{N} |y(k) - \left(\vartheta_\phi^{j-1}\right)^T \phi(k) - \left(\vartheta_\omega^{j-1}\right)^T \tilde{f}(\zeta(k))|^2. \quad (26)$$

Solving (26) corresponds to a PWA identification problem formulated in terms of the minimization of a cost functional, as described in Section 3.1. This is a combinatorial problem that, in order to be solved at the optimum, requires a computational load which grows dramatically with the number of modes s and the number of data N. Section 4.2 will describe how, in the case $s = 2$, it is possible to solve (26) at the optimum by evaluating exhaustively all the possible thresholds of the 2-mode PWA function $f(\cdot)$. Nevertheless, in the rationale of the iterative algorithm, at each step one might be satisfied with finding a PWA map that is not the optimal one, but still improves the cost with respect to the current solution. For this reason, Section 4.3 will describe a suboptimal technique for tackling (26) which makes it possible to limit the computational load at each iteration. The PWA approximation step allows one to obtain an estimate $w^j(k)$ of the unknown signal $w(k)$, by applying the estimated PWA map $f^j(\cdot)$ to the data $z(k)$:

$$w^j(k) = f^j(z(k)), \quad k = 1, \ldots, N. \quad (27)$$

Identification of the linear part

An estimate of the parameter vector $\vartheta = [\ \vartheta_\phi^T\ \vartheta_\omega^T\ ]^T$ in (4) is obtained by solving:

$$(\vartheta_\phi^j, \vartheta_\omega^j) = \arg \min_{(\vartheta_\phi, \vartheta_\omega)} \sum_{k=n+1}^{N} |y(k) - \vartheta_\phi^T \phi(k) - \vartheta_\omega^T \tilde{f}^j(\zeta(k))|^2. \quad (28)$$

Note that, thanks to the choice of the ARX structure to model the linear part $\mathscr{L}$, this step reduces to a classical least squares problem [Ljung (1999)].

Stopping criterion

The identification procedure stops when no significant changes occur to the cost between two consecutive iterations. By letting J_j be the minimum of (28) at iteration j (J_0 is conventionally set to $+\infty$), this criterion is implemented by checking if

$$(J_{j-1} - J_j) \leq \gamma \, J_{j-1}, \tag{29}$$

where $\gamma \in (0, 1)$ is a threshold defined by the user.

The overall iterative procedure for the identification of PWA-LFR models is summarized next.

Algorithm 12.1

(1) Given: ϑ^0, γ, set: $j = 0$, $J_0 = +\infty$.
(2) Set: $j = j + 1$.
(3) *PWA approximation of the nonlinear part*:
Estimate $f^j(\cdot)$ by solving (26).
(4) *Identification of the linear part*:
Estimate ϑ^j by solving (28). Let J^j be the minimum of (28).
(5) If $(J_{j-1} - J_j) > \gamma \, J_{j-1}$ go back to step 2.
(6) Return: $\vartheta = \vartheta^j$, $f(\cdot) = f^j(\cdot)$.

The following proposition formally states the convergence properties of Algorithm 12.1.

Proposition 12.1. *Let* $\{J_j\}_{j=0}^{\infty}$ *be the sequence of cost values generated by Algorithm 12.1. Then,* $J_j \leq J_{j-1}$ *for all* $j = 1, 2, \ldots,$ *and* $\lim_{j\to\infty} J_j$ *exists and is finite.*

Proof. Let J_j^{PWA} be the minimum of (26) at iteration $j = 1, 2, \ldots$. For $j = 2, 3, \ldots,$ $J_j^{PWA} \leq J_{j-1}$ because (26) is minimized with respect to $f(\cdot) \in PWA_s$ and $f^{j-1}(\cdot)$ is a particular choice with cost J_{j-1}. On the other hand, for $j = 1, 2, \ldots,$ $J_j \leq J_j^{PWA}$, because (28) is minimized with respect to $(\vartheta_\phi, \vartheta_\omega)$ and $(\vartheta_\phi^{j-1}, \vartheta_\omega^{j-1})$ is a particular choice with cost J_j^{PWA}. Therefore, $J_j \leq J_{j-1}$, $j = 1, 2, \ldots$. Since $\{J_j\}_{j=0}^{\infty}$ is a non-increasing sequence bounded below by 0 (all J_j are non-negative), $\lim_{j\to\infty} J_j$ exists and is finite. □

It is stressed that Proposition 12.1 does not imply convergence to the global optimum of (25). The cost (25) is non-convex, and therefore the iterative algorithm might get trapped in a local minimum. Clearly, a possible way to get around this problem is to repeat Algorithm 12.1 for different choices of ϑ_0 (e.g. randomly chosen), and take the best solution found.

Note that Proposition 12.1 justifies the termination condition (29). By letting $\bar{J} = \lim_{j\to\infty} J_j$, (29) always becomes true as $J_j \to \bar{J}$, thus guaranteeing the termination of the algorithm.

Remark 12.1. Algorithm 12.1 has the same two-step structure of the algorithm proposed in [Pepona *et al.* (2011)], alternating a bounded-error PWA approximation of the nonlinear part and the identification of the linear part. In that approach, the number of modes of the PWA map need not to be fixed a priori, being automatically estimated from data. However, there are also some drawbacks that have been overcome in the new approach presented in this paper. In particular:

- In [Pepona *et al.* (2011)], the PWA approximation of the nonlinear part is done by recovering an estimate $\hat{w}(k)$ of the internal signal $w(k)$ via a smoothing technique (given the current model of $\mathscr{L}$), and then by fitting a PWA map to the samples $(z(k), \hat{w}(k))$. If the polynomial $G(q)$ in (2a) is not Schur, the smoothing filter may return divergent estimates due to numerical instabilities. Conversely, Algorithm 12.1 does not require $G(q)$ to be Schur.

- The algorithm in [Pepona *et al.* (2011)] might get stuck in cyclic solutions (i.e. the same identified models recur cyclicly at different algorithm iterations), and therefore it might never terminate. The algorithm proposed in this paper always terminates. This is obtained by considering a non-negative cost which is non-increasing at each algorithm iteration.

As mentioned above, the most challenging part in Algorithm 12.1 is the PWA approximation of the nonlinear part via (26). This problem is considered for $s = 2$ in Section 4.2, and for $s > 2$ in Section 4.3.

4.2. *Estimation of PWA maps with two modes*

This section addresses the optimization problem (26) in the case $s = 2$, and shows how it is possible to solve it at the optimum by enumerating all possible thresholds. The motivation for presenting the exhaustive algorithm in the case $s = 2$ is that it requires a reasonable computation time. For instance, the algorithm implemented in Matlab takes about 1 second to solve (26) with $N = 1000$, and about 2.5 seconds with $N = 2000$, on a laptop equipped with a 1.33 GHz processor and 2 Gb RAM. Recall that (26) has to be solved at every iteration of Algorithm 12.1.

Consider the following optimization problem, derived from (26):

$$\hat{f}(\cdot) = \arg \min_{f(\cdot) \in PWA_2} \sum_{k=\bar{n}+1}^{N} |\bar{y}(k) - \vartheta_\omega^T \tilde{f}(\zeta(k))|^2, \tag{30}$$

where $\bar{y}(k) = y(k) - \vartheta_\phi^T \phi(k)$, and $\vartheta_\phi, \vartheta_\omega$ are fixed parameter vectors. It is interesting to note that, rewritten as in (30), the considered optimization problem can be interpreted as the optimal estimation, from input-output data, of a PWA map with two modes, within a Hammerstein system with input $z(k)$ and output $\bar{y}(k)$, whose linear part is described by a known FIR model $G(q)$.

Since $z(k) \in \mathbb{R}$ and $s = 2$, the partition of the domain of any function $f(\cdot) \in PWA_2$ is determined by only one value, namely the threshold $\sigma \in \mathbb{R}$ discriminating the two regions of the PWA map. Since N values $z(k)$ are available, one has $N-1$ possible choices for σ (assuming all $z(k)$ are distinct), namely the centers $\Omega = \{\sigma_1, \ldots, \sigma_{N-1}\}$ of the $N-1$ intervals between consecutive values $z(k)$ along the real axis. Let $\hat{f}_{\sigma_j}(\cdot)$ be the optimizer of (30) among PWA_2 functions with fixed threshold σ_j. Then, the function $\hat{f}(\cdot)$ that solves (30) is the function $\hat{f}_{\sigma_j}(\cdot)$ with smallest cost among all possible choices $\sigma_j \in \Omega$. It is shown next that computing $\hat{f}_{\sigma_j}(\cdot)$ boils down to a standard least squares problem.

Let $\sigma = \sigma_j$, $j = 1, \ldots, N-1$, and define the following quantities:

$$\bar{Y} = [\ \bar{y}(\bar{n}+1)\ \ldots\ \bar{y}(N)\ ]^T \tag{31}$$

$$\Theta_g = \begin{pmatrix} g_{n_g} & \cdots & g_1 & 0 & \ldots & 0 \\ 0 & g_{n_g} & \cdots & g_1 & & \vdots \\ \vdots & & \ddots & & \ddots & \vdots \\ 0 & \ldots & \ldots & g_{n_g} & \cdots & g_1 \end{pmatrix} \tag{32}$$

$$W = [\ w(\bar{n} - n_{k_w} - n_g + 2)\ \ldots\ w(N - n_{k_w})\ ]^T. \tag{33}$$

Then, it is possible to rewrite the cost of (30) as

$$\sum_{k=\bar{n}+1}^{N} |\bar{y}(k) - \vartheta_\omega^T \omega(k)|^2 = \|\bar{Y} - \Theta_g W\|^2, \tag{34}$$

where ϑ_ω is given by (6b) and $\|\cdot\|$ denotes the Euclidean norm of a vector. Define the generic two-mode PWA map as follows:

$$f(z) = \begin{cases} \alpha_1 z + \beta_1 \text{ if } z < \sigma \\ \alpha_2 z + \beta_2 \text{ otherwise,} \end{cases} \tag{35}$$

where α_1, β_1, α_2 and β_2 are the parameters of the two affine modes[b], and the classification functions:

$$\chi_1(k) = \begin{cases} 1 \text{ if } z(k) < \sigma \\ 0 \text{ otherwise} \end{cases} \quad \chi_2(k) = \begin{cases} 1 \text{ if } z(k) \geq \sigma \\ 0 \text{ otherwise.} \end{cases} \tag{36}$$

It turns out that:

$$w(k) = \chi_1(k)[\alpha_1 z(k) + \beta_1] + \chi_2(k)[\alpha_2 z(k) + \beta_2] = \xi(k)^T \theta, \tag{37}$$

where $\xi(k) = [\, \chi_1(k)z(k) \; \chi_1(k) \; \chi_2(k)z(k) \; \chi_2(k) \,]^T$ and $\theta = [\, \alpha_1 \; \beta_1 \; \alpha_2 \; \beta_2 \,]^T$. The vector W can thus be expressed as

$$W = \Xi\, \theta \tag{38}$$

with

$$\Xi = \begin{bmatrix} \xi(\bar{n} - n_{k_w} - n_g + 2)^T \\ \vdots \\ \xi(N - n_{k_w})^T \end{bmatrix}. \tag{39}$$

By substituting (38) into (34), it turns out that the PWA map (35) minimizing (34) is the one defined by the parameters $\hat{\theta} = [\, \hat{\alpha}_1 \; \hat{\beta}_1 \; \hat{\alpha}_2 \; \hat{\beta}_2 \,]^T$ solution of the least squares problem

$$\hat{\theta} = \arg\min_{\theta} \|\bar{Y} - \Theta_g\, \Xi\, \theta\|^2. \tag{40}$$

Summing up, problem (30) requires to solve $N-1$ least squares problems (40). From the computational point of view, if one evaluates the thresholds σ_j in increasing order, only one point $z(k)$ changes classification at each iteration. This implies that only one row of the matrix Ξ must be updated from one iteration to the other, thus reducing the computation time (the matrix Ξ must not be afresh computed at each iteration).

---

[b]The strict inequality in (35) is an arbitrary choice, and can be made non-strict, without affecting the following results.

4.3. Estimation of PWA maps with more than two modes

Section 4.2 illustrated how it is possible to solve the optimization problem (26) in the case $s = 2$ by evaluating exhaustively all possible partitions. For $s > 2$ this approach becomes rapidly impracticable, since the number of possible partitions to be evaluated grows as N^s, where N is the number of data points $z(k)$. A heuristic to tackle (26) in the case $s > 2$ is presented in this section. It is inspired by [Breiman (1993)] and [Pucar and Sjöberg (1998)], and takes the form of an iterative algorithm which, at each iteration, solves a problem of the type of (26) with $f(\cdot) \in PWA_2$, and therefore can be accomplished via the exhaustive approach of Section 4.2.

Consider the following cost, derived from (26):

$$\mathscr{J}(g(\cdot)) = \sum_{k=\bar{n}+1}^{N} |\bar{y}(k) - \vartheta_\omega^T \tilde{g}(\zeta(k))|^2, \tag{41}$$

where $\bar{y}(k) = y(k) - \vartheta_\phi^T \phi(k)$, and $\vartheta_\phi, \vartheta_\omega$ are fixed parameter vectors. Define the optimization problem:

$$\hat{g}(\cdot) = \arg \min_{g(\cdot) \in \mathrm{PWA}_s} \mathscr{J}(g(\cdot)). \tag{42}$$

Note that the PWA maps are denoted by $g(\cdot)$ to avoid confusion with Algorithm 12.1. Assume $g^0(\cdot) \in PWA_s$ is a given initial PWA map, and let:

$$g^h(z) = \begin{cases} \alpha_1^h z + \beta_1^h & \text{if } z < \sigma_1^h \\ \alpha_2^h z + \beta_2^h & \text{if } \sigma_1^h \le z < \sigma_2^h \\ \vdots & \\ \alpha_{s-1}^h z + \beta_{s-1}^h & \text{if } \sigma_{s-2}^h \le z < \sigma_{s-1}^h \\ \alpha_s^h z + \beta_s^h & \text{if } z \ge \sigma_{s-1}^h \end{cases} \tag{43}$$

be the PWA map at iteration h of the iterative algorithm. Moreover, let $\mathscr{J}_h$ be the value of (41) evaluated at $g(\cdot) = g^h(\cdot)$. The iterative algorithm for tackling (42) works as follows.

Algorithm 12.2

(1) At initialization, set $h = 0$ (iteration counter).
(2) Set $h = h + 1$.
(3) Choose the threshold σ_i^{h-1} of $g^{h-1}(\cdot)$ to be updated, with $i \in \{1, \dots, s-1\}$.
(4) Define the two-mode PWA map:

$$\ell_i^{h-1}(z) = \begin{cases} \alpha_i^{h-1} z + \beta_i^{h-1} & \text{if } z < \sigma_i^{h-1} \\ \alpha_{i+1}^{h-1} z + \beta_{i+1}^{h-1} & \text{otherwise.} \end{cases} \tag{44}$$

(5) Let $g_i^{h-1}(z) = g^{h-1}(z) - \ell_i^{h-1}(z)$.
(6) Solve the optimization problem

$$\ell^h(\cdot) = \arg \min_{\ell(\cdot) \in PWA_2} \sum_{t=\bar{n}+1}^{N} |\bar{y}(k) - \vartheta_\omega^T \tilde{g}_i^{h-1}(\zeta(k)) - \vartheta_\omega^T \tilde{\ell}(\zeta(t))|^2 \tag{45}$$

(e.g., by applying the exhaustive approach of Section 4.2).
(7) Let $g^h(z) = g_i^{h-1}(z) + \ell^h(z)$.
(8) If the termination condition is reached, return $g(\cdot) = g^h(\cdot)$, otherwise go back to step 2.

The main property of Algorithm 12.2 is that it generates a sequence of values $\mathscr{J}_h$ which is non-increasing at each iteration. This is a consequence of the fact that (45) is minimized with respect to $\ell(\cdot) \in PWA_2$ (the minimum cost is $\mathscr{J}_h$), and $\ell(\cdot) = \ell_i^{h-1}(\cdot)$ is a possible solution with cost $\mathscr{J}_{h-1}$, therefore $\mathscr{J}_h \leq \mathscr{J}_{h-1}$, $h = 1, 2, \dots$.

In step 3 of Algorithm 12.2, the choice of the threshold to be updated is arbitrary. It could be selected cyclicly, randomly, or by applying a greedy approach, i.e. choosing the threshold to which corresponds the largest improvement of $\mathscr{J}_h$ with respect to $\mathscr{J}_{h-1}$. In step 8, the termination condition is also arbitrary. A termination condition similar to (29) could be implemented, but one could also fix a priori the number of iterations.

When Algorithm 12.2 is applied during the PWA approximation step of the jth iteration of Algorithm 12.1, with $j > 1$, it is invoked with $g^0(\cdot) = f^{j-1}(\cdot)$, $\vartheta_\omega = \vartheta_\omega^{j-1}$ and $\vartheta_\phi = \vartheta_\phi^{j-1}$. It returns $f^j(\cdot) = g^h(\cdot)$, where h is the value of the iteration counter when Algorithm 12.2 is terminated. Note that $J_j^{PWA} = \mathscr{J}_h \leq \mathscr{J}_0 = J_{j-1}$, and hence Proposition 12.1 still holds, even if the optimization problem (26) is not solved at the optimum.

Note that, in order to apply Algorithm 12.2 at the first iteration ($j = 1$) of Algorithm 12.1, an initial PWA approximation $f^0(\cdot) \in PWA_s$ of $\mathscr{N}$ should be provided. One can deal with this in two ways.

- The PWA map $f^0(\cdot) \in PWA_s$ is provided at the initialization of Algorithm 12.1. It could be defined on the basis of the available knowledge about the shape of $\mathscr{N}$. Another option is to let $f^0(\cdot)$ be identically zero, which corresponds to "opening the loop" in Fig. 1 and start from a linear model of the LFR system.
- At the first iteration of Algorithm 12.1, a modified version of Algorithm 12.2 is run. As in [Breiman (1993)], one solves the optimization problem (26) for $s = 2$, and iteratively adds one mode at a time by fitting a two-mode PWA map to the residuals of the currently estimated PWA map (as in steps 6 and 7 of Algorithm 12.2), until the desired number of modes is reached.

5. Applications

Two applications are presented in this section. The first one is a numerical example in which the true system belongs to the class of PWA-LFR models. Since in this case the interconnected system is still piecewise affine (see Section 3), one can compare the parameters of the PWA models (11) identified by different techniques with those of the true system. The second application is the *silverbox* identification problem, a popular "real-life" benchmark in nonlinear system identification [Pintelon and Schoukens (2001); Nolcos (2004)]. By exploiting the knowledge about the internal mechanism generating the data, the proposed iterative algorithm is able to identify PWA-LFR models of the *silverbox* which achieve very good performance both in estimation and in validation.

5.1. *Numerical example*

Consider the following PWA-LFR system:

$$y(k) + a_1 y(k-1) = b_1 u(k-1) + g_1 w(k-2) + g_2 w(k-3) + e(k) \tag{46a}$$

$$z(k) = u(k) - y(k) \tag{46b}$$

$$w(k) = \begin{cases} \theta_1^T \begin{bmatrix} z(k) \\ 1 \end{bmatrix} \text{ if } h^T \begin{bmatrix} z(k) \\ 1 \end{bmatrix} < 0 \\ \theta_2^T \begin{bmatrix} z(k) \\ 1 \end{bmatrix} \text{ if } h^T \begin{bmatrix} z(k) \\ 1 \end{bmatrix} \geq 0, \end{cases} \tag{46c}$$

where $a_1 = -0.4$, $b_1 = 1.2$, $g_1 = 1$, $g_2 = -0.5$, while θ_1, θ_2 and h are reported in Table 1. A data set of $N = 1000$ data points $\{u(k), y(k), z(k)\}$ is generated with $u(k)$ and $e(k)$ uniformly distributed in $[-1, 1]$ and $[-0.25, 0.25]$, respectively. The signal-to-noise ratio is $\frac{\sigma_y}{\sigma_e} \simeq 6.58$, where σ_y and σ_e denote the empirical standard deviations of $y(k)$ and $e(k)$, respectively.

The PWA-LFR identification algorithm is initialized with estimates $a_1^0 = -0.4054$ and $b_1^0 = 1.1531$ obtained by fitting the ARX model

$$y(k) = -a_1^0 y(k-1) + b_1^0 u(k-1) + \varepsilon(k) \tag{47}$$

to the data by least squares. The coefficient g_1^0 is set equal to 1 and kept constant in all iterations, while g_2^0 is set equal to 0. No initial PWA map is needed, since the number of modes of the PWA map is $s = 2$, and the PWA approximation step can thus be accomplished via the exhaustive approach of Section 4.2. By setting $\gamma = 0.0001$ in (29), the algorithm terminates after 16 iterations. The estimates $\hat{a}_1 = -0.4016$, $\hat{b}_1 = 1.1918$ and $\hat{g}_2 = -0.4995$ are finally returned. Concerning the reconstruction of the static nonlinearity (46c), the estimated parameters $\hat{\theta}_1$, $\hat{\theta}_2$ and $\hat{h}$ are reported in Table 1. It is apparent that all the estimated parameters of both the ARX dynamics and the static nonlinearity are very close to the corresponding true values.

Table 1. True and estimated parameters of the PWA nonlinearity in the numerical example.

θ_1	$\hat{\theta}_1$	θ_2	$\hat{\theta}_2$	h	$\hat{h}$
-0.7	-0.6999	0.5	0.4991	1	1
0.25	0.2522	0	0.0052	0	-0.0017

The aim of this example is mainly to show the enhanced results that can be obtained by exploiting the information about the interconnection structure when addressing PWA identification of interconnected systems. For this reason, a PWARX model (11) is fitted directly to the input-output data $\{u(k), y(k)\}$. Note that the system defined by (46) can be equivalently rewritten as a PWARX system with four modes following the procedure described in Example 1, i.e.

$$y(k) = \Theta_{i,j}^T \begin{bmatrix} \boldsymbol{r}(k) \\ 1 \end{bmatrix} + e(k) \quad \text{if } \boldsymbol{r}(k) \in \mathscr{R}_{i,j}, \tag{48}$$

where $\boldsymbol{r}(k) = [\ -y(k-1)\ y(k-2)\ y(k-3)\ u(k-1)\ u(k-2)\ u(k-3)\ ]^T$, $\Theta_{i,j}$ and $\mathscr{R}_{i,j}$ are the parameter vector and the region of validity of each submodel, and the submodel characterized by subscripts (i,j), $i,j = 1,2$, corresponds to the situation when $w(k-2)$ and $w(k-3)$ are generated by the ith and jth mode of (46c), respectively. The PWA-LFR representation (46) of the system is more parsimonious than the PWARX representation (48), as the former requires 7 parameters (a_1, b_1, g_2, θ_1 and θ_2), while the latter 28 ($\Theta_{1,1}$, $\Theta_{1,2}$, $\Theta_{2,1}$ and $\Theta_{2,2}$). For a given finite data set, this is expected to affect the quality of the identification results.

In order to identify (48) directly, the most favorable situation is when the partition of the regressor domain is known. Under this assumption, the PWA system identification problem reduces to a standard least squares problem for each mode (see Section 3.1). Let $\widetilde{\Theta}_{1,1}$, $\widetilde{\Theta}_{1,2}$, $\widetilde{\Theta}_{2,1}$ and $\widetilde{\Theta}_{2,2}$ be the parameter estimates computed in this way. Conversely, the estimates $\hat{\Theta}_{1,1}$, $\hat{\Theta}_{1,2}$, $\hat{\Theta}_{2,1}$ and $\hat{\Theta}_{2,2}$ are reconstructed from the estimated parameters $\hat{a}_1$, $\hat{b}_1$, $\hat{g}_2$, $\hat{\theta}_1$ and $\hat{\theta}_2$ provided by the PWA-LFR identification algorithm. Both $\widetilde{\Theta}_{i,j}$ and $\hat{\Theta}_{i,j}$ are then compared with the corresponding true values $\Theta_{i,j}$: relative estimation errors are reported in Table 2. It turns out that the reconstructed estimates $\hat{\Theta}_{i,j}$ are much closer to the true values than the estimates $\widetilde{\Theta}_{i,j}$ identified directly. This shows that more accurate estimates can be obtained by exploiting the system PWA-LFR structure, rather than identifying a black-box PWARX model of the overall system. It is worthwhile to note that the estimate $\widetilde{\Theta}_{2,2}$ of $\Theta_{2,2}$ is so worse compared to the others since only 36 data points over 1000 are generated by mode $(2,2)$ in the available data set.

Table 2. Comparison of identification results in the numerical example.

mode	$\frac{\|\hat{\Theta}_{i,j}-\Theta_{i,j}\|}{\|\Theta_{i,j}\|} \times 100\%$	$\frac{\|\widetilde{\Theta}_{i,j}-\Theta_{i,j}\|}{\|\Theta_{i,j}\|} \times 100\%$
$(1,1)$	0.5051	1.0260
$(1,2)$	0.5093	6.9529
$(2,1)$	0.6198	3.3254
$(2,2)$	0.5998	25.9544

5.2. *The silverbox benchmark*

The silverbox identification problem is a popular benchmark in nonlinear system identification [Nolcos (2004)], consisting in a nonlinear mechanical resonating system (mass, viscous damping and nonlinear spring). In continuous time, the system can be modelled by the following second order differential equation relating the displacement $y(t)$ to the force $u(t)$:

$$m \frac{d^2y(t)}{dt^2} + \xi \frac{dy(t)}{dt} + \kappa(y(t))\, y(t) = u(t), \tag{49}$$

where m is the mass, ξ is the viscous damping, and the nonlinear spring is described by a static but position-dependent stiffness:

$$\kappa(y) = a + b\, y^2. \tag{50}$$

By substituting (50) into (49), the left- and right-hand side of (49) can be rearranged as follows:

$$m \frac{d^2y(t)}{dt^2} + \xi \frac{dy(t)}{dt} + a\, y(t) = u(t) - b\, y(t)^3, \tag{51}$$

which can be represented in Lur'e form (linear system with static nonlinear feedback) as depicted in Fig. 2, with linear part $\mathscr{L}(s) = 1/(m\, s^2 + \xi\, s + a)$ and static nonlinearity $\mathscr{N}(y) = b\, y^3$.

Several detailed studies of the silverbox benchmark can be found in the literature, using, e.g., best linear approximation based methods [Schoukens *et al.* (2003); Lauwers *et al.* (2008)], Least Squares Support Vector Machines (LS-SVM) [Espinoza *et al.* (2004)], partially linear LS-SVM [Espinoza *et al.* (2005)]. For comparisons of different approaches applied to this benchmark, see [Nolcos (2004)]. Recently, PWA-LFR models of the silverbox have been proposed in [Pepona *et al.* (2011)]. In the same paper, it is also shown that black-box PWA identification algorithms may not provide satisfactory results when applied to the silverbox benchmark, due to difficulties in classifying the available data in a high-dimensional space. The contribution of this section is to show that, by assuming the system structure of Fig. 2, the PWA-LFR identification algorithm presented in this paper is able to identify a PWA-LFR model of the silverbox with better performance both on estimation and validation data than the PWA-LFR models presented

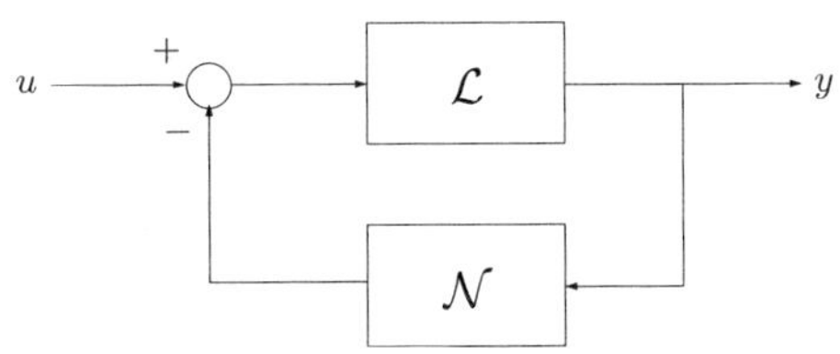

Fig. 2. Lur'e representation of the silverbox benchmark model.

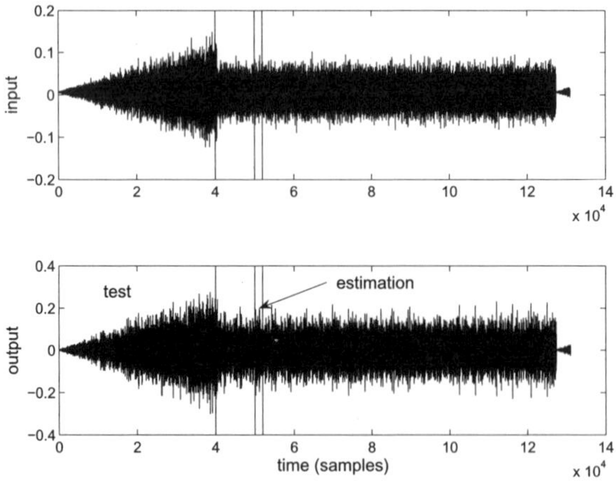

Fig. 3. Available input-output data for the silverbox identification problem, and portions used for estimation and test.

in [Pepona *et al.* (2011)]. Moreover, the estimated PWA-LFR model is ready-to-use for analysis and design purposes, since it allows one to exploit the wide literature on control systems in Lur'e form (see, e.g., the classical textbooks [Šiljak (1969); Narendra and Taylor (1973)]).

The available data set is a sequence $\{u(k), y(k)\}$ of about 130000 inputs and outputs measured from an electrical circuit simulating the nonlinear mechanical resonating system. The input and output sequences are plotted in Fig. 3. The data sets used for estimation and test are indicated in the same figure. The estimation data set is composed by $N = 2000$ data pairs $(u(k), y(k))$ inside the "body of the arrow", namely the samples from 50001 to 52000. As in [Espinoza *et al.* (2004)], the test data set is composed by the first 40000 samples (the "head of the arrow").

In order to apply Algorithm 12.1, the Lur'e structure of Fig. 2 is imposed to model (2) by choosing:

- $G(q) = -B(q)$,
- $n_{k_u} = n_{k_w}$,
- $z(k) = y(k)$.

Moreover, as in the best PWA-LFR model presented in [Pepona *et al.* (2011)], the following settings are made: $n_a = n_b = 4$, $n_{k_u} = 1$ and $s = 5$. Note that, since $s > 2$, Algorithm 12.2 is applied in the PWA approximation

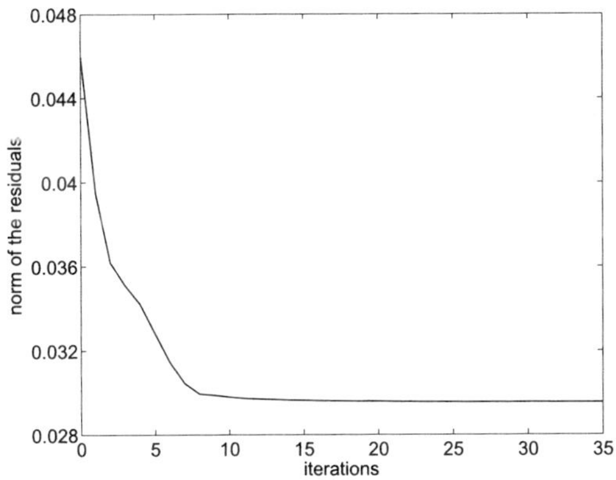

Fig. 4. Decrease of the norm of the residuals with the iterations (first identification run).

step of Algorithm 12.1. To limit the computation time, in Algorithm 12.2 only one iteration is run, in combination with the greedy approach for the choice of the threshold to be updated.

A first identification run is performed by initializing Algorithm 12.1 as described in Section 4.1. In particular, ϑ_{ϕ}^{0} is computed by fitting an ARX model to the input-output data by least squares, while ϑ_{ω}^{0} is chosen according to the constraint $G(q) = -B(q)$. The initial PWA map for Algorithm 12.2 is set equal to zero. By setting $\gamma = 0.0001$, Algorithm 12.1 terminates after 35 iterations. Figure 4 shows the decrease of the norm of the residuals with the iterations, in accordance with Proposition 12.1. The corresponding evolutions of the parameters of the polynomials $A(q)$ and $B(q)$ are plotted in Fig. 5 and Fig. 6, respectively. Finally, Fig. 7 shows the evolution of the estimated PWA map $f(\cdot)$ along the first three iterations of Algorithm 12.1. Quite nicely, already at the third iteration, $f(\cdot)$ looks like the PWA approximation of a cubic polynomial function, which is in agreement with the shape of $\mathcal{N}$ in (51).

Validation of the identified PWA-LFR models is carried out by calculating the fit between the system outputs and the model outputs using test data (i.e. data not used for identification). By letting $\boldsymbol{y} = (y(1), \ldots, y(N))$ be the vector of system outputs, $\bar{y}$ the mean value of $\boldsymbol{y}$, and $\hat{\boldsymbol{y}} = (\hat{y}(1), \ldots, \hat{y}(N))$ the vector of model outputs, the following standard measure of fit is used:

$$\text{FIT} = 100 \times \left(1 - \frac{\|\hat{\boldsymbol{y}} - \boldsymbol{y}\|}{\|\boldsymbol{y} - \bar{y}\|}\right)\%. \tag{52}$$

Figure 8 shows the fit in validation as a function of the PWA-LFR models identified at each iteration of Algorithm 12.1. The PWA-LFR model with best performance in validation is the one identified at iteration 7 (FIT = 95.54%), while the fit of the best model in estimation (iteration 35) is 94.77%. Note that the fit of the best model obtained in [Pepona *et al.* (2011)] is 92.29%.

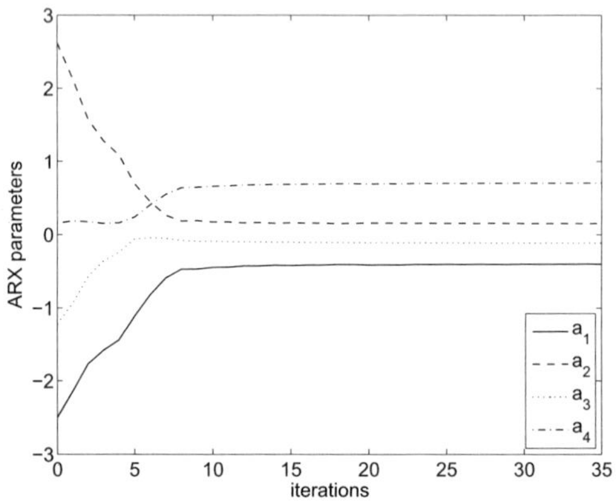

Fig. 5. Evolution of the parameters a_i, $i = 1, \ldots, 4$, of $A(q)$ along the iterations (first identification run).

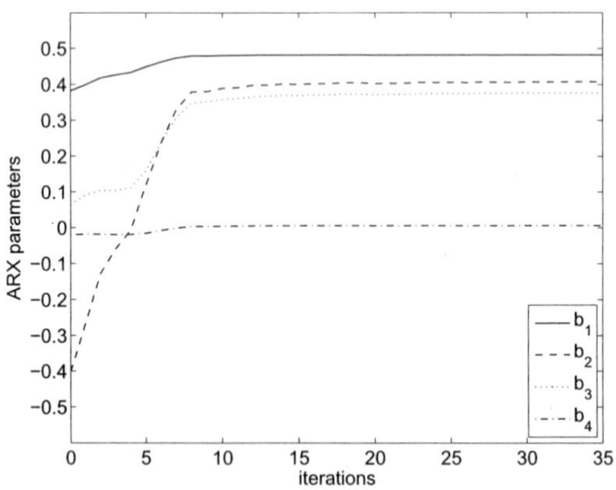

Fig. 6. Evolution of the parameters b_i, $i = 1, \ldots, 4$, of $B(q)$ along the iterations (first identification run).

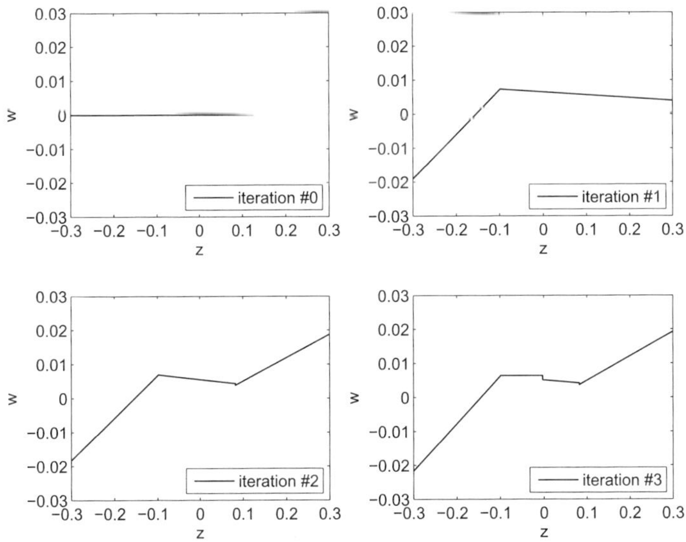

Fig. 7. Evolution of the estimated PWA map $f(\cdot)$ along the first three iterations of Algorithm 12.1 (first identification run).

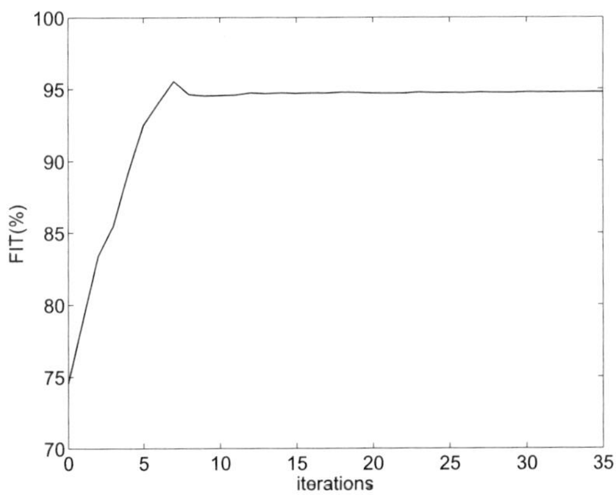

Fig. 8. FIT on test data of the PWA-LFR models identified along the iterations of Algorithm 12.1 (first identification run).

Figure 9 shows the simulation error $y(k) - \hat{y}(k)$ on test data, where $y(k)$ is the system output and $\hat{y}(k)$ is the output of the PWA-LFR model identified at iteration 7 of Algorithm 12.1. It can be observed that the simulation error deteriorates in the final portion of the test data set. This is due to

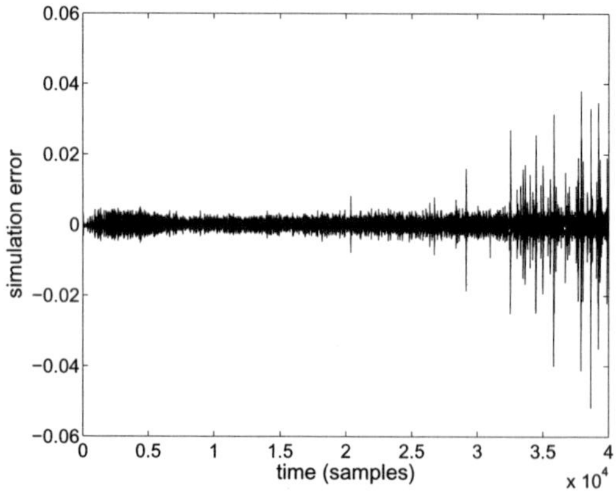

Fig. 9. Simulation error on test data of the PWA-LFR model identified at iteration 7 of Algoritm 12.1 (first identification run).

the fact that the input signal in the test data set (see Fig. 3) excites the PWA-LFR model beyond the range where it was estimated. Nevertheless, the maximum error (in absolute value) is around 0.04 in Fig. 9, while it is around 0.1 for the best PWA-LFR model presented in [Pepona *et al.* (2011)]. This is due to a better approximation of the static nonlinearity obtained by the PWA-LFR model derived in this paper.

A second identification run is performed by initializing Algorithm 12.1 with the best PWA-LFR model presented in [Pepona *et al.* (2011)]. The aim of

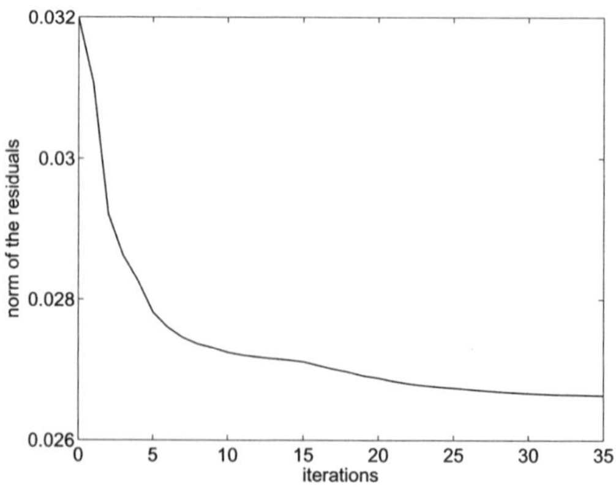

Fig. 10. Decrease of the norm of the residuals with the iterations (second identification run).

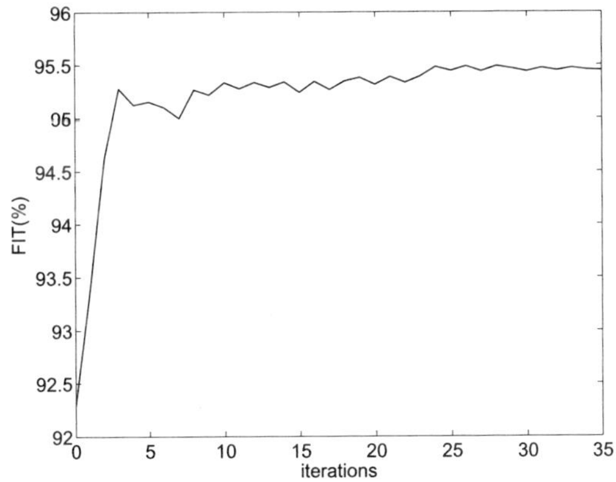

Fig. 11. FIT on test data of the PWA-LFR models identified along the iterations of Algorithm 12.1 (second identification run).

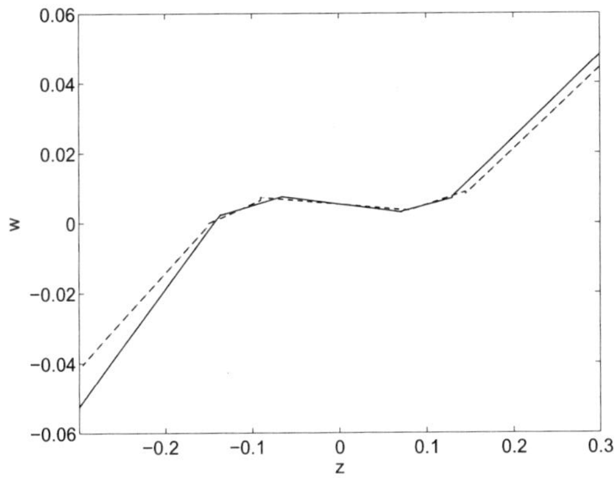

Fig. 12. Comparison between the initial PWA map (dash-dotted) and the final PWA map (solid) in the second identification run.

this trial is to quantify the improvements that Algorithm 12.1 achieves with respect to the previous iterative algorithm. Figure 10 shows the decrease of the norm of the residuals with the iterations, while Fig. 11 shows the fit in validation for each PWA-LFR model identified during the execution of Algorithm 12.1. In both figures, values corresponding to iteration 0 refer to the PWA-LFR model presented in [Pepona *et al.* (2011)]. It is apparent that the iterative procedure implemented in Algorithm 12.1 is able to sensibly enhance the original solution. Note that the norm of the residuals reaches

0.0266 in Fig. 10, while it reaches 0.0295 in Fig. 4. This means that the first identification run got stuck in a local minimum. Nevertheless, the performance in validation of the best models obtained in the first and the second identification run is similar (compare Fig. 8 and Fig. 11). Finally, Fig. 12 compares the plots of the initial and the final PWA map in the second identification run. As can be observed, differences are mostly evident in the slopes of the external modes of the PWA maps.

6. Conclusions

In this paper, an iterative procedure has been proposed for identification of PWA-LFR models, described by the LFR interconnection of an ARX linear model and a static PWA map. Nice features of the proposed approach are a non-increasing identification cost (which guarantees convergence in finite time of the iterative procedure) and the remarkable improvement in the model quality with respect to black-box PWA identification techniques, not taking into account the interconnection structure. It is stressed that, due to the explosion of the number of modes, the black-box approach is often intractable even when dealing with very simple systems, while the proposed procedure allows one to treat models of reasonable size.

Identification of models reflecting the inner structure of interconnected system is a research area still in its infancy. A deep investigation of identifiability issues has not been addressed yet, for a wide variety of model classes. Another key problem is the study of different modeling paradigms for both the linear and the nonlinear blocks. In this respect, the approach proposed in the present paper should be adapted to cope with other linear models, such as ARMAX, OE, or state space models. At the same time, different PWA identification techniques could be employed to estimate the nonlinear block. These topics will be the subject of future research.

Acknowledgements

The authors would like to thank Alessandra Di Tella for her contribution to the applications reported in Section 5, and Johan Schoukens for providing the data of the silverbox benchmark.

References

Bai, E. and Li, D. (2004). Convergence of the iterative Hammerstein system identification algorithm, *IEEE Transactions on Automatic Control* **49**, 11, pp. 1929–1940.

Breiman, L. (1993). Hinging hyperplanes for regression, classification, and function approximation, *IEEE Transactions on Information Theory* **39**, 3, pp. 999–1013.

Espinoza, M., Pelckmans, K., Hoegaerts, L., Suykens, J. A. K. and De Moor, B. (2004). A comparative study of LS-SVM's applied to the silver box identification problem, in *Proceedings of the IFAC Symposium on Nonlinear Control Systems* (Stuttgart, Germany).

Espinoza, M., Suykens, J. A. K. and De Moor, B. (2005). Kernel based partially linear models and nonlinear identification, *IEEE Transactions on Automatic Control* **50**, 10, pp. 1602–1606.

Hsu, K., Claassen, M., Novara, C., Khargonekar, P., Milanese, M. and Poolla, K. (2005a). Nonparametric identification of static nonlinearities in a general interconnected system, in *Proceedings of the IFAC World Congress* (Prague, Czech Republic).

Hsu, K., Poolla, K. and Vincent, T. (2008). Identification of structured nonlinear systems, *IEEE Transactions on Automatic Control* **53**, 11, pp. 2497–2513.

Hsu, K., Vincent, T., Novara, C., Milanese, M. and Poolla, K. (2005b). Identification of nonlinear maps in interconnected systems, in *Proceedings of the IEEE Conference on Decision and Control and European Control Conference* (Seville, Spain), pp. 6430–6435.

Julián, P., Jordan, M. and Desages, A. (1998). Canonical piecewise-linear approximation of smooth functions, *IEEE Transactions on Circuits and Systems I* **45**, 5, pp. 567–571.

Lauwers, L., Schoukens, J., Pintelon, R. and Enqvist, M. (2008). A nonlinear block structure identification procedure using frequency response function measurements, *IEEE Transactions on Instrumentation and Measurement* **57**, 10, pp. 2257–2264.

Lin, J.-N. and Unbehauen, R. (1992). Canonical piecewise-linear approximations, *IEEE Transactions on Circuits and Systems I* **39**, 8, pp. 697–699.

Ljung, L. (1999). *System identification: Theory for the User* (Prentice-Hall, Upper Saddle River, NJ, USA).

Narendra, K. S. and Gallman, P. G. (1966). An iterative method for the identification of nonlinear systems using a Hammerstein model, *IEEE Transactions on Automatic Control* **11**, 3, pp. 546–550.

Narendra, K. S. and Taylor, J. (1973). *Frequency domain methods for absolute stability* (Academic Press, New York, NY, USA).

Nolcos (2004). Identification of nonlinear systems: The silverbox case study, in *Proceedings of the IFAC Symposium Nonlinear Control Systems* (Stuttgart, Germany).

Paoletti, S., Juloski, A. L., Ferrari-Trecate, G. and Vidal, R. (2007). Identification of hybrid systems: A tutorial, *European Journal of Control* **13**, 2–3, pp.

242–260.

Pepona, E., Paoletti, S., Garulli, A. and Date, P. (2011). Identification of piecewise affine LFR models of interconnected systems, *IEEE Transactions on Control Systems Technology* **19**, 1, pp. 148–155.

Pintelon, R. and Schoukens, J. (2001). *System identification : A frequency domain approach* (IEEE Press, Piscataway, NJ, USA).

Previdi, F. and Lovera, M. (2003). Identification of a class of non-linear parametrically varying models, *International Journal Adaptive Control Signal Processing* **17**, 1, pp. 33–50.

Pucar, P. and Sjöberg, J. (1998). On the hinge-finding algorithm for hinging hyperplanes, *IEEE Transactions on Information Theory* **44**, 3, pp. 1310–1319.

Roll, J. (2003). *Local and piecewise affine approaches to system identification*, Ph.D. thesis, Department of Electrical Engineering, Linköping University, Sweden.

Roll, J., Bemporad, A. and Ljung, L. (2004). Identification of piecewise affine systems via mixed-integer programming, *Automatica* **40**, 1, pp. 37–50.

Schoukens, J., Nemeth, J., Crama, P., Rolain, Y. and Pintelon, R. (2003). Fast approximate identification of nonlinear systems, *Automatica* **39**, 7, pp. 1267–1274.

Tóth, R. (2008). *Modeling and identification of linear parameter-varying systems*, Ph.D. thesis, Delft University of Technology, The Netherlands.

Šiljak, D. D. (1969). *Nonlinear systems: Parametric analysis and design* (John Wiley & Sons, New York, NY, USA).

Voros, J. (2003). Modeling and identification of Wiener systems with two-segment nonlinearities, *IEEE Transactions on Control Systems Technology* **11**, 2, pp. 253–257.

Zhou, K., Doyle, J. C. and Glover, K. (1996). *Robust and optimal control* (Prentice Hall, Upper Saddle River, NJ, USA).

Chapter 13

Identification and model (in)validation of switched ARX systems: A moment-based approach

C. Feng[†], N. Ozay[‡], C. M. Lagoa[†] & M. Sznaier[*]

[†]*Department of Electrical Engineering*
The Pennsylvania State University
University Park, PA 16802, USA
feng@psu.edu
lagoa@engr.psu.edu

[‡]*Department of Computing and Mathematical Sciences*
California Institute of Technology, Pasadena, CA 91125, USA
necmiye@caltech.edu

[*]*Electrical and Computer Engineering Department*
Northeastern University, Boston, MA 02115, USA
msznaier@coe.neu.edu

Identification of switched systems has received considerable attention during the past few years. This chapter addresses the problem of identification and (in)validation of switched autoregressive models with external inputs from experimental data. In the first part of the chapter identification of switched linear systems from noisy measurements is considered. Given a finite number of measurements of input/output and a bound on the measurement noise, the objective is to identify a switching sequence and a set of linear submodels that are compatible with the a priori information, while minimizing the number of submodels. A deterministic approach based on convex optimization is proposed. Namely, by recasting the identification problem as polynomial optimization, we develop deterministic algorithms, in which the inherent sparse structure is exploited. A finite dimensional semi-definite problem is then given which is equivalent to the identification problem. Moreover, to address computational complexity issues, an equivalent rank minimization problem subject to deterministic linear matrix inequality constraints is provided, as efficient convex relaxations for rank minimization are available in the literature.

Since the identification problem is generically NP-Hard, the majority of existing algorithms, including the one we proposed, are based on heuristics or relaxations. Therefore, it is crucial to check the validity of the identified models against additional experimental data. The second part of this chapter addresses the problem of model (in)validation for switched linear autoregressive exogenous systems with unknown switches. Necessary and sufficient conditions are obtained for a given model to be (in)validated by the experimental data. In principle, checking these conditions requires solving a sequence of convex optimization problems involving increasingly large matrices. However, as we show, if in the process of solving these problems either a positive solution is found or the so-called flat extension property holds, then the process terminates with a certificate that either the model has been invalidated or that the experimental data is indeed consistent with the model and *a–priori* information. By using duality, the proposed approach exploits the inherently sparse structure of the optimization problem to substantially reduce its computational complexity.

The effectiveness of the proposed methods is illustrated using both academic examples and a non-trivial problem arising in computer vision: activity monitoring.

1. Introduction

Hybrid systems, dynamical systems where continuous and discrete states interact, are ubiquitous in many different contexts (e.g. circuit network, biological systems, systems incorporating logical and continuous elements, manufacturing, automotive, etc.). In addition, these systems can be used to approximate nonlinear dynamics. In this chapter we focus on a class of hybrid systems, switched linear systems where the dynamics switch between a finite number of linear submodels. Switched systems can be considered as a special case of linear parameter varying (LPV) systems with a finite unknown parameter set, each element of which corresponds to a submodel. However, contrary to LPV systems, for switched systems the switches (hence the parameters) are not measurable in real-time in general, calling for the development of a different set of tools for their identification.

During the past few years considerable research efforts have been devoted to the problem of identifying hybrid systems, leading to several methods (see the tutorial paper [Paoletti *et al.* (2007)] for a summary of the main issues and recent developments in the field). In the case where measurements are noise-free, an algebraic procedure, known as Generalized Principal Component Analysis (GPCA), has been proposed in [Ma and Vidal (2005); Vidal *et al.* (2003)] to efficiently solve the problem. The problem can also be formulated as a mixed linear integer optimization problem [Roll *et al.* (2004)] or in terms of linear complementary inequalities [Bemporad *et al.* (2005)], leading to generically NP-hard problems. Alternative methods include clustering based method [Ferrari-Trecate *et al.* (2003); Nakada *et al.* (2005)] where identification is performed on local clusters or those that rely on probabilistic priors [Juloski *et al.* (2005)] which, once again, lead to a combinatorial problem that requires relaxations.

More recently, a greedy algorithm has been proposed to identify the system while minimizing the number of switches [Ozay *et al.* (2008)]. For robust identification of switched affine systems subject to process noise, an efficient moment-based convex approach using convex relaxations on rank minimization has been proposed in [Ozay *et al.* (2009)]. A similar approach was also pursued in [Ozay *et al.* (2010b)] to solve a different problem: segmenting a collection of noisy measurements into subspaces. In the case where measurement noise is considered, approaches based on polynomial optimization and rank minimization are proposed in [Feng *et al.* (2010b,a)] with exploiting the inherent sparse structure. It is known in the literature of polynomial optimization that the sparse structure can be used to significantly reduce computational complexity, e.g. see [Lasserre (2001, 2006); Laurent (2009); Waki *et al.* (2006)].

On the other hand, since the hybrid system identification problem is generically NP-Hard [Roll *et al.* (2004)], the above mentioned identification algorithms are based on heuristics or relaxations. Hence, a crucial step before using the resulting models, is to check their validity against additional experimental data. Model (in)validation of Linear Time Invariant (LTI) systems has been extensively addressed in the past two decades and both time and frequency domain results are available in the literature (see for instance [Poolla *et al.* (1994); Chen (1997); Sánchez Peña and Sznaier (1998)] and references therein). A related line of research is model (in)validation of LPV systems ([Sznaier and Mazzaro (2003); Bianchi and Sánchez Peña (2009)]) where it is assumed that parameter values are measurable during the experiment and used as part of *a posteriori* data during the (in)validation step. Finally, (in)validation of continuous-time nonlinear models was addressed in [Prajna (2006)] using sum of squares methods and barrier functions. In

[Ozay *et al.* (2010a)], an algorithm is proposed for the case where discrete-time switched linear system with unknown switches is considered, with the main difficulty being the combination of noisy measurements and an unknown mode signal.

The main objective of this chapter is to provide efficient algorithms for the identification of switched ARX systems as well as to provide a necessary and sufficient conditions for switch ARX models to be (in)validated by experimental data. In the identification part, we provide an equivalent polynomial optimization problem, which inherently has a sparse structure and satisfies the so-called *running intersection property*. Moreover, it is shown that the sparse polynomial optimization problem can be solved via a *fixed size* semi-definite programming (SDP). Furthermore, for larger size problems, an equivalent rank minimization problem is provided. In formulating this rank minimization problem, inspired by the results in [Ozay *et al.* (2009)], we use a similar approach to isolate the system parameters from the unknown noise terms, and, hence, eliminate them from the decision variables in the formulated rank minimization problem. A major advantage of the approach in this chapter is that the matrix in the objective function is symmetric and of much smaller size. One may note that this feature notably reduces the computational burden in solving convex relaxations of rank minimization problems.

For model (in)validation, given a nominal model and experimental input/output data, we provide certificates for the existence/nonexistence of a switching sequence such that the resulting output sequence interpolates the given experimental data within a given noise bound. The starting point to obtain such certificates is to recast the (in)validation problem as one of checking whether a semialgebraic set is empty. By using a combination of recent results on moments-based sparse polynomial optimization and duality we show that emptiness of this set is equivalent to strict positivity of the solution of a related, convex optimization problem. In principle, checking this condition requires solving a sequence of convex optimization problems involving increasingly large matrices. However, as we show in the sequel, if in the process of solving these problems either a positive solution is found or the rank of certain matrices formed using the solution to the dual problem ceases to increase (the so-called flat extension property), then the process terminates with either an invalidation or a consistency certificate. A salient feature of the proposed approach is its ability to exploit the inherently sparse structure of the optimization problem to substantially reduce its computational complexity.

These results are illustrated both with academic examples and a non-trivial problem arising in computer vision: activity monitoring. Typically, a visual surveillance system captures high volume data streams from multiple cameras. However, interesting (e.g. abnormal) activities are rare. Thus, it is important to be able to automatically eliminate the "normal" behavior and trigger an appropriate response when something potentially interesting or "abnormal" occurs. As we show in Section 5.4, this problem can be recast into a piecewise–linear model invalidation form and solved using the framework developed in this chapter.

The remainder of the chapter is organized as follows. Section 2 introduces the notation used throughout the chapter and presents some background results related to sparse polynomial optimization. The main results are presented in Sections 3 and 4 which are devoted to identification and invalidation problems, respectively. In Section 5, we provide illustrative numerical examples as well as a practical computer vision application. Finally, Section 6 summarizes our results and offers some concluding remarks.

2. Preliminaries

For ease of reference, in this section we summarize the notation used in the chapter and recall some results on sparse polynomial optimization that play a key role in establishing the main results of this chapter. For a more detailed exposition on (sparse) polynomial optimization, the reader is referred to [Lasserre (2001, 2006); Laurent (2009); Waki *et al.* (2006)].

2.1. *Notation*

$\boldsymbol{x}, \boldsymbol{M}$	a vector in $\mathbf{R}^n$ (matrix in $\mathbf{R}^{n\times m}$)
$\mathbf{x}^{\boldsymbol{\alpha}}$	abbreviation for $x_1^{\alpha_1}\cdots x_d^{\alpha_d}$ where d is the dimension of the vector $\mathbf{x}$
$\mathbf{E}_\mu[p(x)]$	the mean value of $p(x)$ w.r.t the probability measure μ on the random variable x
$\mathbf{m}$, $\{m_{\boldsymbol{\alpha}}\}_0^N$	the moment sequence for random variable $\mathbf{x}\in\mathbf{R}^d$ where $m_i = \mathbf{E}_\mu\mathbf{x}^{\boldsymbol{\alpha}}$ for some probability measure μ and $0\le\alpha_1+\ldots+\alpha_d\le N$

$\mathbf{M}_N(\mathbf{m})$	the moment matrix in $\mathbf{R}^{\binom{N+d}{N}\times\binom{N+d}{N}}$ constructed using $\{m_{\boldsymbol{\alpha}}\}_0^{2N}$
$\mathbf{M} \succeq 0$	the matrix $\mathbf{M}$ is positive semi-definite
$\|\mathbf{x}\|_p$	the ℓ_p norm of the vector $\mathbf{x}$
$\mathbf{R}[x_1, \ldots, x_n]$	the ring of polynomials in n variables over $\mathbf{R}$. When n is clear from the context, $\mathbf{R}[\boldsymbol{x}]$ is used.
$\mathbf{N}_n$	positive integers up to n, i.e. $\mathbf{N}_n \doteq \{1, \ldots, n\}$
$\wedge$ ($\vee$)	logical AND (OR)

2.2. *General polynomial optimization*

Consider the following general constrained polynomial optimization problem:

$$p_K^* := \min_{x \in K} p_0(\mathbf{x}) \tag{P1}$$

where $K \subset \mathbf{R}^d$ is a compact semi-algebraic set defined as

$$K \doteq \{\mathbf{x} \colon p_i(\mathbf{x}) \geq 0, i = 1, \ldots L\}$$

where $p_i(x)$ are polynomials with total degree d_i. This problem is usually not convex, and hence, hard to solve in general. Yet, let's consider a related problem in the probability measure space:

$$\tilde{p}_K^* := \min_{\mu \in \mathscr{P}(K)} \int p_0(\mathbf{x})\mu(d\mathbf{x}) := \min_{\mu \in \mathscr{P}(K)} \mathbf{E}_\mu\left[p_0(\mathbf{x})\right] \tag{P2}$$

where $\mathscr{P}(K)$ is the space of Borel probability measures on K. Although (P2) is an infinite dimensional problem, it is, in contrast to (P1), convex. The following result, taken from [Lasserre (2001)], establishes the equivalence between the two problems:

Theorem 1. *Problems* (P1) *and* (P2) *are equivalent in the following sense:*

(1) $\tilde{p}_K^* = p_K^*$.

(2) If $\mathbf{x}^*$ *is a global minimizer of* (P1), *then the Dirac distribution* $\mu^* = \delta_{\mathbf{x}^*}$ *with support on the point* $\mathbf{x}^*$ *is a global minimizer of* (P2).

(3) For every optimal solution μ^* *of* (P2), $p_0(\mathbf{x}) = p_K^*$-μ^* *almost everywhere (i.e.* $p_0(\mathbf{x}) = p_K^*$ *except in a set of* μ^**-measure* 0*).*

One direct consequence of this theorem is that, it is possible to develop a *convergent* sequence of LMI based convex relaxations to problem (P1), where the optimization variables are $m_{\boldsymbol{\alpha}} \doteq \mathbf{E}_\mu \mathbf{x}^{\boldsymbol{\alpha}}$, the moments of the unknown distribution μ. If in addition $p_0(\mathbf{x}) - p_K^*$ has a Sum-of-Squares (SOS) representation on K, i.e.,

$$p_0(\mathbf{x}) - p_K^* = t_0^2(\mathbf{x}) + \sum_{i=1}^{L} p_i(\mathbf{x}) t_i^2(\mathbf{x}) \tag{1}$$

for some polynomial $t_0(\mathbf{x})$ of degree at most N and some polynomials $t_i(\mathbf{x})$ of degree at most $N - d_i/2$, then, it is possible to construct an *equivalent* LMI based convex optimization problem for (P2). To this effect, let

$$\begin{aligned} p_N^* = \min_{\mathbf{m}} & \sum_{\alpha} p_{0,\alpha} m_\alpha \\ \text{s.t. } & \mathbf{M}_N(\mathbf{m}) \succeq 0, \\ & \mathbf{L}_N(p_i \mathbf{m}) \succeq 0, i = 1, \ldots, d, \end{aligned} \tag{2}$$

where $p_{0,\alpha}$ is the coefficient of $\mathbf{x}^\alpha$ in $p_0(\mathbf{x})$, $\mathbf{M}_N(\mathbf{m})$ is the so-called *moment matrix* and $\mathbf{L}_N(p_i\mathbf{m})$ is the so-called *localizing matrix*, both of which are constructed from the truncated moment sequence $\{m_{\boldsymbol{\alpha}}\}_0^{2N}$. For illustration and clarity of exposition, consider the case where $\mathbf{x} \in \mathbf{R}^2$, the moment matrix $\mathbf{M}_N(\mathbf{m})$ is defined as

$$\mathbf{M}_N(\mathbf{m}) = \begin{bmatrix} \mathbf{M}_{0,0}(\mathbf{m}) & \mathbf{M}_{0,1}(\mathbf{m}) & \cdots & \mathbf{M}_{0,N}(\mathbf{m}) \\ \mathbf{M}_{1,0}(\mathbf{m}) & \mathbf{M}_{1,1}(\mathbf{m}) & \cdots & \mathbf{M}_{1,N}(\mathbf{m}) \\ \vdots & \vdots & \ddots & \vdots \\ \mathbf{M}_{N,0}(\mathbf{m}) & \mathbf{M}_{N,1}(\mathbf{m}) & \cdots & \mathbf{M}_{N,N}(\mathbf{m}) \end{bmatrix}$$

where

$$\mathbf{M}_{j,k}(\mathbf{m}) = \begin{bmatrix} m_{j+k,0} & m_{j+k-1,1} & \cdots & m_{j,k} \\ m_{j+k-1,1} & m_{j+k-2,2} & \cdots & m_{j-1,k+1} \\ \vdots & \vdots & \ddots & \vdots \\ m_{k,j} & m_{k-1,j+1} & \cdots & m_{0,j+k} \end{bmatrix}.$$

Note that the size of $\mathbf{M}_N(\mathbf{m})$ is $\binom{d+N}{N} \times \binom{d+N}{N}$. The localizing matrix $\mathbf{L}_{N_i}(p_i\mathbf{m})$ is defined as

$$\mathbf{L}_{N_i}(p_i\mathbf{m})(i,j) = \sum_{\boldsymbol{\alpha}} p_{i,\boldsymbol{\alpha}} m_{(\boldsymbol{\beta}(i,j)+\boldsymbol{\alpha})}$$

where $p_{i,\alpha}$ is the coefficient of $\mathbf{x}^{\boldsymbol{\alpha}}$ in $p_i(\mathbf{x})$, $m(i,j)$ is the entry (i,j) of $\mathbf{M}_N(\mathbf{m})$ and $\boldsymbol{\beta}(i,j)$ is the subscript of $m_{\boldsymbol{\beta}}$. To illustrate, consider $g_1 = a - x_1x_2$, then

$$\mathbf{L}_1(g_1\mathbf{m}) = \begin{bmatrix} a - m_{11} & am_{10} - m_{22} & am_{01} - m_{12} \\ am_{10} - m_{22} & am_{20} - m_{31} & am_{11} - m_{22} \\ am_{01} - m_{12} & am_{11} - m_{22} & am_{02} - m_{13} \end{bmatrix}.$$

Then, according to Theorem 4.2 in [Lasserre (2001)], we have

Theorem 2. *In problem* (P1)

(i) [General Polynomial Optimization] as N increases to infinity,

$$p_N^* \uparrow p_K^*; \tag{3}$$

(ii) [General Polynomial Optimization with SOS] if $p_0(x) - p_K^*$ has the representation (1), *then with a fixed relaxation order* $\bar{N} = \max\{d_0, \ldots, d_L\}$

$$p_{\bar{N}}^* = p_K^*. \tag{4}$$

2.3. *Sparse polynomial optimization*

The previous section describes how to build a finite SDP to solve a polynomial optimization problem given that an SOS representation is known to exist a priori. However, in terms of complexity, it might become computationally intractable if the problem size is large, i.e., the dimension of x is large and/or the degree N is large. Note that the dimension of the moment matrix $\mathbf{M}_N(\mathbf{m})$ is $\binom{d+N}{d}$, which grows polynomially in N *or* d but still very fast, as pointed out in [Lasserre (2001); Laurent (2009); Parrilo (2003)]. On the other hand, however, many polynomial optimization problems encountered in practice have a sparse structure that can be exploited to decrease computational complexity; i.e., the polynomial p_i only contains a small fraction of the overall variables; e.g., see [Waki *et al.* (2006)]. If the set of indices of variables in each p_i satisfies the so-called *running intersection property*, the size of the LMIs can be significantly reduced, i.e., at most $\binom{N+\xi}{N}$ where ξ is the largest number of variables appearing in each polynomial, e.g., see [Lasserre (2006)]. We now state the definition of this property.

Definition 13.1 (Running Intersection Property). *Let I_k, $k = 1, \ldots, \tilde{d}$, be the subsets of variables $X \doteq \{x_1, \ldots, x_d\}$ satisfying $\bigcup_{k=1}^{\tilde{d}} I_k = X$. If*

(i) each constraint polynomial $p_i(\mathbf{x})$ uses only variables in I_k for some k;

(ii) the objective polynomial can be written as $p_0 = p_{0,1} + \cdots + p_{0,l}$ where each $p_{0,i}$ uses only variables in I_k for some k,

then running intersection property *is satisfied in Problem* (P1) *if the collection $\{I_1, \ldots, I_{\tilde{d}}\}$ obeys the following condition:*

$$I_{k+1} \cap \left(\cup_{j=1}^{k} I_j\right) \subseteq I_s \text{ for some } s \leq k, \tag{5}$$

for every $k = 1, \ldots, \tilde{d} - 1$.

Similar to the result stated in Theorem 2, for a sparse polynomial optimization problem that satisfies the *running intersection property* and has a sparse SOS representation on K, one can construct a finite SDP as well. For simplicity, we denote $\mathbf{M}_N(\mathbf{m}, I_k)$ the moment matrix for the reduced variables in the set I_k and denote $\mathbf{M}_{N_i}(p_j\mathbf{m}, I_k)$ the localizing matrix with the reduced variables in I_k. In the spirit of results in [Lasserre (2001, 2006)], we have the following theorem.

Theorem 3 (Sparse Polynomial Optimization with SOS). *Assume that* (P1) *satisfies the Running Intersection Property and let*

$$\begin{aligned} p_N^* = \min_{m} & \sum_{\alpha} p_{0,\alpha} m_\alpha \\ s.t.\ & \mathbf{M}_N(\mathbf{m}, I_k) \succeq 0, \quad k = 1, \ldots, \tilde{d} \\ & \mathbf{L}_{N_i}(p_i\mathbf{m}, I_{k(i)}) \succeq 0, \quad i = 1, \ldots, d, \end{aligned} \tag{6}$$

where $p_i(\mathbf{x})$ contains only variables in $I_{k(i)}$, and $p_0(\mathbf{x})$ has a sparse *SOS representation on K, i.e.,*

$$p_0(\mathbf{x}) - p_K^* = \sum_{k=1}^{L} \left(t_{k,0}^2(\mathbf{x}) + \sum_{i=1}^{d} p_i(\mathbf{x}) t_{k,i}^2(\mathbf{x}) \right), \tag{7}$$

where N_i is the smallest integer no less than $N - d_i/2$, $t_{k,0} \in \mathbf{R}[\mathbf{x}(I_k)]$ is of degree N and $t_{k,0} \in \mathbf{R}[\mathbf{x}(I_k)]$ is of degree N_i. Then,

$$p_N^* = p_K^*. \tag{8}$$

3. System identification

In this section, we proceed by first defining the identification problem for switched ARX systems. We consider multi-input, multi-output (MIMO) switched autoregressive exogenous (SARX) models of the form:

$$\begin{aligned} \boldsymbol{y}_t &= \sum_{k=1}^{n_a} \mathbf{A}_k(\sigma_t)\boldsymbol{y}_{t-k} + \sum_{k=1}^{n_c} \mathbf{C}_k(\sigma_t)\boldsymbol{u}_{t-k} \\ \tilde{\boldsymbol{y}}_t &= \boldsymbol{y}_t + \boldsymbol{\eta}_t \end{aligned} \tag{9}$$

where $\boldsymbol{u}_t \in \mathbf{R}^{n_u}$ is the input, $\tilde{\boldsymbol{y}}_t \in \mathbf{R}^{n_y}$ is the measured output corrupted by the noise $\boldsymbol{\eta}_t \in \mathbf{R}^{n_y}$, and $\sigma_t \in \mathbf{N}_s$ is the discrete mode signal indicating which of the s submodels is active at time t. We do not make any dwell-time assumptions, hence the mode signal σ_t can switch arbitrarily fast among the s submodels G_i, each of which is associated with the set of its coefficient matrices $\{\mathbf{A}_1(i), \ldots, \mathbf{A}_{n_a}(i), \mathbf{C}_1(i), \ldots, \mathbf{C}_{n_c}(i)\}$, $1 \leq i \leq s$.

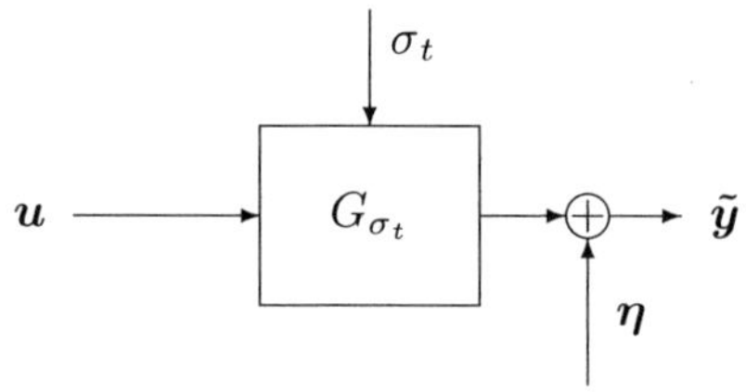

Fig. 1. Problem Setup. The coefficient matrices of the submodels G_i and a bound on the noise are known *a priori*. The experimental data consists of input/output measurements, $\boldsymbol{u}$ and $\tilde{\boldsymbol{y}}$. The mode signal σ_t and noise sequence η are unknown.

3.1. *Problem statement*

Without any additional restrictions, the identification problem admits infinitely many solutions. For example, one can assign a trivial model corresponding to each measurement $\{\mathbf{u}_t, \mathbf{y}_t\}$ at time t. Thus, one needs to add additional constraints or objectives to make the problem well-posed. In this chapter, we aim at minimizing the number of sub-systems and/or minimizing the order of the linear models. For simplicity, let us assume that s, the number of switched sub-systems, is known and (n_a, n_c), the order of the linear models, is also known. This assumption does not imply loss of generality since one can always increase s and/or (n_a, n_c) one by one until a meaningful solution is found. Then, the identification problems for the setup described above and shown in Figure 1 can be formally stated as follows.

Problem 13.1 (Identification). *Given input and output experimental data* $\{\boldsymbol{u}_t, \tilde{\boldsymbol{y}}_t\}_{t=t_0}^{T}$, *an* a priori *bound* ϵ *on noise (i.e.* $\|\boldsymbol{\eta}_t\|_\infty \leq \epsilon$ *for all* t*), the number of sub-models* s *and the order of sub-models* (n_a, n_c)*, find a hybrid linear model of the form (9) that is consistent with all a priori information and the measurement data, or conclude that none exists.*

3.2. *Algebraic reformulation*

We begin by eliminating the discrete variables in Eq. (9) and constructing equivalent polynomial constraints. Rearranging the terms in Eq. (9) yields:

$$\begin{aligned} h_{t,i}(\boldsymbol{\eta}_{t:t-n_a}, \mathbf{A}, \mathbf{C}) &\doteq \mathbf{A}_1(\sigma_t)(\tilde{\boldsymbol{y}}_{t-1} - \boldsymbol{\eta}_{t-1}) + \ldots + \mathbf{A}_{n_a}(\sigma_t)(\tilde{\boldsymbol{y}}_{t-n_a} - \boldsymbol{\eta}_{t-n_a}) \\ &\quad -(\tilde{\boldsymbol{y}}_t - \boldsymbol{\eta}_t) + \mathbf{C}_1(\sigma_t)\boldsymbol{u}_{t-1} + \ldots + \mathbf{C}_{n_c}(\sigma_t)\boldsymbol{u}_{t-n_c} \\ &= \mathbf{0} \end{aligned} \tag{10}$$

which consists of n_y linear equations in $n_a + 1$ unknown noise vectors $\boldsymbol{\eta}_{t:t-n_a}$. Denote each element of $h_{t,i}$ to be $h_{t,i}^{(j)}$, $j = 1, \ldots, n_y$, Eq. (10) is equivalent to

$$[h_{t,i}^{(1)} = 0] \wedge \ldots \wedge [h_{t,i}^{(n_y)} = 0] \tag{11}$$

or algebraically

$$g_{t,\sigma(t)}(\boldsymbol{\eta}_{t:t-n_a}, \mathbf{A}, \mathbf{C}) \doteq \sum_{j=1}^{n_y} \left[h_{t,i}^{(j)}\right]^2 = 0. \tag{12}$$

Note that since the mode signal σ_t is unmeasurable, the actual subsystem G_i that is active at any given time t is unknown. However, since the measured data must be consistent to at least one of the s subsystems, Eq. (12) should hold true for some $i \in \{1, \ldots, s\}$[a]. This condition can be expressed as

$$[g_{t,1} = 0] \vee \ldots \vee [g_{t,s} = 0] \tag{13}$$

or algebraically

$$p_t(\boldsymbol{\eta}_{t:t-n_a}, \mathbf{A}, \mathbf{C}) \doteq \prod_{i=1}^{s} g_{t,i}(\boldsymbol{\eta}_{t:t-n_a}) = 0, \tag{14}$$

---

[a]This idea is similar to the hybrid decoupling constraint proposed in [Ma and Vidal (2005)].

which holds for all t. Then, the identification problem is equivalent to finding some admissible noise $\boldsymbol{\eta}$ and parameters $\mathbf{A}$ and $\mathbf{C}$ such that $p_t = 0$ for all t.

With the above discussion in mind, we consider the following polynomial optimization problem.

Problem 13.2. *Given the number of the sub-systems s and the order of linear submodels (n_a, n_c), find*

$$\begin{aligned} \min_{\boldsymbol{\eta},\mathbf{A},\mathbf{C}} \;& p_{sum}(\boldsymbol{\eta}, \mathbf{A}, \mathbf{C}) \\ s.t. \;& \|\boldsymbol{\eta}_t\|_\infty \le \epsilon \text{ for all } t \in [t_0, T] \end{aligned} \tag{15}$$

where

$$p_{sum}(\boldsymbol{\eta}, \mathbf{A}, \mathbf{C}) = \sum_{t=n_a}^{T} p_t(\boldsymbol{\eta}_{t:t-n_a}, \mathbf{A}, \mathbf{C}).$$

For ease of notation, we also write p_{sum} as

$$p_{\text{sum}} = \sum_{\boldsymbol{\alpha}} p_{\boldsymbol{\alpha}}(\mathbf{A}, \mathbf{C}, \boldsymbol{\eta})^{\boldsymbol{\alpha}}. \tag{16}$$

The equivalence between the above problem and Problem 13.1 is established in the following theorem.

Theorem 4. *Given the number of the sub-systems s and the order of the subsystems (n_a, n_c), if there exists at least one compatible hybrid affine model of the form (9) for Problem 13.1, then there exist some admissible noise $\boldsymbol{\eta}$ and parameters $(\mathbf{A}, \mathbf{C})$ such that the minimum of Problem 13.2 is zero. The converse is also true.*

Proof. If there is a compatible hybrid model, then, given the true values of the parameter and noise, one has $p_t = 0$ for all t. Hence, $p_{\text{sum}} = 0$. Since p_{sum} is a sum-of-squares, p_{sum} is always non-negative. Hence, the minimum of (15) is zero. Conversely, if $p_{\text{sum}} = 0$ for some noise $\boldsymbol{\eta}$ and some parameter $(\mathbf{A}, \mathbf{C})$, then $p_t = 0$ for all t. Since p_t is a product of s polynomials, then at least one of them is equal to zero. Denote the index of one of such polynomials by $\sigma(t)$. Consequently, a system with parameter $(\mathbf{A}, \mathbf{C})$ is a compatible model, and $\sigma(t)$ is the index of the active subsystem at time t. □

3.3. *An equivalent sparse polynomial optimization*

Although Problem 13.1 and Problem 13.2 are equivalent, at first glance it seems that the reformulation is at least equally difficult to solve, since (15) contains system parameters $(\mathbf{A}, \mathbf{C})$ and noise vectors $\boldsymbol{\eta}$ where the number of the noise vectors is equal to the number of measurements. Note that the number of measurements can range in the hundreds, even in a mid-sized identification problem. However, if one carefully checks the structure of the polynomial p_{sum}, it can be found that it has a sparse structure. Note that for each $p_t(\boldsymbol{\eta}_{t:t-n_a}, \mathbf{A}, \mathbf{C})$, it only involves variables $\mathbf{A}, \mathbf{C}$ and $\boldsymbol{\eta}_{t:t-n_a}$. Moreover, the constraints in (15) can be written as polynomial inequalities:

$$f_{t,j}(\eta_t^{(j)}) \doteq \epsilon^2 - \left[\eta_t^{(j)}\right]^2 \geq 0, \quad \forall t \in [0, T], j \in \mathbf{N}_{n_y}. \tag{17}$$

Hence, problem (15) inherently satisfies the *running intersection property* stated in Definition 13.1, with

$$I_k = \{\mathbf{A}, \mathbf{C}, \boldsymbol{\eta}_k, \ldots, \boldsymbol{\eta}_{k-n_a}\}, \quad k = 1, 2, \ldots, T. \tag{18}$$

Moreover, $p_t(\boldsymbol{\eta}_{t:t-n_a}, \mathbf{A}, \mathbf{C})$ defined in (14) is a sum-of-squares polynomial. In particular, if the system model in (9) is single-output, i.e., $n_y = 1$, $p_t(\boldsymbol{\eta}_{t:t-n_a}, \mathbf{A}, \mathbf{C})$ is a square of a single polynomial. Hence, p_{sum} is also a sum of squares, implying that, if the minimum of (15) is zero, $p_{\text{sum}} - p^*_{\text{sum}} = p_{\text{sum}}$ can be represented in the form of (7); i.e., one just needs to take $t_{k,0} = p_k$ and $t_{k,i} = 0$. Consequently, Problem 13.2 can be solved via a fixed and finite sized SDP. This is summarized as follows.

Theorem 5. *Given the number of the sub-systems s and the order of the sub-models (n_a, n_c), consider the following optimization problem*

$$\begin{aligned} p^*_s = \min_{m} \sum_{\alpha} p_\alpha m_\alpha \\ s.t.\ \mathbf{M}_N(\mathbf{m}, I_k) \succeq 0, k = 0, \ldots, T + n_a \\ \mathbf{L}_N(f_{t,j}\mathbf{m}, I_{\beta(t)}) \succeq 0, t \in [0, T], j \in \mathbf{N}_{n_y} \end{aligned} \tag{19}$$

where $N = 2s$; $f_{t,j}$ are defined in (17), $I_k = \{\mathbf{A}, \mathbf{C}, \boldsymbol{\eta}_{k:k+n_a}\}$ is the partial variable set for p_i, where

$$\beta(t) = \begin{cases} t, & \text{for } t = 1, \ldots, T \\ T, & \text{for } t = T + 1, \ldots, T + n_a \end{cases}.$$

Then, if there exists at least one compatible hybrid affine model for Problem 13.1, $p^*_s = 0$ is the optimum of (19). Conversely, if $p^*_s = 0$, *rank* $M_N(m, I_k) =$ *rank* $M_{N_i}(m, I_k)$ *for all* k*, and rank* $M_N(m, I_k \cap I_j) = 1$ *for all pairs* (j, k) *with* $I_k \cap I_j \neq \emptyset$*, then, there exists at least one compatible model.*

Proof. This is a direct consequence of Theorem 3, given the fact that running intersection property holds for the collection of the variable sets I_k defined in (18) and the fact that p_0 has a representation of (7) by taking $t_{k,0} = p_{0,k}$ and $t_{k,i} = 0$. □

Remark 13.1. With the rank conditions

$$\text{rank } M_N(m, I_k) = \text{rank } M_{N_i}(m, I_k), \quad \text{rank } M_N(m, I_k \cap I_j) = 1$$

being satisfied, an optimal solution is achieved [Henrion and Lasserre (2005); Lasserre (2006)] and an algorithm is given in [Henrion and Lasserre (2005)], which can always extract an optimal moment sequence corresponding to a probability measure with point support.

Remark 13.2. In the case where both input noise and output noise are considered, the results above still apply. One just needs to add more variables for input noise terms in the polynomial optimization problem, which has a similar sparse structure as shown above.

The following simple example illustrates that exploiting the sparse structure of the problem substantially reduces computational complexity.

Example 1. Consider a single-input-single-output hybrid system ($n_u = n_y = 1$) with three sub-systems ($s = 3$) of order $(n_a, n_c) = (1, 0)$, and 20 measurements ($T = 20$). Then, in Problem 13.2, one would have to build a moment matrix of size $\Theta_1 = \binom{21+3+6}{6} = 593,775$ and 21 localizing matrices of size $\Theta_2 = \binom{21+3+5}{5} = 118,755$ in the SDP, as solving a general polynomial optimization problem. On the other hand, by taking into account the sparsity, one could build 20 moment matrices of size $\Theta_3 = \binom{2+3+6}{6} = 462$ and 21 localizing matrices of size $\Theta_4 = \binom{2+3+5}{5} = 252$ for constructing an equivalent SDP.

As illustrated above, by taking into account the sparsity, the optimization problem can be solved if its size is relatively small. One important observation is that the complexity is proportional to the number of measurements, i.e. if one fixes the structure of the hybrid system, the maximum size of

LMIs in the SDP remains the same and the number of LMIs increases proportionally. Hence, if the hybrid system is relatively simple, the identification problem can be solved via solving (19) directly. However, if the hybrid system becomes more complex, say, the number of sub-systems increases or the order of the sub-models increases, the problem becomes numerically difficult to solve. For example, in the SISO case where $n_u = n_y = 1$, the size of I_k is equal to $n_\nu = n + 1 + s(n_a + n_c)$, and the maximum size of the moment matrices is equal to $\binom{n_\nu+2s}{2s}$.

To overcome this numerical difficulty, motivated by GPCA related algorithms, we formulate an equivalent rank minimization problem by using quadratic functions of rows of the Veronese matrix to eliminate the system parameters $(\mathbf{A}, \mathbf{C})$ in the optimization problem. This is described in the next section.

3.4. *A "refined" algorithm via rank minimization*

For ease of notation, this section is mainly devoted for SISO hybrid system identification. However, the algorithm proposed can be modified for MIMO systems, as discussed at the end of this section.

It can be seen that the main computational difficulty comes from the fact that all system parameters are a part of the decision variables in the polynomial optimization. Thus, if the structure of the hybrid system is too complex, a large computational effort may be required to solve (19).

In the sequel, we consider the case where the switched ARX system is SISO, and we show that the original identification problem can be reformulated as an LMI constrained rank minimization of a matrix (the Veronese map) whose entries are constructed directly from the noisy measurements. Though the rank minimization problem is in general NP-hard, efficient relaxations are available especially for symmetric positive semi-definite matrices, which is the case encountered here.

We now review the GPCA approach which provides the motivation for the results presented. That is, one can write (14) in a matrix format, in which the noise terms $\boldsymbol{\eta}$ and the parameters $(\mathbf{A}, \mathbf{C})$ are isolated. A similar idea has been used in [Feng *et al.* (2010a)] in the proposed randomized algorithm, where the computational cost is reduced by fixing each part of the variables in the iterations.

First, let us recall the construction of the Veronese matrix. Consider the polynomial equations (14) from time $t = 1$ to T, by collecting all measurements, one can build the noisy Veronese matrix $\mathbf{V}_s$ and polynomial equations as

$$\mathbf{V}_s(\mathbf{r},\boldsymbol{\eta})\mathbf{x} \doteq \begin{bmatrix} \mathbf{v}_s(\mathbf{r}_1,\boldsymbol{\eta}) \\ \vdots \\ \mathbf{v}_s(\mathbf{r}_T,\boldsymbol{\eta}) \end{bmatrix} \mathbf{x} = 0. \tag{20}$$

where $\mathbf{x}$ is a vector with elements being polynomial functions of the system parameters and $\mathbf{r}_t$ is a collection of the measured data from time t to t_{n_a}. Note that the Veronese matrix is only a function of the noise while the null vector is only a function of the system parameters. Thus, in this context, the identification problem is equivalent to finding admissible noise $\boldsymbol{\eta}$ so that $\mathbf{V}_s(\mathbf{r},\boldsymbol{\eta})$ has a non-trivial null space, i.e. finding $\boldsymbol{\eta}$ and a non-zero vector $\mathbf{x}$ so that (20) holds. Once such a null space is found, it can be used to recover the parameters of the sub-systems; e.g. see [Vidal *et al.* (2003)]. This is summarized in the following theorem.

Theorem 6 (Theorem 1, 2 in [Vidal *et al.* (2003)]). *Assume $n_u = n_y = 1$, given the number of the sub-systems s, the order of sub-models (n_a, n_c), if there exists at least one compatible hybrid affine model for Problem 13.1, then there exists some admissible noise $\boldsymbol{\eta}$ such that $\mathbf{V}_s(\mathbf{r},\boldsymbol{\eta})$ is rank deficient; and the system parameters $(\mathbf{A},\mathbf{C})$ can be extracted from its null space.*

A simple example is illustrated on how to construct such a noisy Veronese matrix.

Example 2. For $n_u = n_y = 1$, $s = 2$ and the order $(n_a, n_c) = (1, 1)$, Eq. (14) can be written as

$$\left(y_k + a_1(y_{k-1} - \eta_{k-1}) - c_1 u_k - \eta_k\right)\left(y_k + a_2(y_{k-1} - \eta_{k-1}) - c_2 u_k - \eta_k\right) = 0.$$

Hence, the k-th row of the noisy Veronese matrix $\mathbf{V}_s$ can be written as

$$\mathbf{v}_k(\mathbf{r},\boldsymbol{\eta}) = \begin{pmatrix} (y_k - \eta_k)^2 \\ (y_{k-1} - \eta_{k-1})(y_k - \eta_k) \\ -u_k(y_k - \eta_k) \\ -u_k(y_{k-1} - \eta_{k-1}) \\ (y_{k-1} - \eta_{k-1})^2 \\ u_k^2 \end{pmatrix}^T$$

Remark 13.3. For $n_u = n_y = 1$, the size of the Veronese matrix is $L \times \binom{n_a+n_c+s}{s}$. For each row of the matrix, the highest total degree of the polynomials (in $\boldsymbol{\eta}$) is s, the number of sub-systems.

Given the noisy Veronese matrix constructed above, define

$$\mathbf{Q}(\mathbf{m}) \doteq \sum_{i=1}^{T} \mathbf{E}_\mu\{\mathbf{v}_s(\mathbf{r}_i,\boldsymbol{\eta})^T \mathbf{v}_s(\mathbf{r}_i,\boldsymbol{\eta})\} \tag{21}$$

where $\mathbf{m}$ is the truncated moment sequence corresponding to probability measure μ. Hence, $\mathbf{Q}(m)$ is a symmetric matrix and is linear in $\mathbf{m}$. Now, we are ready to propose the equivalent rank minimization problem.

Problem 13.3. *Given the number of the sub-systems s, the order of sub-models (n_a, n_c), find a rank deficient matrix $\mathbf{Q}(\mathbf{m})$ defined in (21) subject to*

$$\begin{aligned} &\mathbf{M}_N(\mathbf{m}, I_k) \succeq 0, \quad k = 1, \ldots, T + n_a; \\ &\mathbf{L}_N(p_t\mathbf{m}, I_{\beta(t)}) \succeq 0, \quad t \in [1, T]. \end{aligned}$$

Then, we observe that this problem is indeed equivalent to the problem of finding an admissible noise sequence that results in a rank deficient Veronese matrix.

Theorem 7. *If there exists an admissible noise sequence $\boldsymbol{\eta}$ such that the Veronese matrix $\mathbf{V}_s(\mathbf{r},\boldsymbol{\eta})$ is rank deficient, then Problem 13.3 has a feasible solution. Conversely, if Problem 13.3 has a feasible solution $\mathbf{m}$, and if rank $\mathbf{M}_N(\mathbf{m}, I_k) =$ rank $\mathbf{M}_{N-s}(\mathbf{m}, I_k)$ for all k and rank $\mathbf{M}_N(\mathbf{m}, I_k \cap I_j) = 1$ for all pairs (j, k) with $I_k \cap I_j \neq \emptyset$, then, $\mathbf{V}_s(\mathbf{r},\boldsymbol{\eta})$ is rank deficient for some admissible noise $\boldsymbol{\eta}$.*

Proof. First note that there exists some unit vector $\mathbf{x}$ and admissible noise $\boldsymbol{\eta}$ such that

$$\mathbf{V}_s(\mathbf{r},\boldsymbol{\eta})\mathbf{x} = 0$$

if and only if

$$\sum_{i=1}^{T} x^T \mathbf{v}_s(\mathbf{r}_i,\boldsymbol{\eta})^T \mathbf{v}_s(\mathbf{r}_i,\boldsymbol{\eta})\mathbf{x} = 0. \tag{22}$$

Since $\mathbf{v}_s(\mathbf{r}_i, \boldsymbol{\eta})^T \mathbf{v}_s(\mathbf{r}_i, \boldsymbol{\eta})$ is positive semi-definite, it is equivalent to finding a unit vector $\mathbf{x}^*$ that the minimum of the following problem

$$\begin{aligned} \min_{\boldsymbol{\eta}} \; & (\mathbf{x}^*)^T \left[\sum_{i=1}^{T} \mathbf{v}_s(\mathbf{r}_i, \boldsymbol{\eta})^T \mathbf{v}_s(\mathbf{r}_i, \boldsymbol{\eta}) \right] \mathbf{x}^* \\ \text{s.t.} \; & \boldsymbol{\eta} \in K \end{aligned} \tag{23}$$

is zero, where K is defined as

$$K \doteq \{\boldsymbol{\eta} : p_k \doteq \epsilon^2 - \eta_k^2 \geq 0, k = 1, \ldots, T + n_a\}.$$

Note that the objective function in (23) is an SOS in terms of $\boldsymbol{\eta}$ for any fixed $\mathbf{x}$, hence, it can be represented in the form of (7). Moreover, the running intersection property is satisfied for (23). Therefore, by Theorem 3, it follows that the minimum of the following problem

$$\begin{aligned} \min_{\mathbf{m}} \quad & (\mathbf{x}^*)^T \mathbf{Q}(\mathbf{m}) \mathbf{x}^* \\ \text{s.t.} \quad & \mathbf{M}_s(\mathbf{m}, I_k) \succeq 0, \quad k = 1, \ldots, T, \\ & \mathbf{L}_s(p_i \mathbf{m}, I_{\beta(i)}) \succeq 0, \quad i = 1, \ldots, T + n_a, \end{aligned} \tag{24}$$

is zero, where $\mathbf{Q}$ is defined in (21). Since $\mathbf{v}_s(\mathbf{r}_i, \boldsymbol{\eta})^T \mathbf{v}_s(\mathbf{r}_i, \boldsymbol{\eta})$ is positive semi-definite, $\mathbf{Q}(\mathbf{m})$ is positive semi-definite as well.[b] Hence, we have $(\mathbf{x}^*)^T \mathbf{Q}(\mathbf{m}) \mathbf{x}^* = 0$ for some unit vector $\mathbf{x}^*$ if and only if the matrix $\mathbf{Q}(\mathbf{m})$ is rank deficient.

Conversely, if the rank condition holds, one can always extract a moment sequence $\mathbf{m}^*$ corresponding to a probability measure with point support, such that $\mathbf{Q}(\mathbf{m}^*)$ is rank deficient. Hence, there is an admissible noise $\boldsymbol{\eta}$ that $\sum_{i=1}^{T} \mathbf{v}_s(\mathbf{r}_i, \boldsymbol{\eta})^T \mathbf{v}_s(\mathbf{r}_i, \boldsymbol{\eta})$ is rank deficient. This implies the Veronese matrix is rank deficient for the same noise $\boldsymbol{\eta}$, which concludes the proof.

□

Although rank minimization is NP-hard, efficient convex relaxations are available. In particular, good approximate solutions can be obtained by using a log-det heuristic that relaxes rank minimization to a sequence of convex problems, e.g., see [Fazel *et al.* (2003); Ozay *et al.* (2009)]. Furthermore, as stated in Theorem 7, it suffices to find a rank deficient solution. Thus, we use a modification of log-det heuristic that aims at dropping the rank by one, as illustrated in the following.

[b]Note that $\mathbf{Q}(\mathbf{m})$ is p.s.d. since $\mathbf{m}$ satisfies the LMI constraints in (24). This comes directly from the fact that the finite moment condition is the dual formulation of the related SOS problem.

Algorithm 13.1 Drop Rank

Set $X \leftarrow \mathbf{Q}(\mathbf{m})$, $X_0 \leftarrow I$, $k \leftarrow 0$.
repeat
Solve

$$X_{k+1} \leftarrow \arg\min \operatorname{Tr}(X_k + \delta I)^{-1} X \quad \text{s.t.} \quad m \in \mathscr{M}$$

where $\mathscr{M}$ is the feasible set in Problem 13.3.
Decompose the symmetric matrix $X_k = T^{-1}DT$.
Set $\delta \leftarrow \min \operatorname{diag}(D) + \delta_0$.
Set $k \leftarrow k + 1$.
until a convergence criterion is reached.
return X_k

Remark 13.4. Once a rank deficient matrix $\mathbf{Q}(\mathbf{m}^*)$ is found with corresponding moment sequence $\mathbf{m}^*$, a vector in its null space can be easily determined by computing its eigenvectors and eigenvalues. Moreover, by Theorem 7, there exists some admissible noise $\boldsymbol{\eta}$ such that the Veronese matrix $\mathbf{V}_s(\mathbf{r}, \boldsymbol{\eta})$ is rank deficient. One can extract such a noise value $\boldsymbol{\eta}$ from the moment sequence $\mathbf{m}^*$ using the algorithm introduced in [Henrion and Lasserre (2005)]. Once the noise values are estimated, the problem reduces to the noise free case, allowing for computing the system parameters using the procedure introduced in [Ma and Vidal (2005)].

Remark 13.5. In the MIMO cases, one can still build a similar but much more complex noisy Veronese matrix. The algorithm proposed above can be suitably modified to find an admissible noise that leads to a rank deficient Veronese matrix. By plugging the noise estimates, the algebraic solution proposed in [Bako and Vidal (2008)] for MIMO switch ARX system identification can be used to construct the identified model.

4. Model (in)validation

Once a system model is obtained either from first principles or an identification procedure, it must be validated using additional experimental data before it can be used for synthesizing controllers. Since hybrid system identification is generically NP-Hard and most of the identification methods are based on heuristics or relaxations, it is crucial to check the validity of the identified models. In addition, this step can be used to obtain bounds on the modeling error that can be directly incorporated into robust controller design algorithms. In this section, we present a method for model invalidation of switched systems with unknown switches using input/output data.

The model invalidation problem for the setup shown in Figure 1 can be formally stated as follows:

Problem 13.4 (Model (in)validation). Given a nominal hybrid model of the form (9) together with its s submodels $G_1, \ldots, G_s$, an *a priori* bound ϵ on the noise, and experimental data $\{\boldsymbol{u}_t, \tilde{\boldsymbol{y}}_t\}_{t=t_0}^{T}$, determine whether or not the *a priori* information and the *a posteriori* experimental data are consistent, i.e. whether the consistency set

$$\mathscr{T}(\boldsymbol{\eta}, \sigma) = \{\|\boldsymbol{\eta}_t\|_\infty \leq \epsilon, \sigma_t \in \mathbf{N}_s \text{ subject to (9) } \forall t \in [t_0, T]\}$$

is nonempty.

Since $\mathscr{T}(\boldsymbol{\eta}, \sigma)$ contains all possible noise and mode signal sequences that can explain the observed data, clearly establishing that $\mathscr{T}(\boldsymbol{\eta}, \sigma) = \emptyset$ is equivalent to invalidating the model.

4.1. *A convex certificate for (in)validating MIMO SARX models*

Next, we present the main result of this section, showing that a necessary and sufficient condition for the model to be invalidated by the experimental data is strict positivity of the solution to a related convex optimization problem. We begin by constructing a semialgebraic set $\mathscr{T}'(\boldsymbol{\eta})$ and showing its equivalence, in a sense to become clear later, to the consistency set $\mathscr{T}(\boldsymbol{\eta}, \sigma)$. Our derivation follows similar lines as in Section 3.2. We use the equalities (10)-(12)-(14) to define a semialgebraic set and establish its relationship with the consistency set $\mathscr{T}$:

Proposition 13.1. *Let*

$$\mathscr{T}'(\boldsymbol{\eta}) \doteq \left\{ \boldsymbol{\eta} \,\middle|\, \begin{array}{c} f_{t,j}(\eta_t^{(j)}) \geq 0 \; \forall t \in [0, T], j \in \mathbf{N}_{n_y} \\ p_t(\boldsymbol{\eta}_{t:t-n_a}) = 0 \; \forall t \in [n_a, T] \end{array} \right\},$$

where

$$f_{t,j}(\eta_t^{(j)}) \doteq \epsilon^2 - \left[\eta_t^{(j)}\right]^2$$

Then, $\mathscr{T}(\boldsymbol{\eta}, \sigma)$ is empty if and only if $\mathscr{T}'(\boldsymbol{\eta})$ is empty.

Proof. Follows from the construction of $\mathscr{T}'(\boldsymbol{\eta})$ and the fact that $|\eta_t^{(j)}| \leq \epsilon \iff f_{t,j}(\eta_t^{(j)}) \geq 0$ □

At this point, one can use the *Positivstellenstaz* and derive sum of squares certificates for the emptiness of $\mathscr{T}'(\boldsymbol{\eta})$, as proposed in [Parrilo (2003)]. However, as we show next, adopting a dual approach based on the theory of moments allows for exploiting the inherently sparse structure of the problem. In order to pursue this approach, start by considering the following optimization problem:

$$\begin{aligned} o^* = \min_{\boldsymbol{\eta}} & \sum_{t=n_a}^{T} p_t(\boldsymbol{\eta}_{t:t-n_a}) \\ \text{s.t.} & \\ & f_{t,j}(\eta_t^{(j)}) \geq 0 \quad \forall t \in [0, T], j \in \mathbf{N}_{n_y}. \end{aligned} \tag{25}$$

Note that $o^* \geq 0$, since the objective function in (25) is a sum of squares polynomial[c]. Further, if $\mathscr{T}'(\boldsymbol{\eta})$ is non-empty, then there exist a noise sequence $\boldsymbol{\eta}^*$ for which (25) attains its minimum $o^* = 0$. Equivalently, $o^* > 0 \iff \mathscr{T}'(\boldsymbol{\eta}) = \emptyset$.

Proposition 13.2. *Problem* (25) *above satisfies the running intersection property.*

Proof. Consider the $T - n_a + 1$ subsets I_k of the variables $\boldsymbol{\eta}_{0:T}$ where each I_k contains only the variables $\boldsymbol{\eta}_{k:k+n_a}$. One can associate each $f_{t,j}$ with I_0 for $t \leq n_a$ and with I_{t-n_a} for $t > n_a$. The collection I_{kj} formed by repeating each I_k, n_y times, satisfies (5) hence the running intersection property in Definition 13.1 holds. □

Next, we use results from the theory of moments to obtain a convex problem where the objective and constraints have affine (rather than polynomial) dependence on the data and exploits sparsity. Consider the related moments-based optimization:

[c] since it is formed by multiplication and addition of SOS polynomials in (12), and the cone of SOS polynomials is closed under these operations.

$$
\begin{aligned}
d_N^* = \min_{\boldsymbol{m}} & \sum_{t=n_a}^{T} l_t(\boldsymbol{m}_{t-n_a:t}) \\
\text{s.t.} & \\
& \mathbf{M}_N(\boldsymbol{m}_{t-n_a:t}) \succeq 0 \quad \forall t \in [n_a, T] \\
& \mathbf{L}_N(f_{t,j}\boldsymbol{m}_{t-n_a:t}) \succeq 0 \quad \forall t \in [n_a+1, T], j \in \mathbf{N}_{n_y} \\
& \mathbf{L}_N(f_{t,j}\boldsymbol{m}_{0:n_a}) \succeq 0 \quad \forall t \in [0, n_a], j \in \mathbf{N}_{n_y}
\end{aligned} \tag{26}
$$

where each l_t is the linear functional of moments defined as $l_t(\boldsymbol{m}_{t-n_a:t}) \doteq \mathbf{E}\left\{p_t(\boldsymbol{\eta}_{t:t-n_a})\right\}$, and where $\mathbf{M}_N$ and $\mathbf{L}_N$ are the moments and localizing matrices associated with a truncated moments sequence containing terms up to order $2N$ with $N \geq s$. Clearly $d_N^* \geq 0$, since the objective function in (25) is a sum of squares polynomial, and, from Theorem 2, $d_N^* \uparrow o^*$ as $N \to \infty$. These observations lead to the following necessary and sufficient conditions for (in)validation:

Theorem 8. *The following statements are equivalent:*

(i) The consistency set $\mathcal{T}'(\boldsymbol{\eta})$ is empty

(ii) There exists some finite N_o such that $d_{N_o}^* > 0$

(iii) The solution r^* to the following optimization problem is strictly greater than zero:

$$
\begin{aligned}
r^* = \min_{\boldsymbol{m}} & \sum_{t=n_a}^{T} l_t(\boldsymbol{m}_{t-n_a:t}) \\
\text{s.t.} & \\
& \mathbf{M}_s(\boldsymbol{m}_{t-n_a:t}) \succeq 0 \quad \forall t \in [n_a, T] \\
& rank\left[\mathbf{M}_s(\boldsymbol{m}_{t-n_a:t})\right] = 1 \quad \forall t \in [n_a, T] \\
& f_{t,j}(\boldsymbol{m}_t^{(j)}) \succeq 0 \quad \forall t \in [0, T], j \in \mathbf{N}_{n_y}
\end{aligned} \tag{27}
$$

where each of the $T - n_a + 1$ moments sequences $\boldsymbol{m}_{t-n_a:t}$, $t \in [n_a, T]$, contains moments up to order $2s$ (i.e. two times number of submodels) and $\boldsymbol{m}_t^{(j)} \doteq \mathbf{E}(\boldsymbol{\eta}_t^{(j)})$.

Proof. (i)⇔(ii) Recall that $\mathcal{T}'(\boldsymbol{\eta}) = \emptyset \iff o^* > 0$. Since $d_N^* \uparrow o^*$ as $N \to \infty$, if $o^* > 0$, there exist N_o such that $d_{N_o}^* > 0$. On the other hand, if $d_{N_o}^* > 0$ then $o^* > 0$ since $d_{N_o}^* < o^*$. Hence, $\mathcal{T}'(\boldsymbol{\eta})$ is empty.

(i)⇔(iii) To prove this equivalence, we show that r^* in (27) is equal to o^* in (25). Assume η^* is an optimizer of (25), then the moments of the distribution μ^* with point support at η^* is feasible for (27) with the same objective value which implies $r^* \le o^*$. On the other hand, if $\boldsymbol{m}^*$ is a minimizer of (27), the rank condition implies that there is a corresponding measure μ^* with point support, say at η^*. Last constraint in (27) guarantees that η^* is a feasible point of (25), which implies $o^* \le r^*$. Therefore, $r^* = o^*$ from which we conclude that $r^* > 0$ is equivalent to (i). □

Note that by forming a single block diagonal matrix containing all LMI constraints in (26), it can be transformed into the standard dual form of semidefinite programs. That is:

$$\begin{aligned} d_N^* = \inf_{\boldsymbol{m}} \; & \boldsymbol{b}^T \boldsymbol{m} + c_o \\ \text{s.t.} \; & \\ & \sum_{\alpha \in \mathscr{I}} \mathbf{A}_\alpha m_\alpha \succeq \mathbf{C} \end{aligned} \tag{28}$$

where $\mathscr{I}$ is the set of the multi-indexes of all moments that occur in (26) except the zeroth moment $m_{\mathbf{0}} = 1$ which is used to form the constant terms $\mathbf{C}$ and c_o.

Remark 13.6. It is important to highlight the complexity reduction achieved by employing the running intersection property while forming the optimization problem (26). The conventional moment relaxation of order N in [Lasserre (2001)] would require $O((Tn_y + n_y)^{2N})$ variables with a moment matrix of size $\binom{N + Tn_y + n_y}{Tn_y + n_y}$. On the other hand, (26) involves only $O((n_a n_y + n_y)^{2N})$ variables with $T - n_a + 1$ moment matrices of size $\binom{N + n_a n_y + n_y}{n_a n_y + n_y}$ where, in general, the length of the experimental data T is substantially larger than the order of the regressor n_a (i.e. $n_a \ll T$).

4.2. *Numerical considerations*

From Theorem 8 it follows that (in)validation/validation certificates can be obtained, for instance, by solving a sequence of problems with increasing N until either a solution with $d^* > 0$ is found (in which case the data invalidates the model) or condition (iii) fails, in which case the data record is consistent with the model and *a-priori* information. However, a potential numerical difficulty here stems from the fact that in practice $d^* = 0$ is never reached since there is no closed form solution to the problem and numerical solvers terminate when the duality gap drops below a certain level. Thus, from a practical standpoint, more useful conditions can be obtained by examining the primal problem, which can be written as

$$
\begin{aligned}
p_N^* = \max_{\mathbf{X}} &< \mathbf{C}, \mathbf{X} > + c_o \\
\text{s.t.} & \\
& < \mathbf{A}_{\boldsymbol{\alpha}}, \mathbf{X} >= b_{\boldsymbol{\alpha}} \;\; \forall \boldsymbol{\alpha} \in \mathscr{I} \\
& \mathbf{X} \succeq 0
\end{aligned} \tag{29}
$$

where, without loss of generality, $\mathbf{X}$ can be chosen to have the same block-diagonal structure that $\mathbf{A}_{\boldsymbol{\alpha}}$s and $\mathbf{C}$. Note that, from duality, $p_N^* \leq d_N^*$, and hence $p_N^* > 0$ provides an invalidation certificate. In fact if problem (29) is feasible, strong duality holds [Lasserre (2006)]. Thus $p_N^* > 0$ is strictly equivalent to $d_N^* > 0$, leading to the following result:

Proposition 13.3. *The consistency set* $\mathscr{T}'(\boldsymbol{\eta})$ *is empty if and only if* $p_{N_o}^* > 0$ *for some finite* N_o.

Proof. Follows directly from strong duality and Theorem 8. □

Let $\tilde{d}_N$ and $\tilde{p}_N$ be the numerical solutions to problems (26) and (29) respectively, obtained using a standard SDP solver. As noted before, in practice, one always has $\tilde{d}_N > 0$. Thus it can be hard to ascertain whether conditions (ii) or (iii) in Theorem 8 fail. On the other hand, $\tilde{p} > 0$ is an invalidation certificate.

Remark 13.7. A certificate that the existing data record is consistent with the model and *a priori* assumptions can be obtained by resorting to a variation of the so-called flat extension property for sparse polynomial optimization stated in [Lasserre (2006)]. In particular, if for some N, $\tilde{p} \leq 0$ and the dual solution satisfies rank $\left[\mathbf{M}_N(\boldsymbol{m}_{0:n_a}^*)\right] = 1$; rank $\left[\mathbf{M}_N(\boldsymbol{m}_{t-n_a:t-1}^*)\right] = 1$ and rank $\left[\mathbf{M}_N(\boldsymbol{m}_{t-n_a:t}^*)\right] =$ rank $\left[\mathbf{M}_{N-2}(\boldsymbol{m}_{t-n_a:t}^*)\right]$ $\forall t \in [n_a, T]$ where $\boldsymbol{m}^*$ denotes an optimal solution of (26), then this certifies that $o^* = d_N^* = 0$; hence $\mathscr{T}'(\boldsymbol{\eta})$ is not empty.

5. Numerical examples and applications

In this section we illustrate the effectiveness of the proposed methods both using academic examples and a computer vision application. In all cases, we used the moments relaxation corresponding to $N = s$ and the resulting SDP problem was solved using SEDUMI [Sturm (1999)].

5.1. *Identification via sparse polynomial optimization*

In this example, we consider a SISO ($n_u = n_y = 1$) hybrid linear switching system with $s = 3$ and $(n_a, n_c) = (1, 0)$, where the sub-models are

$$\begin{aligned} y_k &= -0.9(y_{k-1} - e_{k-1}) + u_{k-1} + e_k \text{ (Submodel 1)} \\ y_k &= -0.5(y_{k-1} - e_{k-1}) + u_{k-1} + e_k \text{ (Submodel 2)} \\ y_k &= +0.7(y_{k-1} - e_{k-1}) + u_{k-1} + e_k \text{ (Submodel 3)} \end{aligned}$$

and the hybrid system is modeled as (9), i.e.,

$$y_k = -a_1(\sigma_k)(y_{k-1} - e_{k-1}) + u_{k-1} + e_k,$$

where $\sigma_k \in \{1, 2, 3\}$ depending on which sub-model is active at time k. In the simulation, we set $\sigma(k) = 1$ for $k = [1, 5]$, $\sigma(k) = 2$ for $k = [6, 10]$ and $k = [16, 20]$, $\sigma(k) = 3$ for $k = [11, 15]$. The experimental data was obtained with unit step input and with uniformly randomly generated noise bounded by $\bar{e}$. First we set the noise bound $\bar{e} = 0.1$ and then increase the bound by setting $\bar{e} = 0.3$. The identification problems are solved based on Theorem 5 via solving an SDP relaxation built from the equivalent sparse polynomial optimization problem.

The parameter values used in the simulation and their estimated values with respect to different noise bounds are illustrated in Table 1.

Table 1. Estimated and true values of parameters.

		True	$\bar{e} = 0.1$	$\bar{e} = 0.3$
Submodel 1	a_1	0.9000	0.9352	0.5825
Submodel 2	a_1	0.5000	0.4867	0.5825
Submodel 3	a_1	-0.7000	-0.6841	-0.6652

Moreover, the active sub-system at time k can be determined once the sub-models are identified. The results are shown in Figure 2.

Remark 13.8. As shown in Table 1 and Figure 2, if the noise bound is large, there may exist hybrid systems that are consistent with the data and priors and have a smaller number of sub-systems that are the "ground truth". However, it should be emphasized that this does not constitute a failure of the algorithm. Rather, it reflects the fact that when the noise level is large, it can mask the difference between model dynamics.

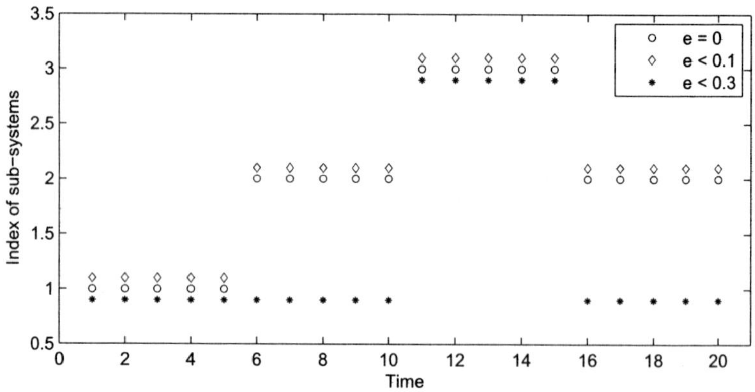

Fig. 2. Identified and true active sub-systems v.s. time.

5.2. *Identification via rank minimization*

In this example, we consider a more complex SISO hybrid system to show the computational efficiency of using Theorem 7 and Algorithm 13.1. It is assumed that there are three sub-systems with the order $(n_a, n_c) = (2, 1)$. That is,

$$y_k = -a_{1,i}(y_{k-1} - e_{k-1}) - a_{2,i}(y_{k-2} - e_{k-2}) + b_{1,i}u_{k-1} + e_k$$

where $i \in \{1, 2, 3\}$ depending on which sub-system is active at time k We take 120 measurement with 11 switches among these three sub-systems. The simulation is run for two different noise levels, i.e. $\bar{e} = 0.1$ and $\bar{e} = 0.3$. The input signal u is uniformly randomly generated between -1 and 1. In Table 2, we present the values of the system parameters identified by the algorithm for the noise levels considered.

Table 2. Estimated and true values of parameters.

		True	Noise $e < 0.1$	Noise $e < 0.3$
Submodel 1	a_1	0.9000	0.9183	0.9586
	a_2	0.1800	0.1736	0.2379
	b_1	0.2000	0.2310	0.2863
Submodel 2	a_1	0.5000	0.5155	0.5494
	a_2	0.0600	0.0564	0.1014
	b_1	1.0000	1.1310	1.0802
Submodel 3	a_1	-1.0000	-0.9716	-0.9618
	a_2	0.3000	0.3144	0.3287
	b_1	0.6000	0.6369	0.6991

Moreover, the active sub-system at time k can be determined once the sub-models are identified. The results are shown in Figure 3.

5.3. *Model (in)validation via sparse polynomial optimization*

We consider the ARX submodels:

$$y_t = 0.2y_{t-1} + 0.24y_{t-2} + 2u_{t-1} \qquad (G_1)$$

$$y_t = -1.4y_{t-1} - 0.53y_{t-2} + u_{t-1} \qquad (G_2)$$

$$y_t = 1.7y_{t-1} - 0.72y_{t-2} + 0.5u_{t-1} \qquad (G_3)$$

and the measurement equation:

$$\tilde{y}_t = y_t + \eta_t. \qquad (30)$$

We ran different sets of simulations investigating different sources of model mismatch. In all cases, we collected input/output data $\{u_t, \tilde{y}_t\}$ for $t \in [0, 96]$ and tried to (in)validate the *a priori* model. In all experiments, when we used data inconsistent with the *a priori* information, $\tilde{p}^*$ turned out to be positive. Hence, we correctly invalidated the model in each of such cases. On the other hand, whenever the *a priori* information was consistent with *a posteriori* data, we had $\tilde{p}^* < 0$.

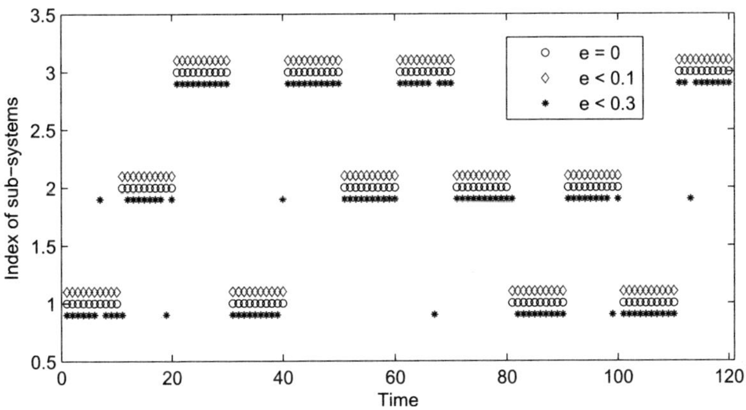

Fig. 3. Identified and true active sub-systems v.s. time.

Example 3. *(Submodel mismatch)* For the first set of experiments, we generated input/output data using different subsets of $\{G_1, G_2, G_3\}$ with a random switching sequence σ_t and with uniform random noise $||\boldsymbol{\eta}_t|| \leq 0.5$. The noise used in the experiments was within the *a priori* noise bound. We used both correct and incorrect *a priori* submodel sets. Hence, the model should be invalid when a submodel that is not contained in the *a priori* submodel set is used in the actual experiment. The results are summarized in Table 3.

Table 3. Invalidation results for example 3. The values of $\tilde{p}$ were respectively $-3.8441e-008$, $-8.2932e-009$, 0.8585, $-5.4026e-008$, $-1.5490e-007$ and 0.7566.

A priori	Actual	Result
G_1, G_2, G_3	G_1, G_2, G_3	not invalidated
G_1, G_2, G_3	G_1, G_2	not invalidated
G_1, G_2	G_1, G_2, G_3	**invalidated**
G_1, G_2	G_2	not invalidated
G_1, G_2, G_3	G_1	not invalidated
G_1, G_2	G_2, G_3	**invalidated**

Example 4. *(Noise bound mismatch)* For this example, we generated input/output data using different subsets of $\{G_1, G_2, G_3\}$ with a random switching sequence σ_t and with uniform random noise $||\boldsymbol{\eta}_t|| \leq \epsilon$. The *a priori* submodel set and the actual submodel set used in the experiment were the same. The source of invalidation was the actual noise level exceeding the *a priori* bound. The results of this set of experiments are summarized in Table 4.

Table 4. Invalidation results for example 4. The values of $\tilde{p}$ were respectively 0.0724, 0.0035, $-5.1810e-007$, 0.0737, 0.0034 and $-1.4930e-007$.

Submodels	A priori ϵ	Actual ϵ	Result
G_1, G_2, G_3	0.5	1	**invalidated**
G_1, G_2, G_3	0.8	1	**invalidated**
G_1, G_2, G_3	1.2	1	not invalidated
G_1, G_2	0.5	1	**invalidated**
G_1, G_2	0.8	1	**invalidated**
G_1, G_2	1.2	1	not invalidated

5.4. *A practical example: Activity monitoring*

In this section we illustrate the application of the proposed model invalidation framework to a non-trivial problem arising in computer vision: activity monitoring. Here we start with a set of dynamic models associated with "normal" behavior. If a person passing in front of the camera exhibits a combination of the normal activities (i.e. his/her behavior can be modeled with a hybrid system that has the normal activity dynamics as submodels), then the activity is considered not interesting. On the other hand, if he/she does something different than what has been encoded in the initial set of normal dynamics, this is an indication of an interesting event. In such cases the model should be invalidated.

We used a training video of walk shown in Fig. 4 to identify an autoregressive model for the dynamics of the center of mass of a person walking. By minimizing the ℓ_∞ norm of process noise via linear programming, we obtained the following model for "walk":

$$\begin{pmatrix} x_t \\ y_t \end{pmatrix} = \begin{pmatrix} 0.4747 & 0.0628 \\ -0.3424 & 1.2250 \end{pmatrix} \begin{pmatrix} x_{t-1} \\ y_{t-1} \end{pmatrix} + \begin{pmatrix} 0.5230 & -0.1144 \\ 0.3574 & -0.2513 \end{pmatrix} \begin{pmatrix} x_{t-2} \\ y_{t-2} \end{pmatrix} \qquad (A_1)$$

where (x_t, y_t) is the normalized coordinate of the center of the person in the tth frame. Another activity that we considered normal is "waiting" which can simply be modeled as:

$$\begin{pmatrix} x_t \\ y_t \end{pmatrix} = \begin{pmatrix} 1 & 0 \\ 0 & 1 \end{pmatrix} \begin{pmatrix} x_{t-1} \\ y_{t-1} \end{pmatrix}. \qquad (A_2)$$

We normalized the image coordinate system so that $(x, y) \in [0, 1] \times [0, 1]$ and set the measurement noise level to $||\boldsymbol{\eta}_t|| \leq 0.04$. This bound together with the submodels (A_1) and (A_2) for "normal" activities constitute the *a priori* information.

As for test purposes, we used three different video sequences. Background subtraction was used to locate the person; and the center of mass was estimated and tracked using the silhouettes. Then the center of mass trajectories were used for model (in)validation. In the first sequence, the person walks, waits and walks again; so the overall activity is normal. In the second sequence, the person runs for which our method found the certificate for invalidity by finding $\tilde{p} > 0$. In the third sequence, the person walks and then starts jumping. Again, our method flagged this abnormal activity by verifying the invalidity of the model. Sample frames from the sequences are shown in Fig. 5 and the results are summarized in Table 5. The primal optimal objective values, $\tilde{p}$, are also reported in Table 5. Since image coordinates are scaled to $[0,1] \times [0,1]$ in this example, higher order polynomials of values within this range results in low values of $\tilde{p}$. However, note that the positive range of $\tilde{p}$ is just a matter of scaling. It is the sign of $\tilde{p}$ that is important. The positivity of $\tilde{p}$ (as well as the existence of any primal feasible point with positive objective value) is a certificate for the invalidity of the model.

Table 5. Invalidation results for activity monitoring. The values of $\tilde{p}$ were, respectively, $-2.3303e-008$, $2.3707e-005$, and $5.0293e-007$.

A priori	Actual	Result
walk, wait	walk, wait	not invalidated
walk, wait	run	**invalidated**
walk, wait	walk, jump	**invalidated**

Fig. 4. Training sequence used in identification of the submodel (A_1) for walking.

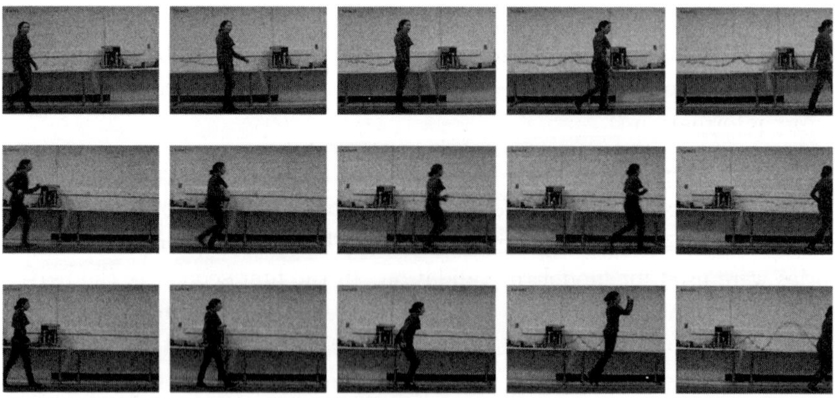

Fig. 5. Top: Walk, wait, walk sequence (not invalidated). Middle: Running sequence (invalidated). Bottom: Walk, jump sequence (invalidated).

6. Concluding remarks

In this chapter, we consider the identification and model invalidation problems for switched ARX systems from noisy experimental input/output data. The proposed approaches start by eliminating the discrete variables corresponding to the mode signal of the switched system and recast the identification and invalidation problems as suitable polynomial optimization problems. These polynomial optimization problems are then converted to semidefinite programs using moments-based convex relaxations. In the case of identification, the relaxation with minimum order is shown to be exact (i.e. equivalent to the original problem) since the objective function has an SOS representation. To address the computational difficulties encountered in complex hybrid systems, a drop rank approach, which can be tackled via efficient convex relaxations for rank minimization, is proposed to solve the original identification problem.

In case of model invalidation, we provide a necessary and sufficient condition that certifies the existence/nonexistence of admissible noise and switching sequences such that the resulting output sequence interpolates the given experimental data within the noise bound. In principle, computing these certificates entails solving a sequence of convex optimization problems involving increasingly large matrices. However, as we show here, if in the process of solving these problems either a positive solution is found or the so-called flat extension property holds, then the process terminates with a certificate that either the model has been invalidated or that the experimental data is indeed consistent with the model and *a–priori* information. In both identification and invalidation problems, further reduction in computational complexity is achieved by utilizing the inherent sparse structure of the corresponding polynomial optimization problems. Finally, the effectiveness of the proposed methods are illustrated using both academic examples and a non-trivial problem arising in computer vision: activity monitoring.

References

Bako, L. and Vidal, R. (2008). Algebraic identification of MIMO SARX models, in M. Egerstedt and B. Mishra (eds.), *Hybrid systems: Computation and control, Lecture Notes in Computer Science*, Vol. 4981 (Springer-Verlag, Berlin, Germany).

Bemporad, A., Garulli, A., Paoletti, S. and Vicino, A. (2005). A bounded-error approach to piecewise affine system identification, *IEEE Transactions on Automatic Control* **50**, 10, pp. 1567–1580.

Bianchi, F. and Sánchez Peña, R. (2009). Robust identification/invalidation in an LPV framework, *International Journal of Robust and Nonlinear Control* **20**, 3, pp. 301–312.

Chen, J. (1997). Frequency-domain tests for validation of linear fractional uncertain models, *IEEE Transactions on Automatic Control* **42**, 6, pp. 748–760.

Fazel, M., Hindi, H. and Boyd, S. (2003). Log-det heuristic for matrix rank minimization with applications to Hankel and euclidean distance matrices, in *Proceedings of the American Conference of Control* (Denver, CO, USA).

Feng, C., Lagoa, C. and Sznaier, M. (2010a). Hybrid system identification via sparse polynomial optimization, in *Proceedings of the American Conference of Control* (Baltimore, MD, USA).

Feng, C., Lagoa, C. M., Ozay, N. and Sznaier, M. (2010b). Hybrid system identification: An SDP approach, in *Proceedings of the Conference on Decision and Control* (Atlanta, GA, USA).

Ferrari-Trecate, G., Muselli, M., Liberati, D. and Morari, M. (2003). A clustering technique for the identification of piecewise affine and hybrid systems, *Automatica* **39**, pp. 205–207.

Henrion, D. and Lasserre, J. B. (2005). Detecting global optimality and extracting solutions in GloptiPoly, Tech. rep., LAAS-CNRS.

Juloski, A., Wieland, S. and Heemels, W. P. M. H. (2005). A Bayesian approach to identification of hybrid systems, *IEEE Transactions on Automatic Control* **50**, 10, pp. 1520–1533.

Lasserre, J. (2001). Global optimization with polynomials and the problem of moments, *SIAM Journal on Optimization* **11**, 3, pp. 796–817.

Lasserre, J. (2006). Convergent SDP-relaxations in polynomial optimization with sparsity, *SIAM Journal on Optimization* **17**, 3, pp. 822–843.

Laurent, M. (2009). Sums of squares, moment matrices and optimization over polynomials, *Emerging applications of algebraic geometry* **149**, pp. 157–270.

Ma, Y. and Vidal, R. (2005). A closed form solution to the identification of yybrid ARX models via the identification of algebraic varieties, in *Proceedings of the Hybrid Systems Computation and Control* (Zurich, Switzerland).

Nakada, H., Takaba, K. and Katayama, T. (2005). Identification of piecewise affine systems based on statistical clustering technique, *Automatica* **41**, 5, pp. 905–913.

Ozay, N., Lagoa, C. and Sznaier, M. (2009). Robust identification of switched affine systems via moments-based convex optimization, in *Proceedings of the Conference on Decision and Control* (Shanghai, China).

Ozay, N., Sznaier, M. and Lagoa, C. (2010a). Model (in)validation of switched ARX systems with unknown switches and its application to activity monitoring, in *Proceedings of the Conference on Decision and Control* (Atlanta, GA, USA).

Ozay, N., Sznaier, M., Lagoa, C. and Camps, O. (2008). A sparsification approach to set membership identification of a class of affine hybrid systems, in *Pro-*

ceedings of the Conference on Decision and Control (Cancún, Mexico).

Ozay, N., Sznaier, M., Lagoa, C. and Camps, O. (2010b). GPCA with denoising: a moment-based convex approach, in *Proceedings of the IEEE Conference on Computer Vision and Pattern Recognition* (San Francisco, CA, USA), pp. 3209–3216.

Paoletti, S., Juloski, A. L., Ferrari-Trecate, G. and Vidal, R. (2007). Identification of hybrid systems: a tutorial, *European Journal of Control* **13**, 2–3, pp. 242–260.

Parrilo, P. A. (2003). Semidefinite programming relaxations for semialgebraic problems, *Mathematical Programming Ser. B* **96**, 2, pp. 293–320.

Poolla, K., Khargonekar, P., Tikku, A., Krause, J. and Nagpal, K. (1994). A time domain approach to model validation, *IEEE Transactions on Automatic Control* **39**, 5, pp. 951–959.

Prajna, S. (2006). Barrier certificates for nonlinear model validation, *Automatica* **42**, 1, pp. 117–126.

Roll, J., Bemporad, A. and Ljung, L. (2004). Identification of piecewise affine systems via mixed-integer programming, *Automatica* **40**, 1, pp. 37–50.

Sánchez Peña, R. and Sznaier, M. (1998). *Robust Systems Theory and Applications* (John Wiley & Sons, Inc., New York City, NY, USA).

Sturm, J. (1999). Using SeDuMi 1.02, a MATLAB toolbox for optimization over symmetric cones, *Optimization Methods and Software* **11–12**, pp. 625–653.

Sznaier, M. and Mazzaro, M. (2003). An LMI approach to control-oriented identification and model (in)validation of LPV systems, *IEEE Transactions on Automatic Control* **48**, 9, pp. 1619–1624.

Vidal, R., Soatto, S., Ma, Y. and Sastry, S. (2003). An algebraic geometric approach to the identification of a class of linear hybrid systems, in *Proceedings of the Conference on Decision and Control* (Maui, HI, USA).

Waki, H., Kim, S., Kojima, M. and Muramatsu, M. (2006). Sums of squares and semidefinite program relaxations for polynomial optimization problems with structured sparsity, *SIAM Journal on Optimization* **17**, 1, pp. 218–242.

Index